MicroRNA as Biomarkers in Cancer Diagnostics and Therapy

MicroRNA as Biomarkers in Cancer Diagnostics and Therapy

Special Issue Editor

Lorenzo F. Sempere

MDPI • Basel • Beijing • Wuhan • Barcelona • Belgrade

MDPI

Special Issue Editor
Lorenzo F. Sempere
Michigan State University,
USA

Editorial Office
MDPI
St. Alban-Anlage 66
4052 Basel, Switzerland

This is a reprint of articles from the Special Issue published online in the open access journal *International Journal of Molecular Sciences* (ISSN 1422-0067) from 2018 to 2019 (available at: https: //www.mdpi.com/journal/ijms/special_issues/miR_biomarkers_cancer)

For citation purposes, cite each article independently as indicated on the article page online and as indicated below:

LastName, A.A.; LastName, B.B.; LastName, C.C. Article Title. *Journal Name* **Year**, *Article Number, Page Range.*

ISBN 978-3-03921-249-1 (Pbk)
ISBN 978-3-03921-250-7 (PDF)

Cover image courtesy of Lorenzo F. Sempere.

Contents

About the Special Issue Editor

Lorenzo F. Sempere is originally from Elche, a sunny city in southeastern Spain with one of largest palm tree groves in the world. He obtained his B.S. in Biochemistry at Universidad Miguel Hernández and trained in the laboratory of Victor Ambros, co-discoverer of microRNAs, at Geisel School of Medicine at Dartmouth College. Dr. Sempere initiated his translational cancer research in the laboratory of Charles Cole, funded by a Susan G. Komen for the Cure postdoctoral fellowship, and continued to climb the academic ladder at Dartmouth in the laboratory of Murray Korc, funded by a Laurie and Paul MacCaskill PanCAN-AACR Career Development award. Dr. Sempere started his independent laboratory as a research track faculty first at Geisel School of Medicine, and then at Van Andel Research Institute. Dr. Sempere joined Michigan State University in 2018 as Assistant Professor of Radiology in the tenure-track system and as faculty member of the campus-wide Precision Health Program. Dr. Sempere has worked in the field of microRNAs since their discovery in 2001. He has experience and expertise as an author, reviewer, and journal editor in diverse areas of microRNA research, including evolutionary and developmental biology, molecular and cellular biology, and immunology and cancer biology.

International Journal of
Molecular Sciences

MDPI

Editorial

Celebrating 25 Years of MicroRNA Research: From Discovery to Clinical Application

Lorenzo F. Sempere

Department of Radiology, Precision Health Program, Michigan State University, East Lansing, MI 48824, USA; semperel@msu.edu; Tel.: +1(517)355-3982

Received: 19 April 2019; Accepted: 21 April 2019; Published: 23 April 2019

In 1993, the Ambros lab reported the cloning and developmental function of lin-4, the first microRNA [1]. This short non-coding RNA was regarded as an oddity of the mighty little roundworm *Caenorhabditis elegans*, controlling gene expression by binding to partially complementary sites on the 3′ UTR of target mRNAs and inhibiting translation [1–3]. The discovery of let-7, also in *C. elegans*, reinforced this molecular oddity and highlighted the power of forward genetics in this model organism [4]. These two small temporal RNAs control developmental timing of larval transition in *C. elegans* and other *Caenorhabditis* species [5]. This restricted classification was short-lived as the evolutionary conservation of let-7 [6] and the shared enzymatic machinery for processing of endogenous short non-coding RNAs and exogenous double-stranded RNA substrates for RNA interference [7] strongly suggested a broadly co-opted regulatory mechanism of short non-coding RNAs in animal evolution. Indeed, let-7 and miR-125 (lin-4 paralog) family members have been implicated in temporal identity in the fly brain, and are likely involved in temporal cell fate decision in vertebrates [5]. Over the last 20 years, microRNAs have been identified in many animal species. Some of these microRNAs are phylum-, order-, genus-, or even species-specific [8–10]. The size of the microRNome and complexity of animal body plans and organ systems suggest a role of microRNAs in cell fate determination and differentiation [8,9]. More than 2000 sequences have been proposed to represent unique microRNA genes in humans with an increasing number of mechanistic roles in developmental, physiological, and pathological processes [10,11].

MicroRNA are short non-coding regulatory RNAs. The mature and biologically active form is about 18–25 nucleotides long. This mature sequence binds to and guides an Ago-containing, RNA-induced, multi-protein silencing complex to partially complementary sites on the mRNA of target genes [12]. Due to this partial complementarity, a single microRNA can regulate the expression of many, if not hundreds, of target genes. Thus, dysregulation of a few key microRNAs can have a profound global effect on gene expression and molecular programs of a cell. Conversely, restoration of baseline microRNA activity can also have profound effects to reverse a pathological process [12,13]. This great potential for clinical intervention captured the interest and imagination of researchers in many fields. However, very few fields have been as prolific as the field of cancer research. This is largely due to early studies by Carlo Croce and colleagues that linked microRNAs with cancer just a couple of years after the discovery of let-7. In 2002, Calin et al. showed that chromosomal deletion of miR-15 and miR-16 was a frequent event in B-cell chronic lymphocytic leukemia [14]. Soon after, Calin et al. also made the tantalizing observation that many microRNA loci are located at fragile sites, breakpoint regions, or frequently altered regions (e.g., deletion or amplification) in the cancer genome [15]. These seminal papers, along with a technological explosion of high-throughput detection platforms, led several groups to extensively profile microRNA expression in healthy and tumor tissues. Altered microRNA expression has been associated with diagnostic and prognostic indicators in many cancer types. Different mechanisms have been reported for this dysregulation, including chromosomal deletion or amplification of a microRNA gene, epigenetic and transcriptional regulation, and mutation in the enzymes responsible for microRNA processing, export, or silencing [12,13]. Despite the overwhelming number of diagnostic and prognostic studies, the current impact of microRNA-based assays is very limited in clinical practice. Some microRNA-based assays have reached the clinic in the form of laboratory developed tests, and several on-going clinical trials propose the use of microRNAs as biomarkers for early disease detection, guiding treatment selection, monitoring disease progression, or other specific clinical endpoints.

This Special Issue celebrates the 25th anniversary of the discovery of the first microRNA. It provides but a glimpse of the large body of literature of microRNA biology in cancer research. This Special Issue contains four original research studies [16–19] and four review papers [20–23] with a disease focus on specific hematologic or solid tumors. Collectively, these papers highlight state-of-the-art approaches and methodologies for microRNA detection in tissue, blood, and other body fluids for biomarker applications from early cancer detection to prognosis and treatment response. These papers also address some of the challenges for clinical implementation. Pezzuto et al. provide a comprehensive review of blood-based detection of cell-free DNA and microRNAs for early detection of hepatitis viruses–related liver cancer [21]. Chang et al. report on a plasma-based 3-microRNA signature for early detection of oral cancer [17]. Konstantinell et al. provide a comprehensive review of extracellular vesicle-based and tissue-based detection of microRNA for diagnostic and prognostic applications in Merkel cell carcinoma [22]. de Oliveira et al. provide a comprehensive review of tissue-based and cell-based detection of microRNAs for diagnostic and prognostic applications in childhood hematological cancers [23]. Drobna et al. describe a methodological advancement of most suitable endogenous microRNA controls for cell-based microRNA detection in T-cell acute lymphoblastic leukemia [18]. Hibner et al. provide a comprehensive review [20], while Moller et al. [19] and Ulivi et al. [16] present specific studies of microRNA biomarkers for diagnostic and prognostic applications in colorectal cancer. Below is a brief summary and highlights of each of these articles.

Liver cancer is the third leading cause of cancer-related death in the world. Hepatocellular carcinoma (HCC) represents 85% of all liver cancer cases and generally has a poor prognosis due to late presentation. Chronic infection of Hepatitis B or Hepatitis C viruses are a major risk factor for development of HCC. Pezzuto et al. review blood-based detection assays that would complement and improve current diagnostic tools based on various imaging modalities and plasma levels of alpha-fetoprotein [21]. Pezzuto et al. describe the utility of detecting circulating cell-free DNA, and cell-free or extracellular vesicle-loaded microRNAs and long non-coding RNAs. While DNA mutations or altered DNA methylation pattern in circulating DNA or altered circulating microRNA levels can detect HCC tumors, altered levels of some microRNAs such as miR-122 can reflect more closely the effect of viral infection on malignant transformation of hepatocytes. Of the discussed blood-based markers, Pezzuto et al. highlight altered circulating levels of miR-122 and let-7, and RASSF1A hypermethylation in circulating free-DNA as the most promising biomarkers for diagnostic application. As in HCC, patients that present with late stage oral cancer, which can be as high as 50% of all cases in some countries, have a much worse prognosis than early stage cases. Oral squamous cell carcinoma (OSCC) is the most common type of oral cancer. Prevalence and risk factors vary with geographic location. In South and Southeast Asia and Taiwan, betel quid chewing is a major risk factor. Chang et al. performed RNA sequencing from plasma of healthy controls, early stage patients with oral leukoplakia (OL, precursor lesion associated with OSCC progression) and late stage patients diagnosed with OSCC to identify differentially expressed microRNAs. Selected microRNAs were further studied in a training set of 72 samples and validated in an independent set of 178 samples of Taiwanese patients. From these analyses, Chang et al. identified a 3-microRNA signature (miR-150-5p, miR-222-3p, miR-423-5p) that could accurately separate OL from OSCC cases and may have clinical utility for early detection of OSCC. As in HCC, viral infection is a major risk factor in Merkel cell carcinoma (MCC), which is a rare but aggressive type of skin cancer. About 80% of MCC tumors are infected with viral DNA of Merkel cell polyomavirus. Konstantinell et al. provide a comprehensive review on the host and viral microRNA expression in MCC tissue samples and present their original data profiling microRNA expression in extracellular vesicles secreted by MCC cell lines [22].

MicroRNA expression and function have been extensively studied during normal hematopoiesis and in several hematological malignancies, including different types of leukemias and lymphomas. Childhood leukemia and lymphomas have certain features that distinguish them from adult counterpart conditions such as mutational spectrum, cell of origin and location, and cellular context and microenvironment. de Oliveira et al. provide a comprehensive review of microRNA dysregulation in childhood leukemia and lymphomas and contrast their differences with adult counterpart conditions [23]. Childhood leukemias and lymphomas represent about 30% and 15% of all pediatric tumors, respectively. de Oliveira et al. systematically review the literature for each leukemia and lymphoma cancer types, with an emphasis on major types, including acute lymphoblastic leukemia (ALL), acute myeloid leukemia and Burkitt lymphoma. In addition to detailed text descriptions, de Oliveira's colorful figures provide informative and concise graphical summaries of microRNA dysregulation associated with diagnostically and therapeutically relevant criteria in each condition such as chromosomal rearrangement and treatment response to a specific drug regimen. While de Oliveira et al. focus their review on microRNA detection in cancer cells or tumor tissues and its diagnostic and prognostic implications, the authors also discuss emerging studies on circulating microRNAs in acute lymphoblastic leukemia and other conditions.

Reliable endogenous controls for normalization of microRNA expression in cells or tissues or for circulating microRNA levels are an important consideration to maximize accuracy of biomarker readout. Identifying such controls is technically challenging because microRNA expression is cell type-, context-, and disease-dependent. Drobna et al. describe a strategy to identify endogenous microRNA controls for adult T-cell ALL using a reverse-transcription quantitative PCR (RT-qPCR) assay [18]. Drobna et al. performed RNA sequencing analysis on sorted T cells from 34 T-cell ALL cases and from bone marrow of five healthy controls. Using an algorithm to identify microRNA with stable expression across the samples, the authors selected 10 microRNAs for further evaluation by RT-qPCR assay; most of these 10 microRNAs had been previously suggested to serve as appropriate controls in other tissue types or disease conditions. Drobna et al. propose three microRNAs (let-7a-5p, miR-16-5p, miR-25-3p) as optimal endogenous controls for evaluation of T-cell ALL samples.

Colorectal cancer (CRC) is the second leading cause of cancer-related death in the world. With over a million new cases and over half a million deaths yearly, clinical management of colorectal cancer is an important and worldwide health problem. Hibner et al. provide a comprehensive review of blood-based and tissue-based studies that utilize a single microRNA or a microRNA signature to find association with diagnostic and prognostic indicators in CRC [20]. Hibner et al. devote individual sections to microRNAs frequently associated with CRC in multiple independent studies, including miR-21, miR-29b, miR-34a, and miR-155. Although these sections focus on diagnostic and prognostic applications, Hibner et al. also report on specific targets of these miRNAs and their potential application for therapeutic intervention. RT-qPCR assay is the preferred method for miRNA expression analysis in most of these CRC studies as well as in other studies described above. The study by Ulivi et al. exemplifies the robustness of RT-qPCR for detection of circulating miRNAs [16]. Ulivi et al. analyzed plasma levels of miR-20b, miR-29b, and miR-155 in a cohort of 52 metastatic CRC cases treated with bevacizumab-containing chemotherapy regimen. Higher circulating levels of these three microRNAs in plasma collected before treatment were associated with longer progression-free and overall survival. The authors only analyzed these microRNAs individually, thus it will be interesting to see if this 3-miRNA signature would provide a stronger prognostic signal. Intriguingly, comparison of circulating levels of these microRNAs before treatment and after 1 month of treatment showed that cases with increased levels of miR-155 after treatment are associated with shorter progression-free and overall survival. The authors discuss mechanisms for the timing and opposite outcome based on circulating miR-155 levels. However, RT-qPCR assay has limitations and is technically challenged to apply for miRNA detection in specific cell types that compose the tumor mass. Møller et al. 2019 describe elegant and robust methodology for in situ co-detection of microRNAs, mRNAs, and non-coding RNA molecules in tumor tissues, combining locked nucleic acid chemistry for microRNA probes and

Int. J. Mol. Sci. **2019**, *20*, 1987

RNAscope® technology for mRNA probes [19]. This in situ co-detection assay enables characterization of RNA expression at single cell resolution providing biologically relevant information of the cell type(s) that present with altered regulation of miR-21 expression in a particular tumor. Previous studies by this group and others have shown that miR-21 expression is predominantly upregulated and carries prognostic value in cancer-associated fibroblasts (CAFs) more than in cancer cells (reviewed in Hibner et al. [20]). Curiously, Møller et al. [19] report upregulation of miR-21 in a discrete set of cancer cells, budding cells, in addition to CAFs in a panel of colorectal cancer cases. Budding cancer cells are single or a small cluster of cells at the invading front that pinched off or detach from the main tumor mass. Co-detection of miR-21 and TNF-α mRNA expression did not indicate a regulatory relation between this pro-fibrotic and pro-survival microRNA and this pro-inflammatory cytokine in budding cancer cells. Nonetheless, upregulation of miR-21 expression suggests a potential role in the survival and/or migration of budding cells.

I would like to thank the authors for their valuable contributions to this Special Issue. I also would like to thank editorial staff members, especially Meredith Liu, and anonymous reviewers who improved the presentation and content of this Special Issue. I hope readers find this Special Issue an accessible reference to keep abreast of recent findings, methodologies, and approaches related to microRNA biology in cancer research and its potential applications in cancer medicine.

Conflicts of Interest: The author declares no conflict of interest.

References

1. Lee, R.C.; Feinbaum, R.L.; Ambros, V. The *C. elegans* heterochronic gene lin-4 encodes small RNAs with antisense complementarity to lin-14. *Cell* **1993**, *75*, 843–854. [CrossRef]
2. Wightman, B.; Ha, I.; Ruvkun, G. Posttranscriptional regulation of the heterochronic gene lin-14 by lin-4 mediates temporal pattern formation in *C. elegans*. *Cell* **1993**, *75*, 855–862. [CrossRef]
3. Moss, E.G.; Lee, R.C.; Ambros, V. The cold shock domain protein LIN-28 controls developmental timing in *C. elegans* and is regulated by the lin-4 RNA. *Cell* **1997**, *88*, 637–646. [CrossRef]
4. Reinhart, B.J.; Slack, F.J.; Basson, M.; Pasquinelli, A.E.; Bettinger, J.C.; Rougvie, A.E.; Horvitz, H.R.; Ruvkun, G. The 21-nucleotide let-7 RNA regulates developmental timing in *Caenorhabditis elegans*. *Nature* **2000**, *403*, 901–906. [CrossRef]
5. Sokol, N.S. Small temporal RNAs in animal development. *Curr. Opin. Genet. Dev.* **2012**, *22*, 368–373. [CrossRef] [PubMed]
6. Pasquinelli, A.E.; Reinhart, B.J.; Slack, F.; Martindale, M.Q.; Kuroda, M.I.; Maller, B.; Hayward, D.C.; Ball, E.E.; Degnan, B.; Muller, P.; et al. Conservation of the sequence and temporal expression of let-7 heterochronic regulatory RNA. *Nature* **2000**, *408*, 86–89. [CrossRef] [PubMed]
7. Grishok, A.; Pasquinelli, A.E.; Conte, D.; Li, N.; Parrish, S.; Ha, I.; Baillie, D.L.; Fire, A.; Ruvkun, G.; Mello, C.C. Genes and mechanisms related to RNA interference regulate expression of the small temporal RNAs that control *C. elegans* developmental timing. *Cell* **2001**, *106*, 23–34. [CrossRef]
8. Sempere, L.F.; Cole, C.N.; McPeek, M.A.; Peterson, K.J. The Phylogenetic Distribution of Metazoan microRNAs: Insights into Evolutionary Complexity and Constraint. *J. Exp. Zool B Mol. Dev. Evol.* **2006**, *306*, 575–588. [CrossRef] [PubMed]
9. Heimberg, A.M.; Sempere, L.F.; Moy, V.N.; Donoghue, P.C.; Peterson, K.J. MicroRNAs and the advent of vertebrate morphological complexity. *Proc. Natl. Acad. Sci. USA* **2008**, *105*, 2946–2950. [CrossRef] [PubMed]
10. Kozomara, A.; Birgaoanu, M.; Griffiths-Jones, S. miRBase: From microRNA sequences to function. *Nucleic Acids Res.* **2019**, *47*, D155–D162. [CrossRef] [PubMed]
11. Fromm, B.; Billipp, T.; Peck, L.E.; Johansen, M.; Tarver, J.E.; King, B.L.; Newcomb, J.M.; Sempere, L.F.; Flatmark, K.; Hovig, E.; Peterson, K.J. A Uniform System for the Annotation of Vertebrate microRNA Genes and the Evolution of the Human microRNAome. *Annu. Rev. Genet.* **2015**, *49*, 213–242. [CrossRef] [PubMed]
12. Di Leva, G.; Garofalo, M.; Croce, C.M. MicroRNAs in cancer. *Annu. Rev. Pathol.* **2014**, *9*, 287–314. [CrossRef]
13. Rupaimoole, R.; Slack, F.J. MicroRNA therapeutics: Towards a new era for the management of cancer and other diseases. *Nat. Rev. Drug Discov.* **2017**, *16*, 203–222. [CrossRef] [PubMed]

14. Calin, G.A.; Dumitru, C.D.; Shimizu, M.; Bichi, R.; Zupo, S.; Noch, E.; Aldler, H.; Rattan, S.; Keating, M.; Rai, K.; et al. Frequent deletions and down-regulation of micro- RNA genes miR15 and miR16 at 13q14 in chronic lymphocytic leukemia. *Proc. Natl. Acad. Sci. USA* **2002**, *99*, 15524–15529. [CrossRef]

15. Calin, G.A.; Sevignani, C.; Dumitru, C.D.; Hyslop, T.; Noch, E.; Yendamuri, S.; Shimizu, M.; Rattan, S.; Bullrich, F.; Negrini, M.; Croce, C.M. Human microRNA genes are frequently located at fragile sites and genomic regions involved in cancers. *Proc. Natl. Acad. Sci. USA* **2004**, *101*, 2999–3004. [CrossRef]

16. Ulivi, P.; Canale, M.; Passardi, A.; Marisi, G.; Valgiusti, M.; Frassineti, G.; Calistri, D.; Amadori, D.; Scarpi, E. Circulating Plasma Levels of miR-20b, miR-29b and miR-155 as Predictors of Bevacizumab Efficacy in Patients with Metastatic Colorectal Cancer. *Int. J. Mol. Sci.* **2018**, *19*, 307. [CrossRef]

17. Chang, Y.; Weng, S.; Yang, S.; Chou, C.; Huang, W.; Tu, S.; Chang, T.; Huang, C.; Jong, Y.; Huang, H. A Three–MicroRNA Signature as a Potential Biomarker for the Early Detection of Oral Cancer. *Int. J. Mol. Sci.* **2018**, *19*, 758. [CrossRef]

18. Drobna, M.; Szarzyńska-Zawadzka, B.; Daca-Roszak, P.; Kosmalska, M.; Jaksik, R.; Witt, M.; Dawidowska, M. Identification of Endogenous Control miRNAs for RT-qPCR in T-Cell Acute Lymphoblastic Leukemia. *Int. J. Mol. Sci.* **2018**, *19*, 2858. [CrossRef] [PubMed]

19. Møller, T.; James, J.P.; Holmstrøm, K.; Sørensen, F.B.; Lindebjerg, J.; Nielsen, B.S. Co-Detection of miR-21 and TNF-α mRNA in Budding Cancer Cells in Colorectal Cancer. *Int. J. Mol. Sci.* **2019**, *20*, 1907. [CrossRef] [PubMed]

20. Hibner, G.; Kimsa-Furdzik, M.; Francuz, T. Relevance of MicroRNAs as Potential Diagnostic and Prognostic Markers in Colorectal Cancer. *Int. J. Mol. Sci.* **2018**, *19*, 2944. [CrossRef]

21. Pezzuto, F.; Buonaguro, L.; Buonaguro, F.; Tornesello, M. The Role of Circulating Free DNA and MicroRNA in Non-Invasive Diagnosis of HBV- and HCV-Related Hepatocellular Carcinoma. *Int. J. Mol. Sci.* **2018**, *19*, 1007.

22. Konstantinell, A.; Coucheron, D.; Sveinbjørnsson, B.; Moens, U. MicroRNAs as Potential Biomarkers in Merkel Cell Carcinoma. *Int. J. Mol. Sci.* **2018**, *19*, 1873. [CrossRef] [PubMed]

23. Carvalho de Oliveira, J.; Molinari Roberto, G.; Baroni, M.; Bezerra Salomão, K.; Alejandra Pezuk, J.; Sol Brassesco, M. MiRNA Dysregulation in Childhood Hematological Cancer. *Int. J. Mol. Sci.* **2018**, *19*, 2688. [CrossRef] [PubMed]

International Journal of
Molecular Sciences

MDPI

Review

The Role of Circulating Free DNA and MicroRNA in Non-Invasive Diagnosis of HBV- and HCV-Related Hepatocellular Carcinoma

Francesca Pezzuto, Luigi Buonaguro, Franco Maria Buonaguro and Maria Lina Tornesello *

Molecular Biology and Viral Oncology Unit, Istituto Nazionale Tumori IRCCS "Fondazione G. Pascale", 80131 Napoli, Italy; francesca.pezzuto1987@gmail.com (F.P.); l.buonaguro@istitutotumori.na.it (L.B.); fm.buonaguro@istitutotumori.na.it (F.M.B.)
* Correspondence: m.tornesello@istitutotumori.na.it; Tel.: +39-081-590-3609

Received: 23 February 2018; Accepted: 24 March 2018; Published: 28 March 2018

Abstract: Hepatocellular carcinoma (HCC) is the third and the fifth leading cause of cancer related death worldwide in men and in women, respectively. HCC generally has a poor prognosis, with a very low 5-year overall survival, due to delayed diagnosis and treatment. Early tumour detection and timely intervention are the best strategies to reduce morbidity and mortality in HCC patients. Histological evaluation of liver biopsies is the gold standard for cancer diagnosis, although it is an invasive, time-consuming and expensive procedure. Recently, the analysis of circulating free DNA (cfDNA) and RNA molecules released by tumour cells in body fluids, such as blood serum, saliva and urine, has attracted great interest for development of diagnostic assays based on circulating liver cancer molecular biomarkers. Such "liquid biopsies" have shown to be useful for the identification of specific molecular signatures in nucleic acids released by cancer cells, such as gene mutations and altered methylation of DNA as well as variations in the levels of circulating microRNAs (miRNAs) and long non-coding RNAs (lncRNAs). Body fluids analysis may represent a valuable strategy to monitor liver disease progression in subjects chronically infected with hepatitis viruses or cancer relapse in HCC treated patients. Several studies showed that qualitative and quantitative assays evaluating molecular profiles of circulating cell-free nucleic acids could be successfully employed for early diagnosis and therapeutic management of HCC patients. This review describes the state of art on the use of liquid biopsy for cancer driver gene mutations, deregulated DNA methylation as well as miRNA levels in HCC diagnosis.

Keywords: liquid biopsy; early diagnosis; circulating free DNA; microRNA; hepatocellular carcinoma; hepatitis B virus; hepatitis C virus; long non coding RNA

---

1. Introduction

Primary liver cancer represents the sixth most common and deadly tumour in the world with 782,000 new cases and 746,000 deaths in 2012 [1]. Hepatocellular carcinoma (HCC) is the major histological subtype accounting for 85% of all liver cancer cases worldwide [2–4]. The major risk factors for the development of HCC are hepatitis B (HBV) and hepatitis C (HCV) chronic infections which have been found to be associated with 56% and 20% of cases, respectively [5]. HBV-related HCC is more frequent (67%) than HCV-related HCC (12%) in less developed countries, while HBV-related HCC is less frequent (23%) than HCV-related tumours (44%) in more developed countries [5]. The HBV and HCV viral proteins along with biological and environmental co-factors promote chronic insult to hepatocytes, accumulation of genetic damages and epigenetic deregulation, which cause over the time, the hepatic damage, cirrhosis, fibrosis and cancer [6].

The diagnosis of liver cancer is generally performed by imaging techniques, such as ultrasonography, computed tomography and magnetic resonance tomography, in combination with the dosage of plasmatic alpha-fetoprotein (AFP) and histological analysis of tissue biopsies [7]. The diagnostic imaging methods have the advantage of not being invasive and the disadvantage of insufficient sensitivity for detection of HCC nodules smaller than one cm [8]. The measurement of AFP in the blood, which is one of the most widely used screening tests to diagnose HCC, has a limited sensitivity and specificity given that some liver nodules may not release AFP, and patients with chronic active hepatitis or liver cirrhosis may have high levels of AFP [9]. To date, liver tissue biopsy is considered the gold standard for HCC diagnosis but has drawbacks of invasiveness, is effective when the nodule has reached considerable dimensions and carries the risk of neoplastic cells diffusion [10].

The treatment options for HCC include surgical resection, transarterial chemoembolization, radiofrequency ablation, high-intensity focused ultrasound, targeted molecular therapy such as sorafenib and more rarely liver transplantation. The success of these treatments could be seriously improved by early cancer detection and effective post-treatment monitoring [11].

Recent studies have shown that specific biomarkers of cancer cells are detectable in the body fluids such as blood serum, urine and saliva, which for this reason have been termed "liquid biopsies". The blood serum contains detectable amounts of circulating free DNA (cfDNA) ranging from 1 to 500 ng/mL, showing the mutational spectrum of the tumour cell DNA [11]. In addition, cfDNA fragments have the same methylation profile as the original tumour DNA, suggesting the possibility of analyzing the cfDNA methylation status for monitoring tumour growth. Many tumour cells, including liver cancer cells, release specific microRNAs (miRNAs) and long non-coding RNAs (lncRNAs) in the bloodstream, either as free molecules or entrapped in vesicles such as exosomes [12–14]. Such molecules may represent important biomarkers of tumour development.

We performed a systematic review of published studies to investigate the state of art on the employment of screening tests based on circulating liver biomarkers for diagnosis and prognosis of HCC associated with different aetiologies (Table 1).

Published data were searched in Medline using the terms ("hepatocellular" OR "Liver" AND "Cancer") AND ("liquid biopsy" OR "blood" OR "plasma" OR "serum" OR "urine") AND ("circulating free DNA") AND ("microRNA OR miRNA") AND ("DNA mutations") AND ("DNA methylation") AND ("microsatellite instability") AND ("microRNA" OR "miRNA") AND ("long non coding RNA" OR "lncRNA") AND ("extracellular vesicles" OR "exosomes"). The search was updated on 28 January 2018.

2. Circulating Free DNA

Circulating free DNA was first described by Mandel and Metais in 1948 [15]. Thirty years later, Leon et al. observed that the amount of cfDNA was higher in cancer patients compared to healthy controls and that its concentration in the serum further increased after radiation therapy [16–18]. Nowadays, it is recognized that cfDNA originates mainly from the activity of macrophages or other scavenger cells which engulf apoptotic and necrotic tumour cells and release digested tumour DNA into the blood stream [19,20]. Because the length of digested DNA molecules is around 160 bp, the recovery and analysis of cfDNA requires highly sensitive techniques [21].

Qualitative and quantitative analysis of cfDNA as a diagnostic and prognostic parameter in cancer patients has been studied by many groups. Piciocchi et al. observed higher levels of cfDNA among patients affected by HCC, cirrhosis and HCV-related chronic hepatitis compared to healthy subjects, and the increase was directly correlated to the disease status and reduced patients' survival [22]. Other studies, however, reported wide inter-subject variations in cfDNA levels, showing sometimes overlapping values between malignant and benign diseases or healthy controls. In addition, patients affected by some non-oncologic pathologies such as autoimmune diseases are also characterized by increased levels of cfDNA in the peripheral blood, making this parameter not specific for cancer diagnosis [23,24].

Table 1. Summary of published articles retrieved from Pubmed on the role of somatic mutations and methylation in non-invasive diagnosis in liver cancer.

DNA Alterations	Gene	Tissue Biopsies N Cases (%)	CfDNA N Cases (%)	Method [2]	Ref.
Single nucleotide mutations	CTNNB1	0	6/48 (12.5)	ddPCR	[25]
	CTNNB1	11/41 (26.8)	4/41 (9.7)	MiSeq	[26]
	TERT promoter	5/41 (12.2)	11/48 (22.9)	ddPCR	[25]
	TERT promoter	29/41 (70.7)	2/41 (4.9)	MiSeq	[26]
	TP53	1/41 (2.4)	7/48 (14.6)	ddPCR	[25]
	TP53	27/41 (65.8)	2/41 (4.9)	MiSeq	[26]
Hypermethylation	APC	NA [1]	49/72 (68.1)	MSRE-qPCR	[27]
	APC	NA	36/98 (36.7)	Methylight	[17]
	BVES	NA	29/98 (29.6)	Methylight	[17]
	ELF	22/34 (64.7)	18/31 (58.1)	MSP	[28]
	GSTP1	NA	40/72 (55.6)	MSRE-qPCR	[27]
	GSTP1	23/34 (67.6)	12/31 (38.7)	MSP	[28]
	GSTP1	23/26 (88.5)	16/32 (50.0)	MSP	[29]
	GSTP1	NA	17/98 (17.3)	Methylight	[17]
	HOXA9	NA	20/98 (20.4)	Methylight	[17]
	P16	16/22 (72.7)	13/22 (59.1)	MSP	[30]
	P16	25/34 (73.5)	13/31 (41.9)	MSP	[28]
	RASSF1A	5/5 (100)	59/63 (93.6)	MSRE, RT-PCR	[31]
	RASSF1A	NA	51/98 (52.0)	Methylight	[17]
	RASSF1A	NA	47/72 (65.3)	MSRE-qPCR	[27]
	RASSF1A	32/34 (94.1)	16/31 (51.6)	MSP	[28]
	RASSF1A	NA	77/105 (73.3)	MSP	[32]
	RASSF1A	37/40 (92.5)	17/40 (42.5)	MSP	[33]
	SFRP1	NA	40/72 (55.6)	MSRE-qPCR	[27]
	SOCS3	23/48 (47.9)	34/119 (28.6)	MSP	[34]
	TGR5	NA	77/160 (48.1)	MSP	[35]
	TIMP3	NA	11/98 (11.2)	Methylight	[17]
Hypomethylation	LINE-1	NA	70/105 (66.7)	MSP	[32]

[1] NA, information not available in the article; [2] ddPCR = digital droplet PCR; MiSeq = next generation sequencing method; MSRE = methylation sensitive restriction enzyme digestion; MSP = methylation specific PCR; Methylight = multiplex PCR assay; qPCR = quantitative PCR; RT-PCR = Real Time PCR.

Moreover, the different methodologies of cfDNA extraction may bias its quantification given that different extraction kits with variable recovery efficiencies can hamper the measurement of the real cfDNA levels in the blood serum [36,37].

Several studies analyzed cfDNA integrity as a parameter of the disease status although with contrasting results [38]. Huang et al. reported low integrity of cfDNA in a cohort of Chinese HCC patients, mainly related to HBV infection, compared to patients with benign liver disease and healthy subjects [39]. Interestingly, the cfDNA integrity test had a sensitivity, specificity and accuracy of 43.4%, 100% and 60%, respectively, in the detection of HCC [39]. The high efficacy of cfDNA integrity as a diagnostic marker was achieved by the improved sensitivity of PCR protocols based on short amplicons targeting the notably short tumour derived DNA fragments [40,41]. Conversely, Wang et al. reported that the increased cfDNA integrity was associated with cancer, and measurement of this parameter may be useful for cancer detection [42]. Accordingly, two other studies observed that cfDNA integrity was significantly higher in HCC patients compared to HBV- and HCV-positive patients and healthy controls [41,43]. Elshimali et al. also observed that cfDNA integrity was associated with tumour size, TNM stage, vascular invasion, lymph node involvement, distant metastasis and poor survival [36].

The majority of cancer types are characterized by distinctive somatic mutations which can be identified in the DNA released by cancer cells and, in combination with the measurement of cfDNA levels, may provide valuable clinical information, Figure 1 [44]. Several methodologies, mainly based on the polymerase chain reaction (PCR) technique, have been used to detect tumour-related known mutations by specific probes in cfDNA including the amplification refractory mutation system (ARMS) PCR, single-strand conformation polymorphism (SSCP), mutant enriched (ME) PCR, mutant allele specific amplification (MASA), pyrophosphorolysis-activated polymerization allele specific (PAP-A) PCR, and restriction fragment length polymorphism (RFLP-PCR) [45]. In addition, novel methods based on digital technology have been introduced in cfDNA analysis such as the droplet digital PCR (ddPCR). This technique is based on a droplet generating system, and BEAMing, involving the use of beads, emulsions, amplification, magnetics, and microfluidics digital PCR [46–49]. All such PCR-based techniques are very sensitive but have the disadvantage of generating false positive results when the target DNA has a low copy number. Next generation sequencing (NGS) is widely used to analyze large genomic regions on cfDNA and to detect, besides the known tumour related mutations, the less common but clinically relevant variations. However, NGS with its high degree of sensitivity may originate false positive results which require careful validation of all steps involved in the experimental procedures including blood collection, cfDNA extraction, library preparation, sequencing and variant callings [50].

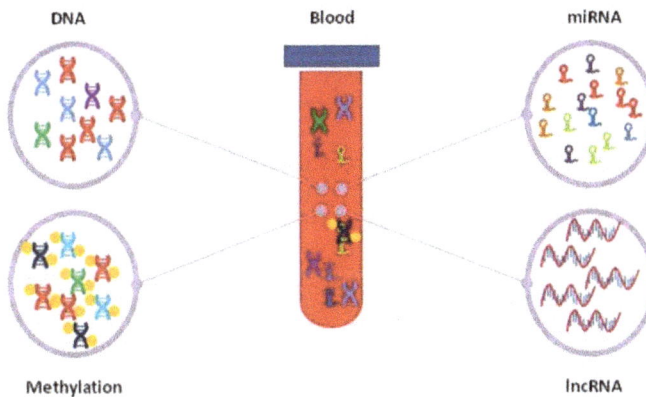

Figure 1. Schematic representation of the liquid biopsy as tool for the analysis of circulating DNAs and RNAs released from apoptotic or necrotic cancer cells into the blood stream.

Many studies have been published on the detection of tumour-specific somatic mutations in cfDNA of various cancer types [51]. A significant association has been reported between tumour stage and cancer-related genetic alterations, such as nucleotide changes in *TP53*, *KRAS*, *APC* and allelic imbalances, in the blood of patients affected by breast, ovarian, pancreatic, colorectal cancer and oral carcinoma as well as HCC [52–54]. Cancer driver mutations in TP53 and CTNNB1 genes as well as in the TERT promoter region have been frequently identified in tumour tissues of HCC patients [55,56]. These mutations have been also detected in the peripheral blood of liver cancer patients. Particularly, Huang et al. analyzed the mutational profile of TP53 (c.747 G > T), CTNNB1 (c.121A > G, c.133 T > C), and TERT promoter (−124 C > T) in 48 HCC cases by digital droplet PCR assay and found that 56.3% of patients had at least one mutation in cfDNA and 22.2% had concordant mutations in tumour DNA and cfDNA [57]. Liao et al. investigated the mutational profile of these three genes in a cohort of Chinese HCC patients and identified TERT, CTNNB1 and TP53 mutations in 4.9%, 9.8% and 4.9%, respectively, of serum samples [26]. Interestingly, one patient had the CTNNB1 mutation (c.122 C > T) in cfDNA but not in the primary tumour DNA, suggesting that circulating DNA fragments originated from different tumour nodules with heterogeneous DNA alterations [26]. However, discordant mutations between the DNA from primary tumour and cfDNA could also indicate the occurrence of false positive results generated by highly sensitive techniques and repeated experiments are needed to rule out such a possibility.

Dietary exposure to aflatoxin B1 (AFB1) in Asia and Africa, in association with HBV infection, has shown to increase the risk of HCC. The AFB1-related HCC patients frequently have distinctive mutations in TP53 gene, such as the G to T transversion at codon 249 causing the arginine substitution to serine (R249S) [58]. Jiao et al. identified the TP53 R249S mutation in 7.3% of HCC from Hispanic patients living in South Texas but not among 218 HCC non-Hispanic patients and not in 96 subjects with advanced fibrosis or cirrhosis living in the same region, suggesting that AFB-1 exposure may have occurred only in the Hispanic population [59]. They observed that patients with TP53 R249S mutations were significantly younger and had a lower overall survival. In Gambia, a country with high exposure to AFB-1, the TP53 R249S mutation has been identified in 35% of HCC biopsies and in 42% of plasma samples from HCC patients with a concordance of 88.5% between tumour tissues and matched plasma [60]. Moreover, Huang et al. [61] studied the intra tumour genetic heterogeneity in relation to the type of mutations identified in cfDNA fragments by analyzing a large panel of mutations in HCC driver genes, comprising TP53, CTNNB1, PIK3CA and ARID1A. They observed that cfDNA might provide a higher genome profiling potential than a single tumour specimen using highly sensitive deep sequencing technology [61]. More recently, Cohen et al. developed a blood assay able to diagnose the most common mutations in eight cancer types, including HCC, through the analysis of circulating proteins, such as CA19-9, HGF, OPN, TIMP-1, CA-125, CEA, MPO and PRL as well as of genetic alterations in cfDNA, such as the mutations in TP53, CTNNB1, CDKN2A, PTEN and KRAS genes [62]. This a combined test, based on Luminex bead immunoassay technology, showed 100% sensitivity for the detection of cancer lesions in the early stages [62]. Cai et al. performed a whole exome sequencing analysis of DNA extracted from paired biopsies and plasma samples of four HCC patients and showed that 96.9% of the tissue mutations could be also detected in cfDNA [63]. Such results strongly suggest that the analysis of cfDNA could overcome tumour heterogeneity with uneven distribution of mutations in different nodules and could allow rapid evaluation of therapeutic responses in the longitudinal monitoring of treated patients [63]. Furthermore, they found that the valine-to-methione substitution at codon 174 in Hck tyrosine kinase, a recurrent metastasis related mutation, could promote the migration and invasion of HCC cells [63].

High levels of HBV DNA in the blood serum have shown to be a strong risk factor for HCC onset. Chen et al. observed that an elevated HBV DNA level ($\geq$10,000 copies/mL) in the serum is a predictor of HCC independently from HBeAg, alanine aminotransferase level and liver cirrhosis [64]. Moreover, circulating HBV DNA has been suggested to be an early indicator of the success or failure of transarterial chemoembolisation [65].

3. DNA Methylation

DNA methylation is one of the most common epigenetic mechanisms used by the cells to control gene expression. It consists of the addition of methyl groups at CpG dinucleotides which are concentrated at specific clusters defined as CpG islands [66]. DNA methylation is usually a repressive mechanism causing specific gene silencing and allele inactivation of the X-chromosome. Aberrant methylation of normally unmethylated 5'-CpG-rich regions in cancer cells leads to the repression of several genes coding for factors involved in DNA damage repair, cell cycle regulation and apoptosis [67]. HBV infection has shown to affect the methylation of several genes including Ras association domain family 1 isoform A (RASSF1A), Glutathione S-Transferase Pi 1 (GSTP1), Cyclin Dependent Kinase Inhibitor 2A (p16[INK4A]), E-cadherin (CDH1) and Cyclin Dependent Kinase Inhibitor 1A (p21[WAF1/CIP1]) genes, while HCV infection has been associated with aberrant methylation of Adenomatous Polyposis Coli (APC), Suppressor of Cytokine Signaling 1 (SOCS-1), Growth Arrest and DNA Damage Inducible Beta (Gadd45β), O-6-Methylguanine-DNA Methyltransferase (MGMT) and Signal Transducer and Activator of Transcription 1 (STAT1) genes [68,69]. Hypermethylation of the RASSF1A gene is frequently observed in HCC [27,33,67]. Chan et al. found the RASSF1A gene hyper methylated in 93% of the sera of HBV-related HCC patients and in 58% of the sera of HBV chronic infected patients suggesting that RASSF1A hypermethylation could represent an early event in HCC pathogenesis [31]. Other studies reported that patients with high RASSF1A methylation at diagnosis or one year after tumour resection show generally poor disease-free survival, suggesting that RASSF1A methylation could be a good cancer prognostic marker [27,32,33]. Conversely, Hui-Chen et al. failed to find RASSF1A gene methylation in plasma of Taiwanese HCC patients although it was hypermethylated in tumour biopsies [70]. Dong et al. reported that several genes, such as RASSF1A, APC, Blood Vessel Epicardial Substance (BVES), Homeobox A9 (HOXA9), GSTP1, and Tissue Inhibitor of Metallopeptidase Inhibitor 3 (TIMP3), were hypermethylated in cancer biopsies of 343 HCC patients but only RASSF1A, BVES, and HOXA9 gene promoters were found significantly hypermethylated also in the sera of these patients [17]. In addition, in this study, the sensitivity of RASSF1A hypermethylation in the serum was higher than AFP ($\geq$20 ng/L) in distinguishing HCC from HBV chronic infected patients [17].

The promoter region of GSTP1, encoding for Glutathione S-transferase P1, has been found to be hyper methylated in about 50% of cancer tissues including HCC [71,72]. The aberrant methylation of GSTP1 has been shown to be associated with HCC progression [1], and to be more frequent in tumours characterized by capsular invasion and metastasis [73]. Several studies suggested GSTP1 methylation as a diagnostic marker for HCC reporting a sensitivity of 50–75% and a specificity of 70–91% with a performance superior to that of APC or RASSF1 genes [28]. The meta-analysis conducted by Liu et al. analyzing the methylation status of GSTP1 in 646 HCC tissues, APC in 592, and SOCS1 in 512 HCC tissues showed a strong correlation between the hypermethylation of such genes and the risk of HCC and suggested such epigenetic alterations as promising biomarkers for HCC development [74]. Huang et al. analyzed the methylation status of GSTP1, RASSF1A, APC and Secreted Frizzled Related Protein 1 (SFRP1) genes in plasma samples of 72 patients with HCC and 37 subjects with benign liver diseases showing that RASSF1A methylation was positively correlated with tumour size, while GSTP1 methylation was associated with elevated AFP levels in the serum, and SFRP1 methylation was more common in females [27]. The authors also found that hypermethylation of all these genes had a sensitivity of 84.7% in the detection of HCC [27]. Wang et al. reported that the methylation status of GSTP1 contributes to hepatic carcinogenesis since this gene has been found hypermethylated in the serum of 50% of HCC patients and in 37.5% of cirrhotic patients [29].

Other hypermethylated genes detected in the plasma of HCC patients are CDKN2A, which encodes for p16, an inhibitor of cyclin D-dependent kinases, and SOCS3, which encodes for the cytokine signaling 3 suppressor [30,34]. Moreover, Han et al. found that G Protein-Coupled Bile Acid Receptor 1 (TGR5), a membrane-bound receptor with a crucial role in regulating bile homeostasis and glucose metabolism, is aberrantly methylated in HCC and could have a diagnostic value of AFP in the discrimination of HCC from HBV chronic infected patients [35]. TGR5 acts as tumour suppressor gene, in fact its activation greatly inhibits the proliferation and migration of human liver cancer cells in vitro while the deficiency of TGR5 enhances chemical-induced liver carcinogenesis [35]. Recently, other genes have been found hypermethylated in HCC, such as AKR1B1, GRASP, MAP9, NXPE3, RSPH9, SPINT2, STEAP4, ZNF154, VIM and FBLN1 genes [75,76]. On the other hand, an elevated level of hypomethylated LINE1 Type Transposase Domain Containing 1 (LINE-1) in the serum has been associated with tumour progression, invasiveness and poor prognosis in HCC patients [77–79]. Liu et al. found that LINE-1 was hypomethylated in 66.7% of sera from HCC patients and was associated with HBsAg positivity, tumour size, AFP levels and poor survival [32]. Importantly, measurement of LINE-1 hypomethylation and RASSF1A promoter hypermethylation was found to be significantly correlated with early recurrence and poor prognosis in HCC patients after curative resection [32].

4. Microsatellite Instability

Microsatellites are short, highly repeated DNA sequences commonly present in the eukaryotic genomes [80]. Loss and length alteration of microsatellite regions are frequent events in the neoplastic process, suggesting their possible employment in the tumour diagnosis. The comparative genomic hybridization (CGH) technique has enabled the study of some microsatellite alterations in HCC genomes such as those affecting chromosome 8p, 17p and 19p deletions, which might cause HCC metastasis [81]. Moreover, two microsatellite markers located on the chromosome 8p, namely D8S258 and D8S264, have been found to be associated with increased cfDNA levels and involved in HCC progression, metastasis and reduced survival [82]. Pang et al. observed microsatellite instability and loss of heterozygosity of D8S277, D8S298, and D8S1771 located on chromosome 8p in the plasma DNA of HCC patients [83].

The analysis of 109 microsatellite markers, representing 24 chromosomal arms, in 21 cases of HCC, six cholangiocarcinoma and 27 chronic hepatitis or cirrhosis cases performed by Chang et al. showed at least one loss of heterozygosity in the cfDNA of approximately 76% of HCC patients. None of the cholangiocarcinoma patients exhibited loss of heterozygosity, suggesting that microsatellite markers might be appropriate for differential diagnosis of primary liver cancers [84]. Interestingly, 71.4% of HCC patients with AFP levels below 20 ng/mL showed loss of heterozygosity in the microsatellite regions, suggesting that this factor is an early marker of tumour development [84].

5. Circulating MicroRNAs

MiRNAs are short non-coding RNAs which regulate gene expression through their binding to the 3′UTR of mRNAs and consequent degradation or translational repression of targeted gene transcripts [85]. Deregulation of miRNAs levels in the cells plays an important role in tumour development.

Numerous miRNAs have shown to be associated with HCC on the basis of their differential expression in tumour versus non-tumour liver tissues such as the miR-122, miR-200a, miR-21, miR-223, let-7f, and miR-155 [85]. The role of circulating miR-122 and let-7 in the early diagnosis of HCC was suggested by the observation that their levels in the sera of HBV positive patients with dysplastic nodules and of early stage HCC patients had a sensitivity comparable to AFP testing [57]. Moreover, the hyper expression of let-7f in the serum has been shown to correlate with tumour size above 5 cm in diameter and with early recurrence [86]. miR223-3p and miR-125b-5p also were evaluated as good biomarkers in HBV-positive HCC [87]. Zheng et al. analyzed the serum levels of miR-125-5p in 120 patients with HCC, 91 with chronic HBV and 164 healthy controls, observing increased expression in liver fibrosis but not in HCC. Low serum levels of miR-125a-5p in HCC patients were correlated with a poor prognosis [88].

miR-122 has been shown to have a major role in HCV-related HCC. Zekri et al, using a panel of miR-122, miR-885-5p, and miR-29b in association with AFP testing, obtained a high diagnostic accuracy for early detection of HCC in a normal population, while using a panel of miR-122, miR-885-5p, miR-221, and miR-22 with AFP, obtained a high diagnostic accuracy for early detection of HCC in cirrhotic patients [89]. In addition, Qu et al., testing for miR-143 and miR-215 in association with AFP, showed a good efficiency in HCC diagnosis [90]. Okajima et al. analyzed the expression of four oncogenic miRNAs, namely miR-151, miR-155, miR-191 and miR-224, in the plasma of 107 HCC patients and 75 healthy volunteers. They observed that miR-224 was highly expressed in HCC tissues and plasma, but the levels decreased significantly following surgery, suggesting that miR-224 reflects tumour dynamic [91]. Similarly, miR-500 has been found to be largely expressed in sera of HCC patients and decreased to normal levels after surgery [92]. The expression profiles of miR-21 showed contrasting results. Ge et al. and Zhuang et al. observed down-regulation of miR-21 in HCC [86,93], while Zhou et al. and Amr et al. found hyper expression of miR-21 in HCC patients [94,95]. Recently, Ding et al. performed a meta-analysis including 24 studies and concluded that the high expression levels of miR-21, as well as miR-122 and miR-199, are highly specific for the diagnosis of HCC [96]. Despite its expression levels, miR-21 has been found to be involved in tumour cell migration, invasion and in metastasis [94,95]. Other miRNAs have been found to be associated with the development of metastasis, including miR-182, which is able to down-regulate metastasis suppressor 1 [97], and miR-331–3P, targeting the PH domain and leucine-rich repeat protein phosphatase [98].

miR-16 is down-regulated in the serum of HCC patients, and the low expression is correlated with some clinical features such as platelets, prothrombin time and bilirubin [86]. miR-30e and miR-223 have also been found at significantly lower levels in the sera of HCC patients compared to chronic liver diseases patients and healthy volunteers [99]. In addition, miR-26a and miR-101 are deregulated in the serum of HCC patients and could be used as biomarkers in combination with AFP testing to obtain a better sensitivity than AFP alone [93]. Yin et al. found that miR-199a-3p have high specificity and good predictive value in the diagnosis of early-stage alcohol-related HCC cases [100]. Zhan et al. found that patients with high levels of circulating miR-210 are resistant to trans-arterial chemoembolization treatment and have generally poor survival [101]. The levels of circulating miR-106b showed high sensitivity and specificity in differentiating HCC patients from chronic liver diseases or healthy subjects, denoting its clinical relevance [12].

6. Long Non-Coding RNA

Long non-coding RNAs (lncRNA) have been defined as transcripts longer than 200 nucleotides that are not translated into proteins and are largely expressed in various tissues [102]. They are also involved in multiple tumour processes including proliferation, apoptosis, invasion and metastasis through chromatin remodelling, epigenetic modifications, and gene regulation. Many previous studies showed that lncRNAs might be used as biomarkers in cancers [11]. Among these, the long intergenic non-protein coding RNA 974 (Linc00974) has been shown to be increased in the serum of HCC patients in comparison to the cytokeratin 19 fragment (CYFRA 21-1) and is useful as a tumour marker to improve the prognosis of HCC patients [103]. In vitro studies showed that Linc00974 causes proliferation and metastasis of HCC cells by interacting with keratin 19 (KRT19) [103]. In addition, the overexpression of lncRNA SPRY4-IT1 has been shown to promote tumour cell proliferation and invasion through the activation of the histone-lysine N-methyltransferase enzyme EZH2 [104]. Accordingly, Jing et al. observed that SPRY4-IT1 levels were significantly upregulated in HCC biopsies compared to the adjacent non-tumour tissues and that the amount of SPRY4-IT1 was significantly higher in the plasma collected in pre-surgery compared to that withdrawn in post-surgery [105]. The lncRNA MALAT1 has been demonstrated to regulate Zinc finger E-box-binding homeobox 1 (ZEB1) expression, promoting HCC development [106]. The evaluation of MALAT1 in peripheral blood and HCC tissues showed that there was a progressive and significant increase of MALAT1 levels in the plasma of patients with increasing severity of disease. On the other hand, plasma MALAT1 levels were significantly low in HCC patients with hepatitis B infection [106]. The circulating lncRNA-CTBP has been shown to have high sensitivity and specificity for discriminating HCC from healthy controls and from cirrhotic patients [107]. Weidong et al. identified three circulating lncRNAs, LINC00152, RP11-160H22.5 and XLOC014172, which, combined with the dosage of AFP, could be potential biomarkers of HCC development both in cirrhotic patients and healthy subjects [108].

7. Extracellular Vesicles

Extracellular vesicles are membrane-derived structures, released by cells into their microenvironment, which are classified into exosomes, microvesicles and apoptotic bodies, based on their biogenesis, size, and membrane markers [109]. Exosomes are the smallest subtype, with a diameter of 100–150 nm, and are formed by the fusion of multivesicular bodies and the plasma membrane [14]. Microvesicles have a larger diameter, approximately of 100–1000 nm, and derive from the cell membrane. Apoptotic bodies have the largest diameter, ranging from 1 to 5 μm, and are formed by the aggregation of apoptotic cells [14,110]. Secretion of extracellular vesicles in the body fluids is a common mechanism of cell homeostasis, thus the vesicles content can reflect the disease-associated cellular changes [13]. An enhancement of extracellular vesicle secretion is frequently observed in the serum of patients affected by alcoholic liver disease or by early stage fibrosis associated with chronic HBV or HCV infection. One of the molecules highly enriched within the extracellular vesicles released by HCC cells is the lncRNA TUC339 [111]. Moreover, cirrhotic patients with chronic HBV or HCV infection have an increased amount of Annexin V+, EpCAM+, ASGPR1+ and CD133+ microvesicles [112]. The number CD4+ and CD8+ microvesicles has also been found to increase in patients with liver diseases due to chronic inflammation and elevated number of T-cells in the injured liver [113]. Since multiple diseases are associated with the activation of inflammatory cells, the quantification of inflammatory cell-derived extracellular vesicles is not specific to liver pathology. However, the detection of asialoglycoprotein receptor 1 (ASGPR1), a hepatocyte-specific receptor, of EpCAM/CD133, markers of liver progenitor cells, and of cytokeratin-18 (CK18), a marker of hepatocytes and cholangiocytes, could help to define the hepatic origin of such vesicles [114].

Exosomes contain a wide range of biological molecules, including proteins, lipids and nucleic acids, which are markers of tumour onset and progression [19]. Tumour cells release numerous exosomes that are involved in intercellular communication, angiogenesis, metastasis, drug and radiotherapy resistance [115]. Kogure et al. identified 134 different types of miRNAs in Hep3B cell line-derived exosomes and found 55 miRNAs were over expressed more than 4-fold in the exosomes compared with the donor cells [116]. The expression levels of 25 of these miRNAs were increased up to 166-fold, 30 miRNAs were decreased up to 113-fold and importantly, 11 miRNAs were only detected in exosomes [116]. Wei et al. identified nine miRNAs differentially expressed in the SMMC-7721 liver cancer cell line expressing VPS4, a protein involved in endosomal transport, and derived exosomes. Particularly, six tumour suppressor miRNAs (miR-122-5p, miR-33a-5p, miR-34a-5p, miR-193a-3p, miR-16-5p, and miR-29b-3p) were significantly up-regulated in exosomes secreted by SMMC-7721 expressing Vps4 versus those produced by SMMC-7721 negative forVps4 [117].

Exosomes have shown to vehicle miRNAs into cells and to alter biological functions by targeting specific genes. Lou et al. found that miR-122 could be transported to HCC cells via exosomes and could regulate the target genes resulting in the improvement of HCC cell sensitivity to chemotherapeutic drugs [118]. Exosomal miR-718 has shown to regulate the homeobox B8 gene expression and to inhibit the differentiation of liver HCC cells [119]. Patients with low numbers of exosomes positive for miR-718 in the serum showed higher probability of tumour recurrence after liver transplantation [119]. Moreover exosomal miRNA content could be useful for the differential diagnosis of liver diseases. In fact, the number of exosomes containing miR-18a, miR-221, miR-222 and miR-224 in the serum of patients with HCC has been found to be significantly higher than that in patients with hepatitis and cirrhosis, whereas the presence of miR-101, miR-106b, miR-122 and miR-195 was found to be significantly reduced in HCC [120]. Shi et al. observed reduced levels of exosomal miR-638 in the serum of HCC patients and a negative association with tumour size, vascular infiltration, TNM stage and poor prognosis [121].

Sohn et al. observed that levels of exosomal miR-18a, miR-221, miR-222 and miR-224 in the serum were significantly higher in patients with HCC than those with chronic hepatitis B or with cirrhosis [120].

Wang et al. [122] analyzed the expression level of exosomal miR-21 in the serum and found significantly higher levels in patients with HCC than in those with chronic hepatitis or healthy volunteers. High levels of miR-21 correlated with cirrhosis and advanced tumour stages. Interestingly, they found high levels of miR-21 both in sera and in exosomes; however, exosomal miR-21 expression showed better sensitivity compared to the circulating free molecules [122]. Exosomal miR-665 levels have been found to be significantly over expressed in tumours with large size (>5 cm), local invasion, advanced clinical stage (stage III/IV) and reduced survival [123]. The expression of miR-939, miR-595 and miR-519d was shown to differentiate cirrhotic patients with and without HCC while miR-939 and miR-595 have been shown to be independent predicting factors for HCC [124].

Exosomal miRNAs are emerging as mediators of the interaction between mast cells and tumour cells. Xiong et al. observed that mast cells are able to block HCC cell metastasis by inhibiting the

ERK1/2 pathway through the transfer of the exosomal miRNAs into HCC cells, thus providing new insights for the biological therapy of HCV-related HCC [125].

Exosomes have the potential to be employed in target therapy. In fact, the transfer of miR-142 and miR-223 from human macrophages to liver cancer cells by exosomes has been shown to inhibit the proliferation of tumour cells [126]. Moreover, Ma et al. showed that bone marrow-derived mesenchymal stem cells showed significant anti-tumour activity after their sensitization with HCC cell-derived exosomes and inhibited the proliferation of HCC cells, suggesting that sensitization with cancer cell-derived exosomes may be a novel therapeutic strategy [1].

8. Conclusions

Several genetic and epigenetic alterations have shown to contribute to tumour development and progression. During neoplastic process, tumour heterogeneity progressively increases, making it extremely difficult to obtain good response to therapeutic treatments. Early diagnosis and dynamic tumour monitoring represent crucial factors to improve the clinical outcome of malignant tumours including HCC. cfDNA, circulating miRNA and epigenetic alterations are a good source of information for tumour diagnosis. Although HCC driver mutations, such as those in TERT promoter, CTNNB1 and TP53 genes, have been widely observed in tissue biopsies, they have been rarely found in cfDNA probably due to the low fraction of circulating mutated molecules or to the lack in sensitivity of most methodologies. Pre-analytical parameters, such as blood storage and processing, also affect cfDNA integrity and recovery yield [36].

Recently, the development of new technologies such as ddPCR, able to detect one mutant copy in a background of 20,000 wild type molecules, is opening new perspectives in the detection of mutant cfDNA [127,128]. The ddPCR method has wide employment also in miRNA detection since it allows absolute miRNA quantification and is not affected by variations caused by samples and PCR amplification efficiency [127].

Some molecular alterations, such as miR-122 and let-7 expression or RASSF1A hypermethylation, provided good diagnostic results when used alone or in combination with AFP dosage [17,57,89]. Conversely, miR-21 levels [86,93–95] and cfDNA integrity [12,42,43,57] show contrasting trends and underlie the need for further investigations.

Although the number of studies evaluating biomarkers in liquid biopsies as diagnostic tools for cancer detection is progressively increasing, very few of them have demonstrated solid diagnostic performance. The Early Detection Research Network, administered by the Cancer Biomarkers Research Group in the Division of Cancer Prevention of the US National Cancer Institute, proposed that the development of biomarkers for cancer diagnosis must undergo five phases, but most of the studies are still in the early phases [129].

Standardization among different laboratories in collecting, storage and analytic methods is a key factor to ensure consistency in clinical application. All the genetic and epigenetic alterations proposed as good tumour markers by the studies described above need further analyses on larger cohorts in order to validate them as HCC diagnostic and prognostic biomarkers. It is likely that in the future some of these biomarkers will be employed, alone or in combination with other already established assays (i.e., AFP), to improve the accuracy in the diagnosis of the medical practice.

Acknowledgments: Francesca Pezzuto is the recipient of a research fellowship awarded by FIRE/AISF ONLUS (Fondazione Italiana per la Ricerca in Epatologia) http://www.fondazionefegato.it/.

Author Contributions: Francesca Pezzuto, Luigi Buonaguro, Franco Maria Buonaguro and Maria Lina Tornesello conceived and wrote the review article.

Conflicts of Interest: The authors have declared no conflicts of interest.

References

1. Ferlay, J.; Soerjomataram, I.; Dikshit, R.; Eser, S.; Mathers, C.; Rebelo, M.; Parkin, D.M.; Forman, D.; Bray, F. Cancer incidence and mortality worldwide: Sources, methods and major patterns in GLOBOCAN 2012. *Int. J. Cancer* **2015**, *136*, E359–E386. [CrossRef] [PubMed]
2. Perz, J.F.; Armstrong, G.L.; Farrington, L.A.; Hutin, Y.J.; Bell, B.P. The contributions of hepatitis B virus and hepatitis C virus infections to cirrhosis and primary liver cancer worldwide. *J. Hepatol.* **2006**, *45*, 529–538. [CrossRef] [PubMed]
3. Okuda, K.; Nakanuma, Y.; Miyazaki, M. Cholangiocarcinoma: Recent progress. Part 1: Epidemiology and etiology. *J. Gastroenterol. Hepatol.* **2002**, *17*, 1049–1055. [CrossRef] [PubMed]
4. Bosetti, C.; Turati, F.; La, V.C. Hepatocellular carcinoma epidemiology. *Best Pract. Res. Clin. Gastroenterol.* **2014**, *28*, 753–770. [CrossRef] [PubMed]
5. Maucort-Boulch, D.; de Martel, C.; Franceschi, S.; Plummer, M. Fraction and incidence of liver cancer attributable to hepatitis B and C viruses worldwide. *Int. J. Cancer* **2018**, *10*. [CrossRef] [PubMed]
6. El-Serag, H.B. Epidemiology of viral hepatitis and hepatocellular carcinoma. *Gastroenterology* **2012**, *142*, 1264–1273. [CrossRef] [PubMed]
7. Tang, A.; Cruite, I.; Mitchell, D.G.; Sirlin, C.B. Hepatocellular carcinoma imaging systems: Why they exist, how they have evolved, and how they differ. *Abdom. Radiol.* **2018**, *43*, 3–12. [CrossRef] [PubMed]
8. Bolondi, L.; Cillo, U.; Colombo, M.; Craxi, A.; Farinati, F.; Giannini, E.G.; Golfieri, R.; Levrero, M.; Pinna, A.D.; Piscaglia, F.; et al. Position paper of the Italian Association for the Study of the Liver (AISF): The multidisciplinary clinical approach to hepatocellular carcinoma. *Dig. Liver Dis.* **2013**, *45*, 712–723. [CrossRef] [PubMed]
9. Gamil, M.; Alboraie, M.; El-Sayed, M.; Elsharkawy, A.; Asem, N.; Elbaz, T.; Mohey, M.; Abbas, B.; Mehrez, M.; Esmat, G. Novel scores combining AFP with non-invasive markers for prediction of liver fibrosis in chronic hepatitis C patients. *J. Med. Virol.* **2018**, *10*. [CrossRef] [PubMed]
10. Attwa, M.H.; El-Etreby, S.A. Guide for diagnosis and treatment of hepatocellular carcinoma. *World J. Hepatol.* **2015**, *7*, 1632–1651. [CrossRef] [PubMed]
11. Tang, J.C.; Feng, Y.L.; Guo, T.; Xie, A.Y.; Cai, X.J. Circulating tumor DNA in hepatocellular carcinoma: Trends and challenges. *Cell Biosci.* **2016**, *6*, 32. [CrossRef] [PubMed]
12. Jiang, L.; Li, X.; Cheng, Q.; Zhang, B.H. Plasma microRNA might as a potential biomarker for hepatocellular carcinoma and chronic liver disease screening. *Tumour. Biol.* **2015**, *36*, 7167–7174. [CrossRef] [PubMed]
13. Raposo, G.; Stoorvogel, W. Extracellular vesicles: Exosomes, microvesicles, and friends. *J. Cell Biol.* **2013**, *200*, 373–383. [CrossRef] [PubMed]
14. Thery, C.; Zitvogel, L.; Amigorena, S. Exosomes: Composition, biogenesis and function. *Nat. Rev. Immunol.* **2002**, *2*, 569–579. [CrossRef] [PubMed]
15. Mandel, P.; Metais, P. Les acides nucléiques du plasma sanguin chez l'homme. *C. R. Seances Soc. Biol. Fil.* **1948**, *142*, 241–243. [PubMed]
16. Leon, S.A.; Shapiro, B.; Sklaroff, D.M.; Yaros, M.J. Free DNA in the serum of cancer patients and the effect of therapy. *Cancer Res.* **1977**, *37*, 646–650. [PubMed]
17. Dong, X.; Hou, Q.; Chen, Y.; Wang, X. Diagnostic Value of the Methylation of Multiple Gene Promoters in Serum in Hepatitis B Virus-Related Hepatocellular Carcinoma. *Dis. Markers* **2017**, *2017*. [CrossRef] [PubMed]
18. Sozzi, G.; Conte, D.; Leon, M.; Ciricione, R.; Roz, L.; Ratcliffe, C.; Roz, E.; Cirenei, N.; Bellomi, M.; Pelosi, G.; et al. Quantification of free circulating DNA as a diagnostic marker in lung cancer. *J. Clin. Oncol.* **2003**, *21*, 3902–3908. [CrossRef] [PubMed]
19. Yin, C.Q.; Yuan, C.H.; Qu, Z.; Guan, Q.; Chen, H.; Wang, F.B. Liquid Biopsy of Hepatocellular Carcinoma: Circulating Tumor-Derived Biomarkers. *Dis. Markers* **2016**, *2016*. [CrossRef] [PubMed]
20. Breitbach, S.; Tug, S.; Simon, P. Circulating cell-free DNA: An up-coming molecular marker in exercise physiology. *Sports Med.* **2012**, *42*, 565–586. [CrossRef] [PubMed]
21. Snyder, M.W.; Kircher, M.; Hill, A.J.; Daza, R.M.; Shendure, J. Cell-free DNA Comprises an In Vivo Nucleosome Footprint that Informs Its Tissues-Of-Origin. *Cell* **2016**, *164*, 57–68. [CrossRef] [PubMed]
22. Piciocchi, M.; Cardin, R.; Vitale, A.; Vanin, V.; Giacomin, A.; Pozzan, C.; Maddalo, G.; Cillo, U.; Guido, M.; Farinati, F. Circulating free DNA in the progression of liver damage to hepatocellular carcinoma. *Hepatol. Int.* **2013**, *7*, 1050–1057. [CrossRef] [PubMed]

23. Truszewska, A.; Foroncewicz, B.; Paczek, L. The role and diagnostic value of cell-free DNA in systemic lupus erythematosus. *Clin. Exp. Rheumatol.* **2017**, *35*, 330–336. [PubMed]

24. Zhang, S.; Lu, X.; Shu, X.; Tian, X.; Yang, H.; Yang, W.; Zhang, Y.; Wang, G. Elevated plasma cfDNA may be associated with active lupus nephritis and partially attributed to abnormal regulation of neutrophil extracellular traps (NETs) in patients with systemic lupus erythematosus. *Intern. Med.* **2014**, *53*, 2763–2771. [CrossRef] [PubMed]

25. Huang, A.; Zhang, X.; Zhou, S.L.; Cao, Y.; Huang, X.W.; Fan, J.; Yang, X.R.; Zhou, J. Detecting Circulating Tumor DNA in Hepatocellular Carcinoma Patients Using Droplet Digital PCR Is Feasible and Reflects Intratumoral Heterogeneity. *J. Cancer* **2016**, *7*, 1907–1914. [CrossRef] [PubMed]

26. Liao, W.; Yang, H.; Xu, H.; Wang, Y.; Ge, P.; Ren, J.; Xu, W.; Lu, X.; Sang, X.; Zhong, S.; et al. Noninvasive detection of tumor-associated mutations from circulating cell-free DNA in hepatocellular carcinoma patients by targeted deep sequencing. *Oncotarget* **2016**, *7*, 40481–40490. [CrossRef] [PubMed]

27. Huang, Z.H.; Hu, Y.; Hua, D.; Wu, Y.Y.; Song, M.X.; Cheng, Z.H. Quantitative analysis of multiple methylated genes in plasma for the diagnosis and prognosis of hepatocellular carcinoma. *Exp. Mol. Pathol.* **2011**, *91*, 702–707. [CrossRef] [PubMed]

28. Huang, W.; Li, T.; Yang, W.; Chai, X.; Chen, K.; Wei, L.; Duan, S.; Li, B.; Qin, Y. Analysis of DNA methylation in plasma for monitoring hepatocarcinogenesis. *Genet. Test. Mol. Biomark.* **2015**, *19*, 295–302. [CrossRef] [PubMed]

29. Wang, J.; Qin, Y.; Li, B.; Sun, Z.; Yang, B. Detection of aberrant promoter methylation of GSTP1 in the tumor and serum of Chinese human primary hepatocellular carcinoma patients. *Clin. Biochem.* **2006**, *39*, 344–348. [CrossRef] [PubMed]

30. Wong, I.H.; Lo, Y.M.; Zhang, J.; Liew, C.T.; Ng, M.H.; Wong, N.; Lai, P.B.; Lau, W.Y.; Hjelm, N.M.; Johnson, P.J. Detection of aberrant p16 methylation in the plasma and serum of liver cancer patients. *Cancer Res.* **1999**, *59*, 71–73. [PubMed]

31. Chan, K.C.; Lai, P.B.; Mok, T.S.; Chan, H.L.; Ding, C.; Yeung, S.W.; Lo, Y.M. Quantitative analysis of circulating methylated DNA as a biomarker for hepatocellular carcinoma. *Clin. Chem.* **2008**, *54*, 1528–1536. [CrossRef] [PubMed]

32. Liu, Z.J.; Huang, Y.; Wei, L.; He, J.Y.; Liu, Q.Y.; Yu, X.Q.; Li, Z.L.; Zhang, J.; Li, B.; Sun, C.J.; et al. Combination of LINE-1 hypomethylation and RASSF1A promoter hypermethylation in serum DNA is a non-invasion prognostic biomarker for early recurrence of hepatocellular carcinoma after curative resection. *Neoplasma* **2017**, *64*, 795–802. [CrossRef] [PubMed]

33. Yeo, W.; Wong, N.; Wong, W.L.; Lai, P.B.; Zhong, S.; Johnson, P.J. High frequency of promoter hypermethylation of RASSF1A in tumor and plasma of patients with hepatocellular carcinoma. *Liver Int.* **2005**, *25*, 266–272. [CrossRef] [PubMed]

34. Wei, L.; Huang, Y.; Zhao, R.; Zhang, J.; Liu, Q.; Liang, W.; Ding, X.; Gao, B.; Li, B.; Sun, C.; et al. Detection of promoter methylation status of suppressor of cytokine signaling 3 (SOCS3) in tissue and plasma from Chinese patients with different hepatic diseases. *Clin. Exp. Med.* **2017**, *18*, 79–87. [CrossRef] [PubMed]

35. Han, L.Y.; Fan, Y.C.; Mu, N.N.; Gao, S.; Li, F.; Ji, X.F.; Dou, C.Y.; Wang, K. Aberrant DNA methylation of G-protein-coupled bile acid receptor Gpbar1 (TGR5) is a potential biomarker for hepatitis B Virus associated hepatocellular carcinoma. *Int. J. Med. Sci.* **2014**, *11*, 164–171. [CrossRef] [PubMed]

36. Elshimali, Y.I.; Khaddour, H.; Sarkissyan, M.; Wu, Y.; Vadgama, J.V. The clinical utilization of circulating cell free DNA (CCFDNA) in blood of cancer patients. *Int. J. Mol. Sci.* **2013**, *14*, 18925–18958. [CrossRef] [PubMed]

37. Page, K.; Guttery, D.S.; Zahra, N.; Primrose, L.; Elshaw, S.R.; Pringle, J.H.; Blighe, K.; Marchese, S.D.; Hills, A.; Woodley, L.; et al. Influence of plasma processing on recovery and analysis of circulating nucleic acids. *PLoS ONE* **2013**, *8*, e77963. [CrossRef] [PubMed]

38. Chen, K.; Zhang, H.; Zhang, L.N.; Ju, S.Q.; Qi, J.; Huang, D.F.; Li, F.; Wei, Q.; Zhang, J. Value of circulating cell-free DNA in diagnosis of hepatocelluar carcinoma. *World J. Gastroenterol.* **2013**, *19*, 3143–3149. [CrossRef] [PubMed]

39. Huang, A.; Zhang, X.; Zhou, S.L.; Cao, Y.; Huang, X.W.; Fan, J.; Yang, X.R.; Zhou, J. Plasma Circulating Cell-free DNA Integrity as a Promising Biomarker for Diagnosis and Surveillance in Patients with Hepatocellular Carcinoma. *J. Cancer* **2016**, *7*, 1798–1803. [CrossRef] [PubMed]

40. Andersen, R.F.; Spindler, K.L.; Brandslund, I.; Jakobsen, A.; Pallisgaard, N. Improved sensitivity of circulating tumor DNA measurement using short PCR amplicons. *Clin. Chim. Acta* **2015**, *439*, 97–101. [CrossRef] [PubMed]

41. Jiang, P.; Chan, C.W.; Chan, K.C.; Cheng, S.H.; Wong, J.; Wong, V.W.; Wong, G.L.; Chan, S.L.; Mok, T.S.; Chan, H.L.; et al. Lengthening and shortening of plasma DNA in hepatocellular carcinoma patients. *Proc. Natl. Acad. Sci. USA* **2015**, *112*, E1317–E1325. [CrossRef] [PubMed]

42. Wang, B.G.; Huang, H.Y.; Chen, Y.C.; Bristow, R.E.; Kassauei, K.; Cheng, C.C.; Roden, R.; Sokoll, L.J.; Chan, D.W.; Shih, I. Increased plasma DNA integrity in cancer patients. *Cancer Res.* **2003**, *63*, 3966–3968. [PubMed]

43. Chen, H.; Sun, L.Y.; Zheng, H.Q.; Zhang, Q.F.; Jin, X.M. Total serum DNA and DNA integrity: Diagnostic value in patients with hepatitis B virus-related hepatocellular carcinoma. *Pathology* **2012**, *44*, 318–324. [CrossRef] [PubMed]

44. Frattini, M.; Gallino, G.; Signoroni, S.; Balestra, D.; Lusa, L.; Battaglia, L.; Sozzi, G.; Bertario, L.; Leo, E.; Pilotti, S.; et al. Quantitative and qualitative characterization of plasma DNA identifies primary and recurrent colorectal cancer. *Cancer Lett.* **2008**, *263*, 170–181. [CrossRef] [PubMed]

45. Polivka, J., Jr.; Pesta, M.; Janku, F. Testing for oncogenic molecular aberrations in cell-free DNA-based liquid biopsies in the clinic: Are we there yet? *Expert Rev. Mol. Diagn.* **2015**, *15*, 1631–1644. [CrossRef] [PubMed]

46. Beaver, J.A.; Jelovac, D.; Balukrishna, S.; Cochran, R.; Croessmann, S.; Zabransky, D.J.; Wong, H.Y.; Toro, P.V.; Cidado, J.; Blair, B.G.; et al. Detection of cancer DNA in plasma of patients with early-stage breast cancer. *Clin. Cancer Res.* **2014**, *20*, 2643–2650. [CrossRef] [PubMed]

47. Higgins, M.J.; Jelovac, D.; Barnathan, E.; Blair, B.; Slater, S.; Powers, P.; Zorzi, J.; Jeter, S.C.; Oliver, G.R.; Fetting, J.; et al. Detection of tumor PIK3CA status in metastatic breast cancer using peripheral blood. *Clin. Cancer Res.* **2012**, *18*, 3462–3469. [CrossRef] [PubMed]

48. Dawson, S.J.; Tsui, D.W.; Murtaza, M.; Biggs, H.; Rueda, O.M.; Chin, S.F.; Dunning, M.J.; Gale, D.; Forshew, T.; Mahler-Araujo, B.; et al. Analysis of circulating tumor DNA to monitor metastatic breast cancer. *N. Engl. J. Med.* **2013**, *368*, 1199–1209. [CrossRef] [PubMed]

49. Yung, T.K.; Chan, K.C.; Mok, T.S.; Tong, J.; To, K.F.; Lo, Y.M. Single-molecule detection of epidermal growth factor receptor mutations in plasma by microfluidics digital PCR in non-small cell lung cancer patients. *Clin. Cancer Res.* **2009**, *15*, 2076–2084. [CrossRef] [PubMed]

50. Malapelle, U.; Pisapia, P.; Rocco, D.; Smeraglio, R.; di Spirito, M.; Bellevicine, C.; Troncone, G. Next generation sequencing techniques in liquid biopsy: Focus on non-small cell lung cancer patients. *Transl. Lung Cancer Res.* **2016**, *5*, 505–510. [CrossRef] [PubMed]

51. Lu, J.L.; Liang, Z.Y. Circulating free DNA in the era of precision oncology: Pre- and post-analytical concerns. *Chronic Dis. Transl. Med.* **2016**, *2*, 223–230. [CrossRef] [PubMed]

52. Pisapia, P.; Pepe, F.; Smeraglio, R.; Russo, M.; Rocco, D.; Sgariglia, R.; Nacchio, M.; De Luca, C.; Vigliar, E.; Bellevicine, C.; et al. Cell free DNA analysis by SiRe® next generation sequencing panel in non small cell lung cancer patients: Focus on basal setting. *J. Thorac. Dis.* **2017**, *9*, S1383–S1390. [CrossRef] [PubMed]

53. Lodrini, M.; Sprussel, A.; Astrahantseff, K.; Tiburtius, D.; Konschak, R.; Lode, H.N.; Fischer, M.; Keilholz, U.; Eggert, A.; Deubzer, H.E. Using droplet digital PCR to analyze MYCN and ALK copy number in plasma from patients with neuroblastoma. *Oncotarget* **2017**, *8*, 85234–85251. [CrossRef] [PubMed]

54. Garcia, J.; Dusserre, E.; Cheynet, V.; Bringuier, P.P.; Brengle-Pesce, K.; Wozny, A.S.; Rodriguez-Lafrasse, C.; Freyer, G.; Brevet, M.; Payen, L.; et al. Evaluation of pre-analytical conditions and comparison of the performance of several digital PCR assays for the detection of major EGFR mutations in circulating DNA from non-small cell lung cancers: The CIRCAN_0 study. *Oncotarget* **2017**, *8*, 87980–87996. [CrossRef] [PubMed]

55. Pezzuto, F.; Buonaguro, L.; Buonaguro, F.M.; Tornesello, M.L. Frequency and geographic distribution of TERT promoter mutations in primary hepatocellular carcinoma. *Infect. Agents Cancer* **2017**, *12*, 27. [CrossRef] [PubMed]

56. Tornesello, M.L.; Buonaguro, L.; Izzo, F.; Buonaguro, F.M. Molecular alterations in hepatocellular carcinoma associated with hepatitis B and hepatitis C infections. *Oncotarget* **2016**, *7*, 25087–25102. [CrossRef] [PubMed]

57. Hung, C.H.; Hu, T.H.; Lu, S.N.; Kuo, F.Y.; Chen, C.H.; Wang, J.H.; Huang, C.M.; Lee, C.M.; Lin, C.Y.; Yen, Y.H.; et al. Circulating microRNAs as biomarkers for diagnosis of early hepatocellular carcinoma associated with hepatitis B virus. *Int. J. Cancer* **2016**, *138*, 714–720. [CrossRef] [PubMed]

58. Tornesello, M.L.; Buonaguro, L.; Tatangelo, F.; Botti, G.; Izzo, F.; Buonaguro, F.M. Mutations in TP53, CTNNB1 and PIK3CA genes in hepatocellular carcinoma associated with hepatitis B and hepatitis C virus infections. *Genomics* **2013**, *102*, 74–83. [CrossRef] [PubMed]

59. Jiao, J.; Niu, W.; Wang, Y.; Baggerly, K.A.; Ye, Y.; Wu, X.; Davenport, D.; Almeda, J.L.; Betancourt-Garcia, M.M.; Forse, R.A.; et al. Prevalence of Aflatoxin-associated TP53R249S mutation in Hepatocellular Carcinoma in Hispanics in South Texas. *Cancer Prev. Res.* **2017**, *11*, 103–112. [CrossRef] [PubMed]

60. Szymanska, K.; Lesi, O.A.; Kirk, G.D.; Sam, O.; Taniere, P.; Scoazec, J.Y.; Mendy, M.; Friesen, M.D.; Whittle, H.; Montesano, R.; et al. Ser-249TP53 mutation in tumour and plasma DNA of hepatocellular carcinoma patients from a high incidence area in the Gambia, West Africa. *Int. J. Cancer* **2004**, *110*, 374–379. [CrossRef] [PubMed]

61. Huang, A.; Zhao, X.; Yang, X.R.; Li, F.Q.; Zhou, X.L.; Wu, K.; Zhang, X.; Sun, Q.M.; Cao, Y.; Zhu, H.M.; et al. Circumventing intratumoral heterogeneity to identify potential therapeutic targets in hepatocellular carcinoma. *J. Hepatol.* **2017**, *67*, 293–301. [CrossRef] [PubMed]

62. Cohen, J.D.; Li, L.; Wang, Y.; Thoburn, C.; Afsari, B.; Danilova, L.; Douville, C.; Javed, A.A.; Wong, F.; Mattox, A.; et al. Detection and localization of surgically resectable cancers with a multi-analyte blood test. *Science* **2018**. [CrossRef] [PubMed]

63. Cai, Z.X.; Chen, G.; Zeng, Y.Y.; Dong, X.Q.; Lin, M.J.; Huang, X.H.; Zhang, D.; Liu, X.L.; Liu, J.F. Circulating tumor DNA profiling reveals clonal evolution and real-time disease progression in advanced hepatocellular carcinoma. *Int. J. Cancer* **2017**, *141*, 977–985. [CrossRef] [PubMed]

64. Chen, C.J.; Yang, H.I.; Su, J.; Jen, C.L.; You, S.L.; Lu, S.N.; Huang, G.T.; Iloeje, U.H. Risk of hepatocellular carcinoma across a biological gradient of serum hepatitis B virus DNA level. *JAMA* **2006**, *295*, 65–73. [CrossRef] [PubMed]

65. Su, Y.W.; Huang, Y.W.; Chen, S.H.; Tzen, C.Y. Quantitative analysis of plasma HBV DNA for early evaluation of the response to transcatheter arterial embolization for HBV-related hepatocellular carcinoma. *World J. Gastroenterol.* **2005**, *11*, 6193–6196. [CrossRef] [PubMed]

66. Barros, S.P.; Offenbacher, S. Epigenetics: Connecting environment and genotype to phenotype and disease. *J. Dent. Res.* **2009**, *88*, 400–408. [CrossRef] [PubMed]

67. Schagdarsurengin, U.; Wilkens, L.; Steinemann, D.; Flemming, P.; Kreipe, H.H.; Pfeifer, G.P.; Schlegelberger, B.; Dammann, R. Frequent epigenetic inactivation of the RASSF1A gene in hepatocellular carcinoma. *Oncogene* **2003**, *22*, 1866–1871. [CrossRef] [PubMed]

68. Rongrui, L.; Na, H.; Zongfang, L.; Fanpu, J.; Shiwen, J. Epigenetic mechanism involved in the HBV/HCV-related hepatocellular carcinoma tumorigenesis. *Curr. Pharm. Des.* **2014**, *20*, 1715–1725. [CrossRef] [PubMed]

69. Iyer, P.; Zekri, A.R.; Hung, C.W.; Schiefelbein, E.; Ismail, K.; Hablas, A.; Seifeldin, I.A.; Soliman, A.S. Concordance of DNA methylation pattern in plasma and tumor DNA of Egyptian hepatocellular carcinoma patients. *Exp. Mol. Pathol.* **2010**, *88*, 107–111. [CrossRef] [PubMed]

70. Wu, H.C.; Yang, H.I.; Wang, Q.; Chen, C.J.; Santella, R.M. Plasma DNA methylation marker and hepatocellular carcinoma risk prediction model for the general population. *Carcinogenesis* **2017**, *38*, 1021–1028. [CrossRef] [PubMed]

71. Lambert, M.P.; Paliwal, A.; Vaissiere, T.; Chemin, I.; Zoulim, F.; Tommasino, M.; Hainaut, P.; Sylla, B.; Scoazec, J.Y.; Tost, J.; et al. Aberrant DNA methylation distinguishes hepatocellular carcinoma associated with HBV and HCV infection and alcohol intake. *J. Hepatol.* **2011**, *54*, 705–715. [CrossRef] [PubMed]

72. Gurioli, G.; Martignano, F.; Salvi, S.; Costantini, M.; Gunelli, R.; Casadio, V. GSTP1 methylation in cancer: A liquid biopsy biomarker? *Clin. Chem. Lab. Med.* **2018**. [CrossRef] [PubMed]

73. Qu, Z.; Jiang, Y.; Li, H.; Yu, D.C.; Ding, Y.T. Detecting abnormal methylation of tumor suppressor genes GSTP1, P16, RIZ1, and RASSF1A in hepatocellular carcinoma and its clinical significance. *Oncol. Lett.* **2015**, *10*, 2553–2558. [CrossRef] [PubMed]

74. Liu, M.; Cui, L.H.; Li, C.C.; Zhang, L. Association of APC, GSTP1 and SOCS1 promoter methylation with the risk of hepatocellular carcinoma: A meta-analysis. *Eur. J. Cancer Prev.* **2015**, *24*, 470–483. [CrossRef] [PubMed]

75. Yamada, N.; Yasui, K.; Dohi, O.; Gen, Y.; Tomie, A.; Kitaichi, T.; Iwai, N.; Mitsuyoshi, H.; Sumida, Y.; Moriguchi, M.; et al. Genome-wide DNA methylation analysis in hepatocellular carcinoma. *Oncol. Rep.* **2016**, *35*, 2228–2236. [CrossRef] [PubMed]

76. Holmila, R.; Sklias, A.; Muller, D.C.; Degli, E.D.; Guilloreau, P.; Mckay, J.; Sangrajrang, S.; Srivatanakul, P.; Hainaut, P.; Merle, P.; et al. Targeted deep sequencing of plasma circulating cell-free DNA reveals Vimentin and Fibulin 1 as potential epigenetic biomarkers for hepatocellular carcinoma. *PLoS ONE* **2017**, *12*, e0174265. [CrossRef] [PubMed]

77. Zhou, J.; Shi, Y.H.; Fan, J. Circulating cell-free nucleic acids: Promising biomarkers of hepatocellular carcinoma. *Semin. Oncol.* **2012**, *39*, 440–448. [CrossRef] [PubMed]

78. Ramzy, I.I.; Omran, D.A.; Hamad, O.; Shaker, O.; Abboud, A. Evaluation of serum LINE-1 hypomethylation as a prognostic marker for hepatocellular carcinoma. *Arab J. Gastroenterol.* **2011**, *12*, 139–142. [CrossRef] [PubMed]

79. Tangkijvanich, P.; Hourpai, N.; Rattanatanyong, P.; Wisedopas, N.; Mahachai, V.; Mutirangura, A. Serum LINE-1 hypomethylation as a potential prognostic marker for hepatocellular carcinoma. *Clin. Chim. Acta* **2007**, *379*, 127–133. [CrossRef] [PubMed]

80. Nawroz, H.; Koch, W.; Anker, P.; Stroun, M.; Sidransky, D. Microsatellite alterations in serum DNA of head and neck cancer patients. *Nat. Med.* **1996**, *2*, 1035–1037. [CrossRef] [PubMed]

81. Zhang, L.H.; Qin, L.X.; Ma, Z.C.; Ye, S.L.; Liu, Y.K.; Ye, Q.H.; Wu, X.; Huang, W.; Tang, Z.Y. Allelic imbalance regions on chromosomes 8p, 17p and 19p related to metastasis of hepatocellular carcinoma: Comparison between matched primary and metastatic lesions in 22 patients by genome-wide microsatellite analysis. *J. Cancer Res. Clin. Oncol.* **2003**, *129*, 279–286. [PubMed]

82. Ren, N.; Qin, L.X.; Tu, H.; Liu, Y.K.; Zhang, B.H.; Tang, Z.Y. The prognostic value of circulating plasma DNA level and its allelic imbalance on chromosome 8p in patients with hepatocellular carcinoma. *J. Cancer Res. Clin. Oncol.* **2006**, *132*, 399–407. [CrossRef] [PubMed]

83. Pang, J.Z.; Qin, L.X.; Ren, N.; Ye, Q.H.; Ying, W.D.; Liu, Y.K.; Tang, Z.Y. Microsatellite alterations of circulating DNA in the plasma of patients with hepatocellular carcinoma. *Zhonghua Yi Xue Za Zhi* **2006**, *86*, 1662–1665. [PubMed]

84. Chang, Y.C.; Ho, C.L.; Chen, H.H.; Chang, T.T.; Lai, W.W.; Dai, Y.C.; Lee, W.Y.; Chow, N.H. Molecular diagnosis of primary liver cancer by microsatellite DNA analysis in the serum. *Br. J. Cancer* **2002**, *87*, 1449–1453. [CrossRef] [PubMed]

85. Romano, G.; Veneziano, D.; Acunzo, M.; Croce, C.M. Small non-coding RNA and cancer. *Carcinogenesis* **2017**, *38*, 485–491. [CrossRef] [PubMed]

86. Ge, W.; Yu, D.C.; Li, Q.G.; Chen, X.; Zhang, C.Y.; Ding, Y.T. Expression of serum miR-16, let-7f, and miR-21 in patients with hepatocellular carcinoma and their clinical significances. *Clin. Lab.* **2014**, *60*, 427–434. [CrossRef] [PubMed]

87. Giray, B.G.; Emekdas, G.; Tezcan, S.; Ulger, M.; Serin, M.S.; Sezgin, O.; Altintas, E.; Tiftik, E.N. Profiles of serum microRNAs; miR-125b-5p and miR223-3p serve as novel biomarkers for HBV-positive hepatocellular carcinoma. *Mol. Biol. Rep.* **2014**, *41*, 4513–4519. [CrossRef] [PubMed]

88. Zheng, J.; Zhou, Z.; Xu, Z.; Li, G.; Dong, P.; Chen, Z.; Lin, D.; Chen, B.; Yu, F. Serum microRNA-125a-5p, a useful biomarker in liver diseases, correlates with disease progression. *Mol. Med. Rep.* **2015**, *12*, 1584–1590. [CrossRef] [PubMed]

89. Zekri, A.N.; Youssef, A.S.; El-Desouky, E.D.; Ahmed, O.S.; Lotfy, M.M.; Nassar, A.A.; Bahnassey, A.A. Serum microRNA panels as potential biomarkers for early detection of hepatocellular carcinoma on top of HCV infection. *Tumour. Biol.* **2016**, *37*, 12273–12286. [CrossRef] [PubMed]

90. Qu, K.Z.; Zhang, K.; Li, H.; Afdhal, N.H.; Albitar, M. Circulating microRNAs as biomarkers for hepatocellular carcinoma. *J. Clin. Gastroenterol.* **2011**, *45*, 355–360. [CrossRef] [PubMed]

91. Okajima, W.; Komatsu, S.; Ichikawa, D.; Miyamae, M.; Kawaguchi, T.; Hirajima, S.; Ohashi, T.; Imamura, T.; Kiuchi, J.; Arita, T.; et al. Circulating microRNA profiles in plasma: Identification of miR-224 as a novel diagnostic biomarker in hepatocellular carcinoma independent of hepatic function. *Oncotarget* **2016**, *7*, 53820–53836. [CrossRef] [PubMed]

92. Yamamoto, Y.; Kosaka, N.; Tanaka, M.; Koizumi, F.; Kanai, Y.; Mizutani, T.; Murakami, Y.; Kuroda, M.; Miyajima, A.; Kato, T.; et al. MicroRNA-500 as a potential diagnostic marker for hepatocellular carcinoma. *Biomarkers* **2009**, *14*, 529–538. [CrossRef] [PubMed]

93. Zhuang, C.; Jiang, W.; Huang, D.; Xu, L.; Yang, Q.; Zheng, L.; Wang, X.; Hu, L. Serum miR-21, miR-26a and miR-101 as potential biomarkers of hepatocellular carcinoma. *Clin. Res. Hepatol. Gastroenterol.* **2016**, *40*, 386–396. [CrossRef] [PubMed]

94. Zhou, L.; Yang, Z.X.; Song, W.J.; Li, Q.J.; Yang, F.; Wang, D.S.; Zhang, N.; Dou, K.F. MicroRNA-21 regulates the migration and invasion of a stem-like population in hepatocellular carcinoma. *Int. J. Oncol.* **2013**, *43*, 661–669. [CrossRef] [PubMed]

95. Amr, K.S.; Ezzat, W.M.; Elhosary, Y.A.; Hegazy, A.E.; Fahim, H.H.; Kamel, R.R. The potential role of miRNAs 21 and 199-a in early diagnosis of hepatocellular carcinoma. *Gene* **2016**, *575*, 66–70. [CrossRef] [PubMed]

96. Ding, Y.; Yan, J.L.; Fang, A.N.; Zhou, W.F.; Huang, L. Circulating miRNAs as novel diagnostic biomarkers in hepatocellular carcinoma detection: A meta-analysis based on 24 articles. *Oncotarget* **2017**, *8*, 66402–66413. [CrossRef] [PubMed]

97. Wang, J.; Li, J.; Shen, J.; Wang, C.; Yang, L.; Zhang, X. MicroRNA-182 downregulates metastasis suppressor 1 and contributes to metastasis of hepatocellular carcinoma. *BMC Cancer* **2012**, *12*, 227. [CrossRef] [PubMed]

98. Chang, R.M.; Yang, H.; Fang, F.; Xu, J.F.; Yang, L.Y. MicroRNA-331-3p promotes proliferation and metastasis of hepatocellular carcinoma by targeting PH domain and leucine-rich repeat protein phosphatase. *Hepatology* **2014**, *60*, 1251–1263. [CrossRef] [PubMed]

99. Bhattacharya, S.; Steele, R.; Shrivastava, S.; Chakraborty, S.; Di Bisceglie, A.M.; Ray, R.B. Serum miR-30e and miR-223 as Novel Noninvasive Biomarkers for Hepatocellular Carcinoma. *Am. J. Pathol.* **2016**, *186*, 242–247. [CrossRef] [PubMed]

100. Yin, J.; Hou, P.; Wu, Z.; Wang, T.; Nie, Y. Circulating miR-375 and miR-199a-3p as potential biomarkers for the diagnosis of hepatocellular carcinoma. *Tumour. Biol.* **2015**, *36*, 4501–4507. [CrossRef] [PubMed]

101. Zhan, M.; Li, Y.; Hu, B.; He, X.; Huang, J.; Zhao, Y.; Fu, S.; Lu, L. Serum microRNA-210 as a predictive biomarker for treatment response and prognosis in patients with hepatocellular carcinoma undergoing transarterial chemoembolization. *J. Vasc. Interv. Radiol.* **2014**, *25*, 1279–1287. [CrossRef] [PubMed]

102. Wong, C.M.; Tsang, F.H.; Ng, I.O. Non-coding RNAs in hepatocellular carcinoma: Molecular functions and pathological implications. *Nat. Rev. Gastroenterol. Hepatol.* **2018**, *15*, 137–151. [CrossRef] [PubMed]

103. Tang, J.; Zhuo, H.; Zhang, X.; Jiang, R.; Ji, J.; Deng, L.; Qian, X.; Zhang, F.; Sun, B. A novel biomarker Linc00974 interacting with KRT19 promotes proliferation and metastasis in hepatocellular carcinoma. *Cell Death Dis.* **2014**, *5*, e1549. [CrossRef] [PubMed]

104. Zhou, M.; Zhang, X.Y.; Yu, X. Overexpression of the long non-coding RNA SPRY4-IT1 promotes tumor cell proliferation and invasion by activating EZH2 in hepatocellular carcinoma. *Biomed. Pharmacother.* **2017**, *85*, 348–354. [CrossRef] [PubMed]

105. Jing, W.; Gao, S.; Zhu, M.; Luo, P.; Jing, X.; Chai, H.; Tu, J. Potential diagnostic value of lncRNA SPRY4-IT1 in hepatocellular carcinoma. *Oncol. Rep.* **2016**, *36*, 1085–1092. [CrossRef] [PubMed]

106. Konishi, H.; Ichikawa, D.; Yamamoto, Y.; Arita, T.; Shoda, K.; Hiramoto, H.; Hamada, J.; Itoh, H.; Fujita, Y.; Komatsu, S.; et al. Plasma level of metastasis-associated lung adenocarcinoma transcript 1 is associated with liver damage and predicts development of hepatocellular carcinoma. *Cancer Sci.* **2016**, *107*, 149–154. [CrossRef] [PubMed]

107. El-Tawdi, A.H.; Matboli, M.; Shehata, H.H.; Tash, F.; El-Khazragy, N.; Azazy, A.; Abdel-Rahman, O. Evaluation of Circulatory RNA-Based Biomarker Panel in Hepatocellular Carcinoma. *Mol. Diagn. Ther.* **2016**, *20*, 265–277. [CrossRef] [PubMed]

108. Yuan, W.; Sun, Y.; Liu, L.; Zhou, B.; Wang, S.; Gu, D. Circulating LncRNAs Serve as Diagnostic Markers for Hepatocellular Carcinoma. *Cell. Physiol. Biochem.* **2017**, *44*, 125–132. [CrossRef] [PubMed]

109. Lambrecht, J.; Verhulst, S.; Mannaerts, I.; Reynaert, H.; van Grunsven, L.A. Prospects in non-invasive assessment of liver fibrosis: Liquid biopsy as the future gold standard? *Biochim. Biophys. Acta* **2018**. [CrossRef] [PubMed]

110. Sato-Kuwabara, Y.; Melo, S.A.; Soares, F.A.; Calin, G.A. The fusion of two worlds: Non-coding RNAs and extracellular vesicles—Diagnostic and therapeutic implications (Review). *Int. J. Oncol.* **2015**, *46*, 17–27. [CrossRef] [PubMed]

111. Kogure, T.; Yan, I.K.; Lin, W.L.; Patel, T. Extracellular Vesicle-Mediated Transfer of a Novel Long Noncoding RNA TUC339: A Mechanism of Intercellular Signaling in Human Hepatocellular Cancer. *Genes Cancer* **2013**, *4*, 261–272. [CrossRef] [PubMed]

112. Julich-Haertel, H.; Urban, S.K.; Krawczyk, M.; Willms, A.; Jankowski, K.; Patkowski, W.; Kruk, B.; Krasnodebski, M.; Ligocka, J.; Schwab, R.; et al. Cancer-associated circulating large extracellular vesicles in cholangiocarcinoma and hepatocellular carcinoma. *J. Hepatol.* **2017**, *67*, 282–292. [CrossRef] [PubMed]

113. Kornek, M.; Lynch, M.; Mehta, S.H.; Lai, M.; Exley, M.; Afdhal, N.H.; Schuppan, D. Circulating microparticles as disease-specific biomarkers of severity of inflammation in patients with hepatitis C or nonalcoholic steatohepatitis. *Gastroenterology* **2012**, *143*, 448–458. [CrossRef] [PubMed]

114. Rautou, P.E.; Bresson, J.; Sainte-Marie, Y.; Vion, A.C.; Paradis, V.; Renard, J.M.; Devue, C.; Heymes, C.; Letteron, P.; Elkrief, L.; et al. Abnormal plasma microparticles impair vasoconstrictor responses in patients with cirrhosis. *Gastroenterology* **2012**, *143*, 166–176. [CrossRef] [PubMed]

115. Pan, J.H.; Zhou, H.; Zhao, X.X.; Ding, H.; Li, W.; Qin, L.; Pan, Y.L. Role of exosomes and exosomal microRNAs in hepatocellular carcinoma: Potential in diagnosis and antitumour treatments (Review). *Int. J. Mol. Med.* **2018**, *41*, 1809–1816. [CrossRef] [PubMed]

116. Kogure, T.; Lin, W.L.; Yan, I.K.; Braconi, C.; Patel, T. Intercellular nanovesicle-mediated microRNA transfer: A mechanism of environmental modulation of hepatocellular cancer cell growth. *Hepatology* **2011**, *54*, 1237–1248. [CrossRef] [PubMed]

117. Wei, J.X.; Lv, L.H.; Wan, Y.L.; Cao, Y.; Li, G.L.; Lin, H.M.; Zhou, R.; Shang, C.Z.; Cao, J.; He, H.; et al. Vps4A functions as a tumor suppressor by regulating the secretion and uptake of exosomal microRNAs in human hepatoma cells. *Hepatology* **2015**, *61*, 1284–1294. [CrossRef] [PubMed]

118. Lou, G.; Song, X.; Yang, F.; Wu, S.; Wang, J.; Chen, Z.; Liu, Y. Exosomes derived from miR-122-modified adipose tissue-derived MSCs increase chemosensitivity of hepatocellular carcinoma. *J. Hematol. Oncol.* **2015**, *8*, 122. [CrossRef] [PubMed]

119. Sugimachi, K.; Matsumura, T.; Hirata, H.; Uchi, R.; Ueda, M.; Ueo, H.; Shinden, Y.; Iguchi, T.; Eguchi, H.; Shirabe, K.; et al. Identification of a bona fide microRNA biomarker in serum exosomes that predicts hepatocellular carcinoma recurrence after liver transplantation. *Br. J. Cancer* **2015**, *112*, 532–538. [CrossRef] [PubMed]

120. Sohn, W.; Kim, J.; Kang, S.H.; Yang, S.R.; Cho, J.Y.; Cho, H.C.; Shim, S.G.; Paik, Y.H. Serum exosomal microRNAs as novel biomarkers for hepatocellular carcinoma. *Exp. Mol. Med.* **2015**, *47*, e184. [CrossRef] [PubMed]

121. Shi, M.; Jiang, Y.; Yang, L.; Yan, S.; Wang, Y.G.; Lu, X.J. Decreased levels of serum exosomal miR-638 predict poor prognosis in hepatocellular carcinoma. *J. Cell. Biochem.* **2017**. [CrossRef] [PubMed]

122. Wang, H.; Hou, L.; Li, A.; Duan, Y.; Gao, H.; Song, X. Expression of serum exosomal microRNA-21 in human hepatocellular carcinoma. *Biomed. Res. Int.* **2014**, *2014*. [CrossRef] [PubMed]

123. Qu, Z.; Wu, J.; Wu, J.; Ji, A.; Qiang, G.; Jiang, Y.; Jiang, C.; Ding, Y. Exosomal miR-665 as a novel minimally invasive biomarker for hepatocellular carcinoma diagnosis and prognosis. *Oncotarget* **2017**, *8*, 80666–80678. [CrossRef] [PubMed]

124. Fornari, F.; Ferracin, M.; Trere, D.; Milazzo, M.; Marinelli, S.; Galassi, M.; Venerandi, L.; Pollutri, D.; Patrizi, C.; Borghi, A.; et al. Circulating microRNAs, miR-939, miR-595, miR-519d and miR-494, Identify Cirrhotic Patients with HCC. *PLoS ONE* **2015**, *10*, e0141448. [CrossRef] [PubMed]

125. Xiong, L.; Zhen, S.; Yu, Q.; Gong, Z. HCV-E2 inhibits hepatocellular carcinoma metastasis by stimulating mast cells to secrete exosomal shuttle microRNAs. *Oncol. Lett.* **2017**, *14*, 2141–2146. [CrossRef] [PubMed]

126. Aucher, A.; Rudnicka, D.; Davis, D.M. MicroRNAs transfer from human macrophages to hepato-carcinoma cells and inhibit proliferation. *J. Immunol.* **2013**, *191*, 6250–6260. [CrossRef] [PubMed]

127. Hindson, C.M.; Chevillet, J.R.; Briggs, H.A.; Gallichotte, E.N.; Ruf, I.K.; Hindson, B.J.; Vessella, R.L.; Tewari, M. Absolute quantification by droplet digital PCR versus analog real-time PCR. *Nat. Methods* **2013**, *10*, 1003–1005. [CrossRef] [PubMed]

128. Pinheiro, L.B.; Coleman, V.A.; Hindson, C.M.; Herrmann, J.; Hindson, B.J.; Bhat, S.; Emslie, K.R. Evaluation of a droplet digital polymerase chain reaction format for DNA copy number quantification. *Anal. Chem.* **2012**, *84*, 1003–1011. [CrossRef] [PubMed]

129. Srivastava, S. Cancer biomarker discovery and development in gastrointestinal cancers: Early detection research network—A collaborative approach. *Gastrointest. Cancer Res.* **2007**, *1*, S60–S63. [PubMed]

International Journal of
Molecular Sciences

MDPI

Article

A Three–MicroRNA Signature as a Potential Biomarker for the Early Detection of Oral Cancer

Yi-An Chang [1,2,†], Shun-Long Weng [3,4,5,†], Shun-Fa Yang [6,7], Chih-Hung Chou [1,8],
Wei-Chih Huang [1,8], Siang-Jyun Tu [8], Tzu-Hao Chang [9], Chien-Ning Huang [7,10],
Yuh-Jyh Jong [1,11,12] and Hsien-Da Huang [1,8,*]

[1] Department of Biological Science and Technology, National Chiao Tung University,
    Hsinchu City 300, Taiwan; b8703126@gmail.com (Y.-A.C.); chchou23@gmail.com (C.-H.C.);
    loveariddle.bi96g@g2.nctu.edu.tw (W.-C.H.); yjjongnctu@gmail.com (Y.-J.J.)
[2] Department of Medical Research, Hsinchu Mackay Memorial Hospital, Hsinchu City 300, Taiwan
[3] Department of Medicine, Mackay Medical College, New Taipei City 252, Taiwan; 4467@mmh.org.tw
[4] Department of Obstetrics and Gynaecology, Hsinchu Mackay Memorial Hospital, Hsinchu City 300, Taiwan
[5] Mackay Junior College of Medicine, Nursing and Management College, Taipei 112, Taiwan
[6] Department of Medical Research, Chung Shan Medical University Hospital, Taichung 402, Taiwan;
    ysf@csmu.edu.tw
[7] Institute of Medicine, Chung Shan Medical University, Taichung 402, Taiwan; cshy049@csh.org.tw
[8] Institute of Bioinformatics and Systems Biology, National Chiao Tung University, Hsinchu City 300, Taiwan;
    mist0205@gmail.com
[9] Graduate Institute of Biomedical Informatics, Taipei Medical University, Taipei 110, Taiwan;
    kevinchang@tmu.edu.tw
[10] Department of Internal Medicine, Division of Endocrinology and Metabolism,
    Chung Shan Medical University Hospital, Taichung 402, Taiwan
[11] Graduate Institute of Clinical Medicine, College of Medicine, Kaohsiung Medical University,
    Kaohsiung 807, Taiwan
[12] Departments of Pediatrics and Laboratory Medicine, Kaohsiung Medical University Chung-Ho Memorial Hospital,
    Kaohsiung 807, Taiwan
*   Correspondence: bryan@mail.nctu.edu.tw; Tel.: +886-3-571-2121 (ext. 56952)
†   These authors contributed equally to this work.

Received: 18 January 2018; Accepted: 3 March 2018; Published: 7 March 2018

Abstract: Oral squamous cell carcinoma (OSCC) is often diagnosed at a late stage and may be malignantly transformed from oral leukoplakia (OL). This study aimed to identify potential plasma microRNAs (miRNAs) for the early detection of oral cancer. Plasma from normal, OL, and OSCC patients were evaluated. Small RNA sequencing was used to screen the differently expressed miRNAs among the groups. Next, these miRNAs were validated with individual samples by quantitative real-time polymerase chain reaction (qRT-PCR) assays in the training phase ($n = 72$) and validation phase ($n = 178$). The possible physiological roles of the identified miRNAs were further investigated using bioinformatics analysis. Three miRNAs (miR-222-3p, miR-150-5p, and miR-423-5p) were identified as differentially expressed among groups; miR-222-3p and miR-423-5p negatively correlated with T stage, lymph node metastasis status, and clinical stage. A high diagnostic accuracy (Area under curve = 0.88) was demonstrated for discriminating OL from OSCC. Bioinformatics analysis reveals that miR-423-5p and miR-222-3p are significantly over-expressed in oral cancer tissues and involved in various cancer pathways. The three-plasma miRNA panel may be useful to monitor malignant progression from OL to OSCC and as potential biomarkers for early detection of oral cancer.

Keywords: miRNA; biomarker; oral cancer; leukoplakia; early diagnosis

---

1. Introduction

Oral squamous cell carcinoma (OSCC) is the most common type (84–97%) of oral cancer [1]. In South and Southeast Asia and Taiwan, the major risk factor is betel quid chewing. The five-year survival rates for early- and late-stage oral cancer are approximately 82% and 20%, respectively [1]. Unfortunately, around 50% of oral cancer patients present at an advanced stage (TNM III or IV) [2,3], signifying the importance of early diagnosis. Malignant transformation of mucosal lesions predispose to oral cancer. The World Health Organization (WHO) defined these lesions as "potentially malignant disorders (PMD)". In Taiwan, oral leukoplakia (OL) is the most common oral precancerous lesion [4] and usually transforms into OSCC after five years. Betel quid chewing, alcohol consumption and smoking habits have been indicated to increase the risk of malignant transformation [5,6]. The rate of dysplastic or malignant transformation is between 15.6% and 39.2% [7–9]. Thus, the assessment and follow-up of OL should be focused for the early detection of OSCC.

MicroRNAs (miRNAs) are a large family of about 22-nucleotide-long, non-coding, single-stranded RNA molecules that interact with target sequences to degrade or repress translation [9]. They have also been documented to have roles in all of the cancer hallmarks [10] by acting as oncogenes or tumor suppressor genes. Recent studies have revealed the association of miRNA deregulation and its role in OSCC. The most reported possible players in the tumorigenesis of OSCC were miR-21, miR-221, miR-184, miR-133a, miR-375, and let-7b [11]. Besides, over expression of miR-21, miR-181, and miR-345 is associated with the malignant transformation of OL [12]. Three-miRNA signatures (miR-129-5p, miR-339-5p, and miR-31-3p) have been identified as mediators in the initiation and progression of non-malignant to aggressive type of OL [13]. These studies were based on tissue expression, which might not be a practical tool for clinical diagnosis.

Circulating miRNA is an ideal biomarker for the diagnosis and assessment of disease progression and metastasis [14], due to its stability in the extracellular environment and accessibility in various body fluids [15,16]. In previous studies, circulating miRNAs were identified by comparing normal patients to those with OSCC. However, if PMD patients are not included, miRNAs might not help to monitor the transformation from PMD to OSCC, thus missing out on the opportunity to diagnose early OSCC. Moreover, the reference miRNAs for normalization of quantitative real-time polymerase chain reaction (qRT-PCR) data used in past studies, such as RNU-6B, have been variably expressed as in serum and plasma [17]; this might question the reliability of qPCR data.

In the present study, we aimed to identify circulating miRNAs as potential biomarkers for the early detection of OSCC. Moreover, suitable reference miRNA for OL and OSCC populations were also investigated. We collected plasma from normal, OL, and OSCC patients and designed a three phase study to investigate potential biomarkers. In the screening phase, miRNA expression was profiled by small RNA sequencing platform in order to identify differentially expressed miRNAs and reference miRNAs. Subsequently, we used qRT-PCR assays to confirm miRNA expression in individual samples and refined the number of miRNAs. Potential miRNA biomarkers were accessed by an independent cohort in the validation phase. Our results provided a three-miRNA panel for discrimination between normal, OL, and OSCC patients.

2. Results

2.1. Characteristics of Study Subjects

We recruited 250 participants (including 72 and 178 participants in training and validation phases, respectively) (Table 1). All of the OSCC patients were free from distant metastasis. There was no age or gender distribution difference between the screening/training phase ($p = 0.960, 0.453$, respectively) and validation phase ($p = 0.130, 0.877$, respectively).

Table 1. Clinical characteristics of study subjects.

Variables	Screening and Training Phase (*n* = 72)			Validation Phase (*n* = 178)		
	Normal (%)	OL (%)	OSCC (%)	Normal (%)	OL (%)	OSCC (%)
Number	20	20	32	50	46	82
Age (mean ± SD)	52.05 ± 12.78	52.20 ± 12.53	52.20 ± 9.03	52.86 ± 14.06	48.35 ± 12.11	53.79 ± 11.25
Sex						
Male	20 (100.0)	18 (90.0)	31 (96.8)	48 (96.0)	44 (95.6)	80 (97.5)
Female	0 (0.0)	2 (10.0)	1 (3.2)	2 (4.0)	2 (4.4)	2 (2.5)
Smoking						
Non-smoker	4 (20.0)	6 (30.0)	1 (3.2)	2 (4.0)	6 (13.0)	7 (8.5)
Former smoker	9 (45.0)	11 (55.0)	12 (37.5)	15 (30.0)	12 (26.1)	16 (19.5)
Current smoker	7 (35.0)	3 (15.0)	19 (59.3)	33 (66.0)	28 (60.9)	59 (72.0)
BQ chewing						
Non-BQ	11 (55.0)	6 (30.0)	3 (9.4)	27 (54.0)	12 (26.1)	7 (8.5)
Former BQ chewing	6 (30.0)	10 (50.0)	25 (78.1)	15 (30.0)	28 (60.9)	63 (76.8)
Current BQ-chewing	3 (15.0)	4 (20.0)	4 (12.5)	8 (16.0)	6 (13.0)	12 (14.7)
Alcohol consumption						
Non-drinker	5 (25.0)	9 (45.0)	11 (34.4)	20 (40.0)	20 (43.5)	22 (26.8)
Former drinker	10 (50.0)	9 (45.0)	13 (40.6)	22 (44.0)	23 (50.0)	36 (43.9)
Current drinker	5 (25.0)	2 (10.0)	8 (25.0)	8 (16.0)	3 (6.5)	24 (29.3)
Stage						
I			14 (43.8)			32 (39.0)
II			0 (0.0)			15 (18.3)
III			0 (0.0)			11 (13.4)
IV			18 (56.2)			24 (29.3)
T stage						
T1			14 (43.8)			33 (40.2)
T2			3 (9.4)			24 (29.3)
T3			0 (0.0)			4 (4.9)
T4			15 (46.8)			21 (25.6)
N stage						
N0			19 (59.4)			60 (73.2)
N1			5 (15.6)			11 (13.4)
N2			8 (25.0)			11 (13.4)

Abbreviations: OL = oral leukoplakia; OSCC = oral squamous cell carcinoma; BQ = betel quid; SD = standard deviation.

2.2. Expression Profiling of miRNAs by Small RNA-seq

As illustrated in Figure 1, small RNA-seq analysis was performed on three pooled samples to identify the differentially expressed miRNA. NGS raw data was uploaded and submitted to a public repository Gene Expression Omnibus (GEO) database (GSE104440). Small RNA reads were highly qualified (Figure 2) for subsequent analysis. Differentially expressed miRNA (Figure 3A) were identified according to the criteria detailed in Figure 1. Total of 14 miRNAs (Figure 3B) were found to be deregulated between normal/OL, OL/OSCC, or OSCC/normal groups. Among the three sets of comparisons, nine candidate miRNA were deregulated in any two sets and were selected for subsequent confirmation.

We tried to identify potential reference miRNA across normal, OL, and OSCC patients for relative quantification by qRT-PCR. Potential reference miRNA were identified if they met the criteria, as illustrated in Figure 1. Four miRNA (miR-130b-3p, miR-221-3p, miR-101-3p, and miR-16-5p) were selected for further analysis.

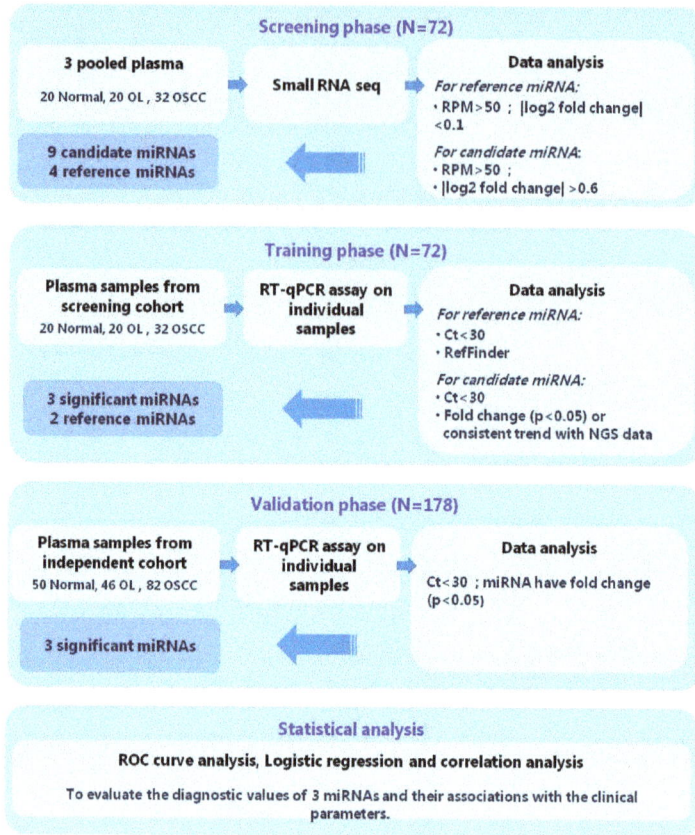

Figure 1. Study design. Abbreviations: RPM = reads per million; qRT-PCR = quantitative reverse transcription polymerase chain reaction; ROC = receiver operating characteristics.

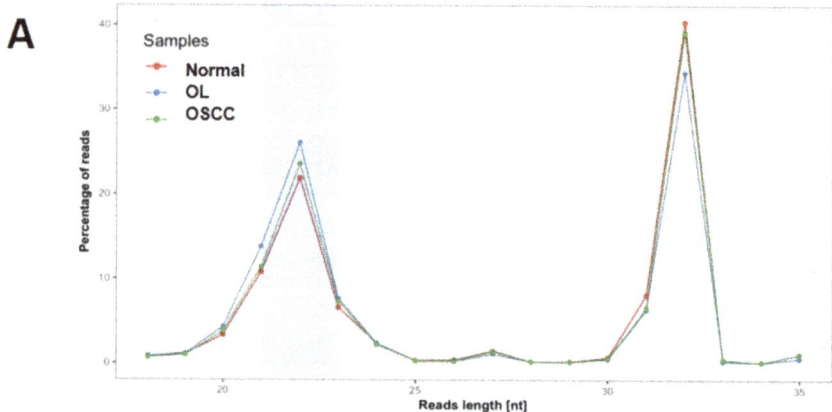

Figure 2. *Cont.*

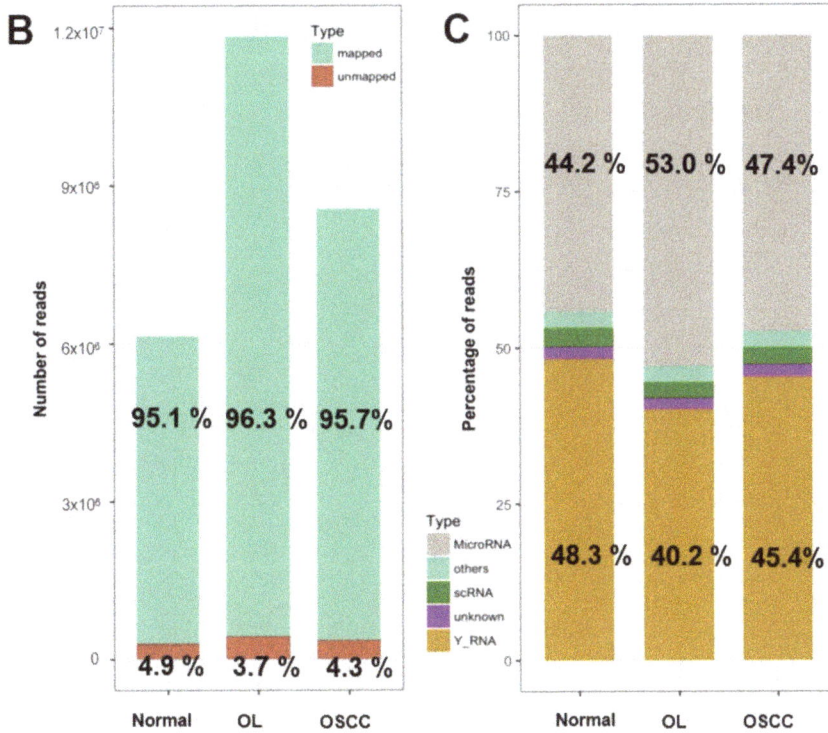

Figure 2. Quality of small RNA sequencing data. (**A**) Read length distribution indicates that all the samples with a peak in read length 21–23 of miRNA length. (**B**) More than 95% reads were mapped to reference genome. (**C**) More than 44% of mapped reads were miRNAs.

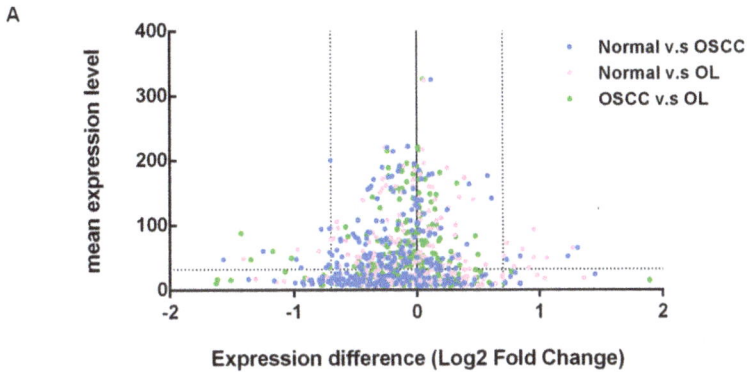

Figure 3. *Cont.*

B

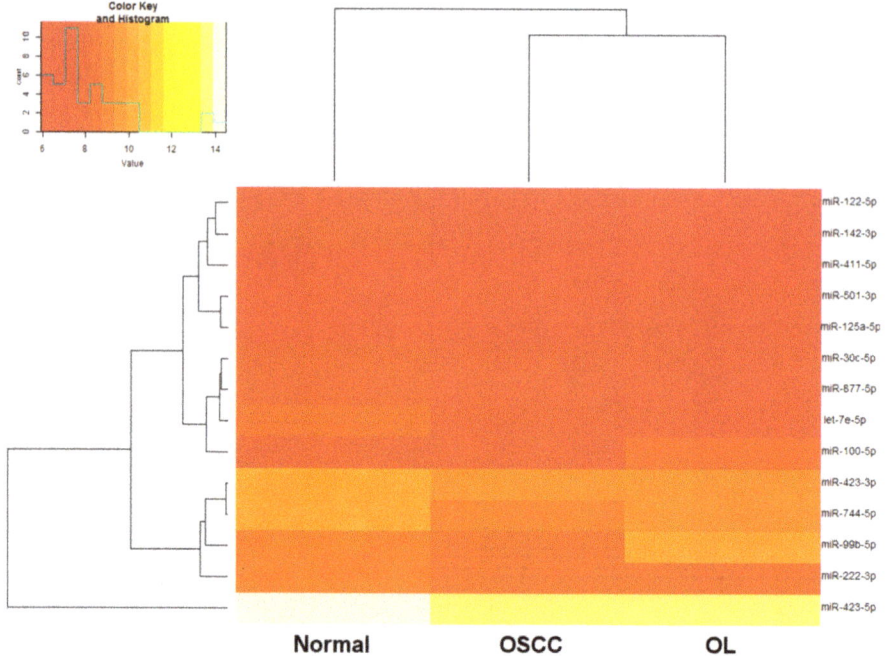

Figure 3. miRNA profiling by small RNA sequencing. (**A**) Each point indicates the expression difference of a single miRNA between specified groups. The dotted line represents the cut-off value of expression difference and expression level while the solid line indicates no discrepant expression among groups. (**B**) Fourteen deregulated miRNAs. The miRNA expression level (log2 transformed RPM value) is presented. The color legend from red to white means the expression value from 6 to 14, and the light blue line presents a histogram of the expression value.

2.3. Selection of Suitable Reference miRNAs

The expression levels of four reference miRNAs were confirmed by qRT-PCR in 72 individual samples (Figure 1). Four miRNAs were detected among 72 samples with median C_t <30 and were considered suitable. By analyzing the C_t values, miR-101-3p revealed differential expression among groups ($p = 0.001$) and was excluded (Supplementary Table S1). Stability was investigated by RefFinder, which is an online tool integrating four programs (http://150.216.56.64/referencegene.php); lower values indicate greater stability. As a result, miR-130b-3p and miR-221-3p were combined and selected as the reference miRNA set (Supplementary Table S1).

2.4. Investigation of the Six Candidate miRNAs in Training Phase

Three out of nine candidate miRNAs were excluded, owing to the relatively low expression levels ($C_t > 30$) in pooled samples. The remaining six miRNAs (miR-let-7e-5p, miR-222-3p, miR-423-5p, miR-150-5p, miR-125a-5p, and miR-100-5p) were investigated in individual samples.

The relative expression level of each miRNA among groups was obtained by normalization with reference miRNA set (mean C_t) by comparative C_t method. Significant miRNAs were chosen according to the criteria listed in Figure 1. No significantly different miRNA levels of miR-125a-5p and miR-100-5p were observed (Figure 4A) between groups. In addition, miR-let-7e-5p displayed an inconsistent trend with NGS profiling data (Supplementary Figure S1). Taken together, three miRNAs (miR-222-3p, miR-423-5p, and miR-150-5p) were considered significant for the next validation.

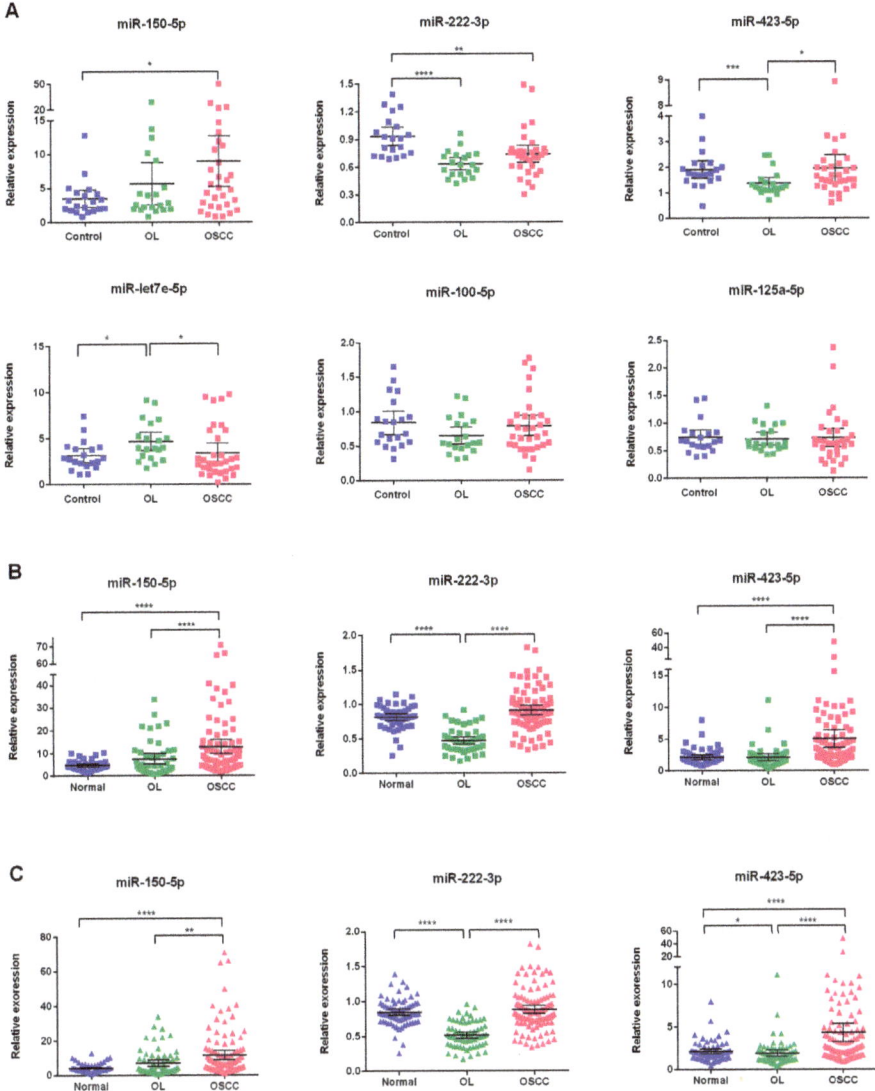

Figure 4. Expression of candidate miRNA in different data sets. (**A**) Abundance of miR-let-7e-5p, miR-125a-5p, miR-100-5p, miR-150-5p, miR-222-3p, and miR-423-5p in plasma of subjects in training phase (*n* = 72). Expression of significant miRNA (miR-222-3p, miR-423-5p, and miR-150-5p) in plasma from validation phase (*n* = 178) (**B**) and from all subjects (*n* = 250) (**C**). Significance of two-sided *p*-values is indicated as follows: * *p* < 0.05, ** *p* < 0.01, *** *p* < 0.001, and **** *p* < 0.001 (Mann–Whitney test).

2.5. Validation of Three Significant miRNA with an Independent Cohort

An independent cohort was used to validate the expression levels of the three miRNAs (Figure 1). We found that miR-222-3p was found to be significantly down regulated in OL patients when compared to that in normal and OSCC patients ($p < 0.0001$), whereas miR-423-5p and miR-150-5p increased in OSCC patients as compared to those in normal and OL patients ($p < 0.001$, Figure 4B). Similar results were observed when analyzing all of the samples (Figure 4C) from training and validation phases. These results reveal that the three miRNA could be potential biomarkers for the diagnosis of OL and OSCC.

2.6. Correlations between miRNA Signature and Clinical Parameters

Spearman rank analysis showed that miR-222-3p and miR-423-5p negatively correlated with clinical stage, lymph node metastasis status, and T stage (Table A1). For OSCC patients, miR-222-3p, and miR-423-5p significantly down regulated when tumors spread to lymph node ($p = 0.026, 0.019$, respectively), and gradually declined with tumor progression (Figure 5A). Although miR-150-5p did not correlate with the node metastasis and tumor progression, decreased the expression level at late-stage tumor was observed (Figure 5A). These findings imply that miR-222 and miR-423-5p could be predictors for tumor progression.

In non-cancer patients, betel quid chewing was shown to have negative correlation with miR-222-3p level (Table A1). A significant difference was observed in miR-222-3p level between non-chewers and former chewers ($p = 0.001$); long-term smokers (>10 years); and, non-smokers ($p = 0.032$) (Figure 5B). However, the miRNA abundance neither correlated with (Table A1) nor showed difference among patients with different drinking habits.

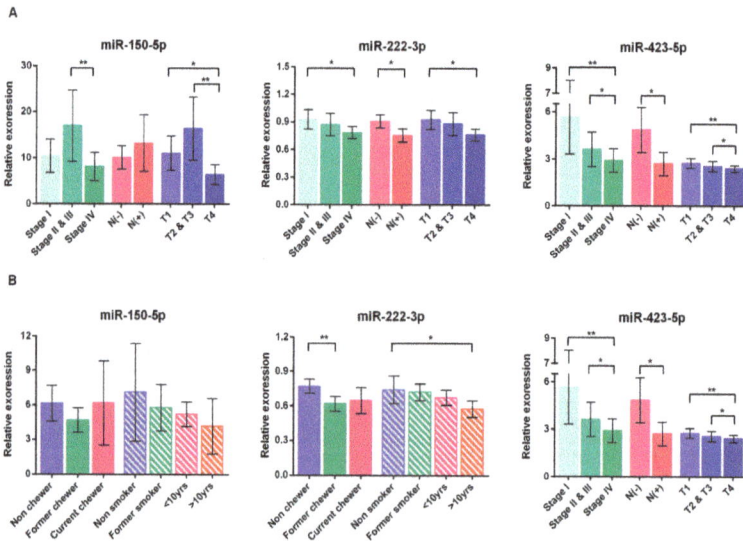

Figure 5. Expression of miR-222-3p, miR-423-5p, and miR-150-5p in different groups of patients. (**A**) OSCC patients ($n = 114$) were categorized according to clinical stage, lymph node metastasis status indicates the presence [N(+)] and absence [N(−)] of metastasis and T stage. (**B**) Normal and oral leukoplakia (OL) patients ($n = 136$) were classified by the habit of smoking and betel nut chewing. Current smokers were further divided into two groups according to the history of smoking. Significance of two-sided *p*-values is indicated as follows: * $p < 0.05$, ** $p < 0.01$ (Mann–Whitney test).

2.7. Logistic Regression Analysis of miRNA Biomarkers

For all of the patients, regression analysis was conducted to determine the diagnostic efficacy of three miRNA signatures. In model 1, we set the normal group as the reference category. The relative risks (RRs) of these miRNAs for OL were shown in Table 2. These results reveal that miR-222-3p and miR-150-5p were independently associated with OL, whereas miR-150-5p and miR-423-5p associated with OSCC. Similarly, we found that all three miRNAs were significant independent predictors of OSCC when the OL group was treated as the reference category (Table 2). In addition, multivariate logistic regression analysis adjusted for sex, age, smoking, and betel quid chewing habits were also performed. As with univariate logistic regression analysis, these independent associations still remained significant (Table 2).

Furthermore, logistic regression analysis was used to conduct a risk score analysis to identify the best combinations of miRNAs to predict OL and OSCC. The combination of miR-222-3p and miR-150-5p, miR-150-5p, and miR-423-5p, produced the best model to predict OL and OSCC, respectively. A combination of the three miRNAs enabled distinction between OL and OSCC (Table 2).

Table 2. Regression analysis.

	Univariate		Mulitivariate	
	Relative Risk	*p*-Value	**Relative Risk**	*p*-Value
Model 1				
OL				
miR-222-3p	0.205 (0.123−0.344)	<0.001	0.212 (0.127−0.357)	<0.001
miR-150-5p	1.114 (1.027−1.210)	0.010	1.124 (1.032−1.223)	0.007
miR-423-5p	0.880 (0.673−1.150)	0.349	0.897 (0.682−1.180)	0.437
miR panel [a]	1.348 (1.233−1.474)	<0.001	1.361 (1.238−1.496)	<0.001
OSCC				
miR-222-3p	1.038 (0.858−1.256)	0.699	1.114 (0.898−1.383)	0.324
miR-150-5p	1.189 (1.083−1.306)	<0.001	1.198 (1.079−1.330)	0.001
miR-423-5p	1.466 (1.182−1.817)	<0.001	1.599 (1.238−2.066)	<0.001
miR panel [b]	1.377 (1.198−1.584)	<0.001	1.386 (1.189−1.615)	<0.001
Model 2				
OSCC				
miR-222-3p	2.915 (2.087−4.072)	<0.001	3.014 (2.102−4.321)	<0.001
miR-150-5p	1.038 (1.003−1.075)	0.035	1.048 (1.007−1.091)	0.020
miR-423-5p	1.581 (1.238−2.021)	<0.001	1.601 (1.236−2.075)	<0.001
miR panel [c]	1.455 (1.308−1.617)	<0.001	1.448 (1.292−1.623)	<0.001

The reference category was normal and OL patients in mode 1 and model 2, respectively. [a] miR222-3p and miR-150-5p; [b] miR-150-5p and miR-423-5p; [c] miR-222-3p, miR-150-5p, and miR-423-5p.

2.8. Diagnostic Performance of miRNA Signature

Receiver operating characteristic (ROC) analysis was applied on the miRNAs in all of the subjects (Figure 6) after being adjusted by the multivariate model. We also evaluated the diagnostic value of the combined miRNA panel. The combination of miRNAs panel increased the AUC when individually when compared with any of the miRNAs (Figure 6). The area under curve (AUC) for combined miRNA panel was 0.959 (miR-150-5p/miR-222-3p, 95% CI, 0.927–0.991, *p* < 0.0001) and 0.749 (miR-150-5p/miR-423-5p, 95% CI, 0.678–0.819, *p* < 0.0001) for OL and OSCC patients, respectively. The three miRNA combined panel for distinguishing OL from OSCC patients yielded an AUC value of 0.916 (95% CI, 0.874–0.957, *p* < 0.0001).

We observed lower miRNA expression at different stages of OSCC; therefore, we considered whether these miRNAs may discriminate between OL and early (stage I) OSCC. The three-miRNA panel yielded an AUC value of 0.917 (95% CI, 0.861–0.973, *p* < 0.0001) (Figure 6). Accordingly, these data suggest that different combinations of the three miRNAs serve as potential biomarkers for

OL and OSCC. Importantly, the three-miRNA panel helped detection of transformation from OL to early malignancy.

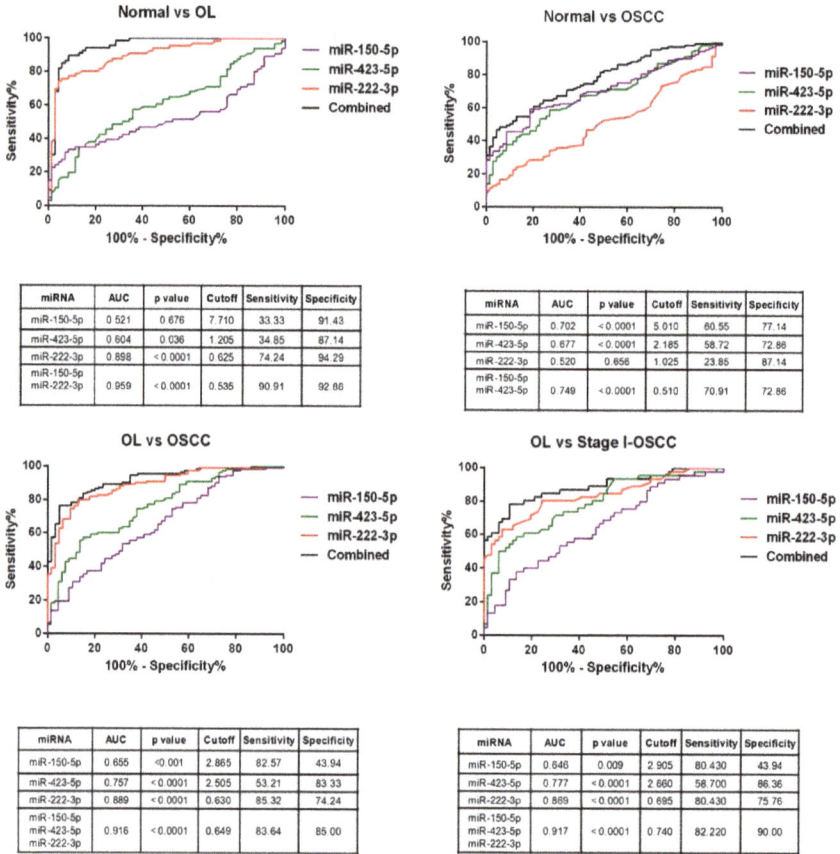

Normal vs OL

miRNA	AUC	p value	Cutoff	Sensitivity	Specificity
miR-150-5p	0.521	0.676	7.710	33.33	91.43
miR-423-5p	0.604	0.036	1.205	34.85	87.14
miR-222-3p	0.898	<0.0001	0.625	74.24	94.29
miR-150-5p miR-222-3p	0.959	<0.0001	0.535	90.91	92.86

Normal vs OSCC

miRNA	AUC	p value	Cutoff	Sensitivity	Specificity
miR-150-5p	0.702	<0.0001	5.010	60.55	77.14
miR-423-5p	0.677	<0.0001	2.185	58.72	72.86
miR-222-3p	0.520	0.656	1.025	23.85	87.14
miR-150-5p miR-423-5p	0.749	<0.0001	0.510	70.91	72.86

OL vs OSCC

miRNA	AUC	p value	Cutoff	Sensitivity	Specificity
miR-150-5p	0.655	<0.001	2.865	82.57	43.94
miR-423-5p	0.757	<0.0001	2.505	53.21	83.33
miR-222-3p	0.889	<0.0001	0.630	85.32	74.24
miR-150-5p miR-423-5p miR-222-3p	0.916	<0.0001	0.649	83.64	85.00

OL vs Stage I-OSCC

miRNA	AUC	p value	Cutoff	Sensitivity	Specificity
miR-150-5p	0.646	0.009	2.905	80.430	43.94
miR-423-5p	0.777	<0.0001	2.660	58.700	86.36
miR-222-3p	0.889	<0.0001	0.695	80.430	75.76
miR-150-5p miR-423-5p miR-222-3p	0.917	<0.0001	0.740	82.220	90.00

Figure 6. ROC analysis for individual miRNA and combined panel. ROC curve was generated by analyzing samples from all patients. The combined panels were generated by linear combination of values of each miRNA. AUC = area under the ROC curve. *p*-Values were calculated using Mann–Whitney test.

2.9. Bioinformatics Analysis of miR-222-3p, miR-423-5p, and miR-150-5p

Circulating miRNA could originate from tumor cells. To elucidate this, we analyzed the miRNA expression profiles of solid tissue of head and neck cancer from The Cancer Genome Atlas (TCGA, http://cancergenome.nih.gov/). Tissue miR-222-3p and miR-423-5p levels were significantly up-regulated in OSCC patients (Supplementary Figure S2) when compared to those in normal ($p < 0.0001$). Furthermore, tissue miR-222-3p was down regulated in OSCC with lymph node metastasis, similar to the levels observed in OSCC plasma. Although tissue miR-150-5p was not found to be differently expressed between normal and OSCC patients, a gradual decrease was observed in miR-150-5p with tumor growth (Supplementary Figure S2).

Putative predicted target genes of miRNA and experimentally validated miRNA-gene interactions were included for pathway enrichment analysis. The identified pathways (Table 3) reveal the involvement of these miRNAs in cancer-related pathways, such as Wnt, PI3K-Akt, MAPK, and Ras signaling pathway. Among these pathways, Wnt signaling pathway, significant in head and neck cancer, was the most enriched pathway. IPA analysis also indicated that Wnt signaling was the one of the most enriched pathways (Supplementary Figure S3) and cancer was the top enriched disease (Supplementary Table S2). These results provided a clue of the roles played by these miRNAs in OSCC.

Table 3. The top 10 most enriched KEGG pathway with the targetome of three miRNAs.

Term	No. of Genes	*p*-Value	Fold Enrichment
Wnt signaling pathway	16	4.69×10^{-6}	4.22
Pathways in cancer	26	6.07×10^{-5}	2.41
Hepatitis B	14	1.55×10^{-4}	3.51
Axon guidance	13	1.70×10^{-4}	3.72
Sphingolipid signaling pathway	11	1.58×10^{-3}	3.33
HTLV-I infection	17	1.66×10^{-3}	2.42
PI3K-Akt signaling pathway	20	2.71×10^{-3}	2.11
Proteoglycans in cancer	14	3.19×10^{-3}	2.55
FoxO signaling pathway	11	3.57×10^{-3}	2.99
Rap1 signaling pathway	14	4.79×10^{-3}	2.42
MicroRNAs in cancer	17	4.84×10^{-3}	2.17

KEGG: Kyoto Encyclopedia of Genes and Genomes.

3. Discussion

To date, histopathology remains as the golden standard for reporting cancer risk of PMD. The invasiveness of histopathology leads to poor compliance for patients and is impossible to be used for monitoring the disease progression. Thus, non-invasive tools, such as tolonium chloride or toluidine blue dye, Oral CDx brush biopsy and latest optical systems (e.g., Vizilite and Velscope) were developed to detect precancer lesions [18]. Unfortunately, morphological finding only indicates the malignant potential (dysplasia) of a given lesion at that time, whereas subtle molecular changes can be detected before the morphological changes. Therefore, molecular biomarkers for detection of oral cancers have been developed and extended to point of care tests (e.g., IL-8 and IL-8 mRNA) [19,20].

This study aims to identify plasma miRNAs as biomarkers for early detection of OSCC. We found that miR-130b-3p and miR-221-3p were the most suitable reference miRNAs in this population. The expression levels of three miRNAs, miR-222-3p, miR-150-5p, and miR-423-3p, were found to be different between groups. For non-cancer patients, miR-222-3p correlated with betel chewing, whereas miR-222-3p and miR-423-5p were associated with tumor progression and lymph node metastasis. Among these miRNAs, the combination of miR-150-5p/miR-222-3p and miR-150-5p/miR-423-5p best discriminated normal from OL and OSCC, respectively. Importantly, we demonstrated that a three-miRNA panel can be used in OL patients for early detection of OSCC.

Several studies demonstrated the differential miRNA levels between the plasma of OSCC and normal [21–23]; Yang et al., 2011a [24]. However, little is known about circulating miRNA in OL. Only one study [25] revealed that salivary miR-31 was lower in OL compared to that in OSCC. The miRNA alteration we observed in this study differs from that in previous reports. Recent studies have indicated some pre-analytical and analytical factors causing these problems, e.g., sample type, extraction methods, and measurement platforms [26,27]. Most studies used RNU-6 and miR-16 as reference genes for the relative quantification of target miRNA. Unfortunately, RNU-6 unstably expressed in plasma and serum [17], whereas miR-16 was affected by hemolysis [28]. Instead of using RNU-6 and miR-16, we identified miR-221-3p and miR-130b-3p as suitable miRNA for our study. These factors might contribute to the different findings we observed.

In the screening phase, we used a small RNA sequencing platform than using microarray, which is widely used in previous studies. With this technique, it is possible to profile miRNA without knowing the sequence of miRNA beforehand and is powerful for miRNA discovery [29]. The application of NGS to measure miRNAs in serum/plasma is still in its early phase. To our knowledge, there was only one study utilized NGS strategy to profile plasma miRNA in the field of oral oncology and identified plasma miRNA biomarkers to monitor OSCC recurrence in patients after surgery [30].

It was indicated that miR-222 was co-transcribed in a cluster with miR-221, and the expression of these two miRNAs were shown to be highly correlated in OSCC [31]. Most recently, down regulation of miR-221/222 was shown to promote apoptosis in OSCC cells [32]. The expression levels of reference miRNA sets, miR-130b, and miR-221-3p, were used to calculate the relative expression level of miR-222-3p in this study. Therefore, not only the correlations between miR-221-3p and miR-222-3p but also the independence of the reference miRNA sets were examined. We found that the reference miRNA sets were not correlated with any of the three identified miRNAs. Moreover, the high correlated expression level between miR-221 and miR-222 demonstrated in previous tissue or cell line studies were not observed in the present study ($\varrho = -0.221, p = 0.061$).

The limitation of this study is the limited source of tissue samples, especially for OL and early stage OSCC. Therefore, we investigated the tissue expression levels of the three identified miRNAs by analyzing the HNSC data set from TCGA. Among them, miR-222-3p was confirmed to have higher expression level in OSCC. A previous study indicated increased miR-222 expression that was found in 40% OSCC and was correlated with tumor growth [31]. We found that tissue and plasma miR-222-3p was down regulated in OSCC if lymph node metastasis were present. A previous study also demonstrated that miR-222-3p contributed to metastasis in tongue cancer by targeting matrix metalloproteinase 1 and manganese superoxide dismutase 2. Ectopic transfection of miR-222-3p resulted in aberrant decrease in cell invasion and migration [33]. Other studies suggested that miR-222 affected cell growth, invasive and apoptotic abilities by targeting to PUMA in OSCC [34,35]. Our analysis also revealed that tissue miR-150-5p declined with tumor progression. The expression of vascular endothelial growth factor A, which is the target gene of miR-150-5p [36], was significantly associated with the tumor stage [37]. In OSCC, increased expression of miR-423-5p was demonstrated in plasma and tissues; however, another study reported a down-regulation in miR-423 in OSCC tissues [38]. Nonetheless, these findings suggest that the identified miRNA in this study acts as an oncomiR during tumor development.

Among the three identified miRNAs, we observed the expressions of miR-222-3p and miR-150-5p did not match the results from TCGA analysis. Up-regulation of miR-222-3p in OSCC was found in tissue but not in plasma when compared to normal. On the other hand, the up-regulation of miR-150-5p in OSCC was only found in plasma. Regarding the consistency of expression level between circulating miRNA and tissue miRNA, only a limited number of studies addressed this issue. Findings were controversial: some researchers described a similar trend of alteration, both in circulating and tissue miRNAs [39,40], whereas others observed the inconsistency between cellular miRNA and circulating miRNA [41–44]. Moreover, Pigati et al. [45] suggested the existence of a cellular selection mechanism for miRNA release and indicated that the extracellular and cellular miRNA profiles differ.

Pathway enrichment analysis also revealed possible functions of the identified miRNAs. Our results suggest that Wnt signaling pathway was the most enriched pathway. The deregulation of Wnt signaling pathway promoted the development and progression of oral cancer; it is also associated with prognosis in OSCC. Of note, β-catenin, which is a downstream mediator of Wnt signaling pathway, was demonstrated to be involved in oral malignant transformation. In dysplastic oral tissues or cancer tissues, β-catenin translocated form membrane to cytoplasm or nucleus [46]. A recent study revealed up-regulation in MAPK, ERK, JNK, IL-6/STAT3, WNT, TGFβ, and glucocorticoid receptor signaling to be the possible driving force behind the early stages of OSCC tumorigenesis [47]. Taken together, our results suggest that the three identified miRNAs might play important roles in the early stage OSCC development. However, the underlying mechanisms are beyond the scope of this study and will be further investigated in the future.

In conclusion, we identified three plasma miRNA for the detection of OL and OSCC by integration of small RNA sequencing and qRT-PCR platforms. By different combination of these three miRNA, OL and OSCC could be diagnosed. The three-miRNA panel demonstrated a high diagnostic value for discriminating OL from OSCC, and could be useful in the follow-up of OL and early detection of OSCC. Our results provide the basis of applying circulating miRNA to monitor malignant transformation of OL and could be extended to other PMD to benefit OSCC patients.

4. Materials and Methods

4.1. Clinical Samples

Two hundred and fifty patients (70 normal, 66 OL, and 114 OSCC) at the Chung Shan Medical University Hospital, Taiwan, were recruited between 2013 and 2016. Ethics approval for this study was obtained from the Institutional Review Board of Chung Shan Medical University Hospital (CSMUH No.: CS13214-1, 28 November 2014). Patients' blood was drawn within two weeks after the diagnosis was confirmed. The whole blood samples were collected in EDTA tubes from each patient after obtaining written informed consent. Plasma was separated by centrifuged at $3000\times g$ within two hours after the blood was drawn. RNA extraction from clinical samples was described in the Supplementary Materials and Methods.

4.2. Small RNA Library Preparation and Sequencing

Library was constructed by TruSeq Small RNA Preparation Kits (Illumina Inc., San Diego, CA, USA), according to the manufacturer's instructions. Library was size selected by 6% TBE PAGE gels to remove excess adapter dimmers. The final library size was confirmed by Agilent tape station 2200 (HSD1000 assay). Subsequently, indexed libraries were quantified by KAPA Library Quantification Kit (Kapa Biosystems, Wilmington, MA, USA), and 2 nM of library sample was subjected to NextSeq 500 (Illumina) for cluster generation and sequencing.

4.3. Small RNA Sequencing Analysis

Reads of small RNA sequencing were trimmed and processed before mapping to human genome. Detailed information is provided in the Supplementary Materials and Methods.

4.4. miRNA Quantification by qRT-PCR Assays

Plasma miRNA was reverse transcribed using miScript II RT Kit (Qiagen) according to the manufacturer's manual. Subsequent qPCR quantification was performed using miScript SYBR Green PCR Kit (Qiagen) on Rotor-Gene Q (Qiagen) instrument. Each sample was analyzed in triplicate. The *C. elegans* synthetic mir-39 spike-in control was used to normalize and evaluate technical variation in RNA extraction experiment as previously described [48].

4.5. Statistical Analysis

Differences in clinical characteristics among patients were compared using χ^2 test and one-way ANOVA for categorical and continuous variables, respectively. The Mann–Whitney U test was used to compare different miRNAs levels between groups, and data were presented as means $\pm$ 95% confidence interval (CI). Spearman rank correlation test was used to determine the association between miRNAs and clinical parameters. ROC curves of individual miRNAs were constructed to obtain the optimal cutoff for the detection of OL and OSCC. The risk score analysis is described in Supplementary Materials and Methods. Statistical analysis was performed using the GraphPad Prism 6 (GraphPad Software, Inc., La Jolla, CA, USA) or SPSS software version 22.0, (SPSS Inc., Chicago, IL, USA). A p-value < 0.05 was considered statistically significant.

4.6. Bioinformatics Analysis

The miRNA expression profiles of solid tissue from head and neck cancer were analyzed and compared to those of plasma in our observations. In addition, pathway enrichment analysis was conducted to discover potential functional roles of identified miRNAs. Further information is provided in the Supplementary Materials and Methods.

Supplementary Materials: Supplementary materials can be found at www.mdpi.com/1422-0067/19/3/758/s1.

Acknowledgments: We would like to acknowledge Chun-Yi Chuang (ENT physician) and Chih-Yu Peng (oral surgeon) for their contributions in collecting the clinical samples and providing clinical assistance. This work was supported by the Ministry of Health and Welfare (MOHW-106-TDU-B-212-144005, MOHW-107-TDU-B-212-114025) and the Ministry of Science and Technology, Taiwan (MOST-106-2319-B-400-001 and MOST-106-2633-B-009-001).

Author Contributions: Yi-An Chang, Shun-Long Weng and Shun-Fa Yang conceived and planned the experiments. Yi-An Chang carried out the experiments and statistical analysis. Chih-Hung Chou, Wei-Chih Huang, Siang-Jyun Tu and Tzu-Hao Chang planned and carried out the bioinformatics analysis. Shun-Fa Yang and Chien-Ning Huang contributed to sample preparation. Yi-An Chang and Shun-Long Weng took the lead in writing the manuscript. Yuh-Jyh Jong helped supervise the project. Hsien-Da Huang was in charge of overall direction. All authors provided critical feedback and helped shape the research, analysis and manuscript.

Conflicts of Interest: The authors declare no conflict of interest.

Abbreviations

OSCC	Oral squamous cell carcinoma
OL	Oral leukoplakia
miRNAs	MicroRNAs
qRT-PCR	Quantitative real-time polymerase chain reaction
PMD	Potentially malignant disorders
ROC	Receiver operating characteristic
AUC	Area under curve
TCGA	The Cancer Genome Atlas

Appendix A

Table A1. Spearman correlation analysis.

Variables	mir-222-3p		mir-423-5p		mir-150-5p	
	ϱ	*p*-Value	ϱ	*p*-Value	ϱ	*p*-Value
Age	0.105	NS	0.037	NS	−0.052	NS
Gender	0.039	NS	0.004	NS	0.103	NS
Betel chewing status	−0.241	0.005	−0.149	NS	−0.099	NS
Smoking status	−0.161	NS	0.034	NS	0.001	NS
Alcohol status	−0.121	NS	0.086	NS	−0.178	NS
Clinical stage	−0.201	0.032	−0.237	0.011	−0.116	NS
T stage	−0.220	NS	−0.276	0.003	−0.156	NS
Lymph node metastasis	−0.222	0.018	−0.220	0.019	0.012	NS

NS: Not significant.

References

1. Corso, G.D.; Villa, A.; Tarsitano, A.; Gohel, A. Current trends in oral cancer: A systematic review. *Cancer Cell Microenviron.* **2016**, *3*. [CrossRef]
2. Lo, W.L.; Kao, S.Y.; Chi, L.Y.; Wong, Y.K.; Chang, R.C. Outcomes of oral squamous cell carcinoma in Taiwan after surgical therapy: Factors affecting survival. *J. Oral Maxillofac. Surg.* **2003**, *61*, 751–758. [CrossRef]
3. Chen, Y.K.; Huang, H.C.; Lin, L.M.; Lin, C.C. Primary oral squamous cell carcinoma: An analysis of 703 cases in southern Taiwan. *Oral Oncol.* **1999**, *35*, 173–179. [CrossRef]

4. Kao, S.Y.; Chu, Y.W.; Chen, Y.W.; Chang, K.W.; Liu, T.Y. Detection and screening of oral cancer and pre-cancerous lesions. *J. Chin. Med. Assoc.* **2009**, *72*, 227–233. [CrossRef]

5. Jeng, J.H.; Chang, M.C.; Hahn, L.J. Role of areca nut in betel quid-associated chemical carcinogenesis: Current awareness and future perspectives. *Oral Oncol.* **2001**, *37*, 477–492. [CrossRef]

6. Yeh, C.Y.; Lin, C.L.; Chang, M.C.; Chen, H.M.; Kok, S.H.; Chang, S.H.; Kuo, Y.S.; Hahn, L.J.; Chan, C.P.; Lee, J.J.; et al. Differences in oral habit and lymphocyte subpopulation affect malignant transformation of patients with oral precancer. *J. Formos. Med. Assoc.* **2016**, *115*, 263–268. [CrossRef] [PubMed]

7. Neville, B.W.; Day, T.A. Oral cancer and precancerous lesions. *CA Cancer J. Clin.* **2002**, *52*, 195–215. [CrossRef] [PubMed]

8. Mortazavi, H.; Baharvand, M.; Mehdipour, M. Oral potentially malignant disorders: An overview of more than 20 entities. *J. Dent Res. Dent Clin. Dent Prospect.* **2014**, *8*, 6–14.

9. Zhang, B.; Pan, X.; Cobb, G.P.; Anderson, T.A. microRNAs as oncogenes and tumor suppressors. *Dev. Biol.* **2007**, *302*, 1–12. [CrossRef] [PubMed]

10. Hanahan, D.; Weinberg, R.A. Hallmarks of cancer: The next generation. *Cell* **2011**, *144*, 646–674. [CrossRef] [PubMed]

11. Gorenchtein, M.; Poh, C.F.; Saini, R.; Garnis, C. MicroRNAs in an oral cancer context—from basic biology to clinical utility. *J. Dent. Res.* **2012**, *91*, 440–446. [CrossRef] [PubMed]

12. Cervigne, N.K.; Reis, P.P.; Machado, J.; Sadikovic, B.; Bradley, G.; Galloni, N.N.; Pintilie, M.; Jurisica, I.; Perez-Ordonez, B.; Gilbert, R.; et al. Identification of a microRNA signature associated with progression of leukoplakia to oral carcinoma. *Hum. Mol. Genet.* **2009**, *18*, 4818–4829. [CrossRef] [PubMed]

13. Maimaiti, A.; Abudoukeremu, K.; Tie, L.; Pan, Y.; Li, X. MicroRNA expression profiling and functional annotation analysis of their targets associated with the malignant transformation of oral leukoplakia. *Gene* **2015**, *558*, 271–277. [CrossRef] [PubMed]

14. Krysan, K.; Kusko, R.; Grogan, T.; O'Hearn, J.; Reckamp, K.L.; Walser, T.C.; Garon, E.B.; Lenburg, M.E.; Sharma, S.; Spira, A.E.; et al. PGE2-driven expression of c-Myc and oncomiR-17-92 contributes to apoptosis resistance in NSCLC. *Mol. Cancer Res.* **2014**, *12*, 765–774. [CrossRef] [PubMed]

15. Mitchell, P.S.; Parkin, R.K.; Kroh, E.M.; Fritz, B.R.; Wyman, S.K.; Pogosova-Agadjanyan, E.L.; Peterson, A.; Noteboom, J.; O'Briant, K.C.; Allen, A.; et al. Circulating microRNAs as stable blood-based markers for cancer detection. *Proc. Natl. Acad. Sci. USA* **2008**, *105*, 10513–10518. [CrossRef] [PubMed]

16. Ng, E.K.; Chong, W.W.; Jin, H.; Lam, E.K.; Shin, V.Y.; Yu, J.; Poon, T.C.; Ng, S.S.; Sung, J.J. Differential expression of microRNAs in plasma of patients with colorectal cancer: A potential marker for colorectal cancer screening. *Gut* **2009**, *58*, 1375–1381. [CrossRef] [PubMed]

17. Wang, K.; Yuan, Y.; Cho, J.H.; McClarty, S.; Baxter, D.; Galas, D.J. Comparing the MicroRNA spectrum between serum and plasma. *PLoS ONE* **2012**, *7*, e41561. [CrossRef] [PubMed]

18. Messadi, D.V. Diagnostic aids for detection of oral precancerous conditions. *Int. J. Oral Sci.* **2013**, *5*, 59–65. [CrossRef] [PubMed]

19. Khan, R.S.; Khurshid, Z.; Asiri, F.Y.I. Advancing Point-of-Care (PoC) Testing Using Human Saliva as Liquid Biopsy. *Diagnostics* **2017**, *7*, 39. [CrossRef] [PubMed]

20. Torrente-Rodriguez, R.M.; Campuzano, S.; Ruiz-Valdepenas Montiel, V.; Gamella, M.; Pingarron, J.M. Electrochemical bioplatforms for the simultaneous determination of interleukin (IL)-8 mRNA and IL-8 protein oral cancer biomarkers in raw saliva. *Biosens. Bioelectron.* **2016**, *77*, 543–548. [CrossRef] [PubMed]

21. Lu, Y.C.; Chang, J.T.; Huang, Y.C.; Huang, C.C.; Chen, W.H.; Lee, L.Y.; Huang, B.S.; Chen, Y.J.; Li, H.F.; Cheng, A.J. Combined determination of circulating miR-196a and miR-196b levels produces high sensitivity and specificity for early detection of oral cancer. *Clin. Biochem.* **2015**, *48*, 115–121. [CrossRef] [PubMed]

22. Liu, C.J.; Tsai, M.M.; Tu, H.F.; Lui, M.T.; Cheng, H.W.; Lin, S.C. miR-196a overexpression and miR-196a2 gene polymorphism are prognostic predictors of oral carcinomas. *Ann. Surg. Oncol.* **2013**, *20*, S406–S414. [CrossRef] [PubMed]

23. Hung, P.S.; Liu, C.J.; Chou, C.S.; Kao, S.Y.; Yang, C.C.; Chang, K.W.; Chiu, T.H.; Lin, S.C. miR-146a enhances the oncogenicity of oral carcinoma by concomitant targeting of the IRAK1, TRAF6 and NUMB genes. *PLoS ONE* **2013**, *8*, e79926. [CrossRef] [PubMed]

24. Liu, C.J.; Kao, S.Y.; Tu, H.F.; Tsai, M.M.; Chang, K.W.; Lin, S.C. Increase of microRNA miR-31 level in plasma could be a potential marker of oral cancer. *Oral Dis.* **2010**, *16*, 360–364. [CrossRef] [PubMed]

25. Liu, C.J.; Lin, S.C.; Yang, C.C.; Cheng, H.W.; Chang, K.W. Exploiting salivary miR-31 as a clinical biomarker of oral squamous cell carcinoma. *Head Neck* **2012**, *34*, 219–224. [CrossRef] [PubMed]

26. Tiberio, P.; Callari, M.; Angeloni, V.; Daidone, M.G.; Appierto, V. Challenges in using circulating miRNAs as cancer biomarkers. *Biomed Res. Int.* **2015**, *2015*, 731479. [CrossRef] [PubMed]

27. He, Y.; Lin, J.; Kong, D.; Huang, M.; Xu, C.; Kim, T.K.; Etheridge, A.; Luo, Y.; Ding, Y.; Wang, K. Current State of Circulating MicroRNAs as Cancer Biomarkers. *Clin. Chem.* **2015**, *61*, 1138–1155. [CrossRef] [PubMed]

28. Kirschner, M.B.; Kao, S.C.; Edelman, J.J.; Armstrong, N.J.; Vallely, M.P.; van Zandwijk, N.; Reid, G. Haemolysis during sample preparation alters microRNA content of plasma. *PLoS ONE* **2011**, *6*, e24145. [CrossRef] [PubMed]

29. Moldovan, L.; Batte, K.E.; Trgovcich, J.; Wisler, J.; Marsh, C.B.; Piper, M. Methodological challenges in utilizing miRNAs as circulating biomarkers. *J. Cell Mol. Med.* **2014**, *18*, 371–390. [CrossRef] [PubMed]

30. Yan, Y.; Wang, X.; Veno, M.T.; Bakholdt, V.; Sorensen, J.A.; Krogdahl, A.; Sun, Z.; Gao, S.; Kjems, J. Circulating miRNAs as biomarkers for oral squamous cell carcinoma recurrence in operated patients. *Oncotarget* **2017**, *8*, 8206–8214. [CrossRef] [PubMed]

31. Yang, C.J.; Shen, W.G.; Liu, C.J.; Chen, Y.W.; Lu, H.H.; Tsai, M.M.; Lin, S.C. miR-221 and miR-222 expression increased the growth and tumorigenesis of oral carcinoma cells. *J. Oral Pathol. Med.* **2011**, *40*, 560–566. [CrossRef] [PubMed]

32. Zhou, L.; Jiang, F.; Chen, X.; Liu, Z.; Ouyang, Y.; Zhao, W.; Yu, D. Downregulation of miR-221/222 by a microRNA sponge promotes apoptosis in oral squamous cell carcinoma cells through upregulation of PTEN. *Oncol. Lett.* **2016**, *12*, 4419–4426. [CrossRef] [PubMed]

33. Liu, X.; Yu, J.; Jiang, L.; Wang, A.; Shi, F.; Ye, H.; Zhou, X. MicroRNA-222 regulates cell invasion by targeting matrix metalloproteinase 1 (MMP1) and manganese superoxide dismutase 2 (SOD2) in tongue squamous cell carcinoma cell lines. *Cancer Genom. Proteom.* **2009**, *6*, 131–139.

34. Jiang, F.; Zhao, W.; Zhou, L.; Liu, Z.; Li, W.; Yu, D. MiR-222 targeted PUMA to improve sensitization of UM1 cells to cisplatin. *Int. J. Mol. Sci.* **2014**, *15*, 22128–22141. [CrossRef] [PubMed]

35. Jiang, F.; Zhao, W.; Zhou, L.; Zhang, L.; Liu, Z.; Yu, D. miR-222 regulates the cell biological behavior of oral squamous cell carcinoma by targeting PUMA. *Oncol. Rep.* **2014**, *31*, 1255–1262. [CrossRef] [PubMed]

36. Yu, Z.Y.; Bai, Y.N.; Luo, L.X.; Wu, H.; Zeng, Y. Expression of microRNA-150 targeting vascular endothelial growth factor-A is downregulated under hypoxia during liver regeneration. *Mol. Med. Rep.* **2013**, *8*, 287–293. [CrossRef] [PubMed]

37. Naruse, T.; Kawasaki, G.; Yanamoto, S.; Mizuno, A.; Umeda, M. Immunohistochemical study of VEGF expression in oral squamous cell carcinomas: Correlation with the mTOR-HIF-1alpha pathway. *Anticancer Res* **2011**, *31*, 4429–4437. [PubMed]

38. Roy, R.; Singh, R.; Chattopadhyay, E.; Ray, A.; Sarkar, N.D.; Aich, R.; Paul, R.R.; Pal, M.; Roy, B. MicroRNA and target gene expression based clustering of oral cancer, precancer and normal tissues. *Gene* **2016**, *593*, 58–63. [CrossRef] [PubMed]

39. Brase, J.C.; Johannes, M.; Schlomm, T.; Falth, M.; Haese, A.; Steuber, T.; Beissbarth, T.; Kuner, R.; Sultmann, H. Circulating miRNAs are correlated with tumor progression in prostate cancer. *Int. J. Cancer* **2011**, *128*, 608–616. [CrossRef] [PubMed]

40. Zhu, C.; Ren, C.; Han, J.; Ding, Y.; Du, J.; Dai, N.; Dai, J.; Ma, H.; Hu, Z.; Shen, H.; et al. A five-microRNA panel in plasma was identified as potential biomarker for early detection of gastric cancer. *Br. J. Cancer* **2014**, *110*, 2291–2299. [CrossRef] [PubMed]

41. Cabibi, D.; Caruso, S.; Bazan, V.; Castiglia, M.; Bronte, G.; Ingrao, S.; Fanale, D.; Cangemi, A.; Calo, V.; Listi, A.; et al. Analysis of tissue and circulating microRNA expression during metaplastic transformation of the esophagus. *Oncotarget* **2016**, *7*, 47821–47830. [CrossRef] [PubMed]

42. Jo, P.; Azizian, A.; Salendo, J.; Kramer, F.; Bernhardt, M.; Wolff, H.A.; Gruber, J.; Grade, M.; Beissbarth, T.; Ghadimi, B.M.; et al. Changes of Microrna Levels in Plasma of Patients with Rectal Cancer during Chemoradiotherapy. *Int. J. Mol. Sci.* **2017**, *18*, 1140. [CrossRef] [PubMed]

43. Molina-Pinelo, S.; Suarez, R.; Pastor, M.D.; Nogal, A.; Marquez-Martin, E.; Martin-Juan, J.; Carnero, A.; Paz-Ares, L. Association between the miRNA signatures in plasma and bronchoalveolar fluid in respiratory pathologies. *Dis. Markers* **2012**, *32*, 221–230. [CrossRef] [PubMed]

44. Brunet Vega, A.; Pericay, C.; Moya, I.; Ferrer, A.; Dotor, E.; Pisa, A.; Casalots, A.; Serra-Aracil, X.; Oliva, J.C.; Ruiz, A.; et al. microRNA expression profile in stage III colorectal cancer: Circulating miR-18a and miR-29a as promising biomarkers. *Oncol. Rep.* **2013**, *30*, 320–326. [CrossRef] [PubMed]

45. Pigati, L.; Yaddanapudi, S.C.; Iyengar, R.; Kim, D.J.; Hearn, S.A.; Danforth, D.; Hastings, M.L.; Duelli, D.M. Selective release of microRNA species from normal and malignant mammary epithelial cells. *PLoS ONE* **2010**, *5*, e13515. [CrossRef] [PubMed]

46. Shiah, S.G.; Shieh, Y.S.; Chang, J.Y. The Role of Wnt Signaling in Squamous Cell Carcinoma. *J. Dent. Res.* **2016**, *95*, 129–134. [CrossRef] [PubMed]

47. Makarev, E.; Schubert, A.D.; Kanherkar, R.R.; London, N.; Teka, M.; Ozerov, I.; Lezhnina, K.; Bedi, A.; Ravi, R.; Mehra, R.; et al. In silico analysis of pathways activation landscape in oral squamous cell carcinoma and oral leukoplakia. *Cell Death Discov.* **2017**, *3*, 17022. [CrossRef] [PubMed]

48. Kroh, E.M.; Parkin, R.K.; Mitchell, P.S.; Tewari, M. Analysis of circulating microRNA biomarkers in plasma and serum using quantitative reverse transcription-PCR (qRT-PCR). *Methods* **2010**, *50*, 298–301. [CrossRef] [PubMed]

International Journal of
Molecular Sciences

MDPI

Review

MicroRNAs as Potential Biomarkers in Merkel Cell Carcinoma

Aelita Konstantinell, Dag H. Coucheron, Baldur Sveinbjørnsson and Ugo Moens *

Molecular Inflammation Research Group, Department of Medical Biology, The Arctic University of Norway, N-9037 Tromsø, Norway; aelita.konstantinell@uit.no (A.K.); dag.coucheron@uit.no (D.H.C.); baldursveinbjornsson@uit.no (B.S.)
* Correspondence: ugo.moens@uit.no; Tel.: +47-77-644-622

Received: 10 May 2018; Accepted: 22 June 2018; Published: 26 June 2018

Abstract: Merkel cell carcinoma (MCC) is a rare and aggressive type of skin cancer associated with a poor prognosis. This carcinoma was named after its presumed cell of origin, the Merkel cell, which is a mechanoreceptor cell located in the basal epidermal layer of the skin. Merkel cell polyomavirus seems to be the major causal factor for MCC because approximately 80% of all MCCs are positive for viral DNAs. UV exposure is the predominant etiological factor for virus-negative MCCs. Intracellular microRNA analysis between virus-positive and virus-negative MCC cell lines and tumor samples have identified differentially expressed microRNAs. Comparative microRNA profiling has also been performed between MCCs and other non-MCC tumors, but not between normal Merkel cells and malignant Merkel cells. Finally, Merkel cell polyomavirus encodes one microRNA, but its expression in virus-positive MCCs is low, or non-detectable or absent, jeopardizing its biological relevance in tumorigenesis. Here, we review the results of microRNA studies in MCCs and discuss the potential application of microRNAs as biomarkers for the diagnosis, progression and prognosis, and treatment of MCC.

Keywords: exosomes; extracellular microRNA; large T-antigen; protein-miRNA complex; small t-antigen

---

1. MicroRNAs

MicroRNAs (miRNAs) are ~18–24 nucleotides long, non-coding RNA molecules encoded by the genomes of viruses, protists, plants and animals [1]. The human genome may code for more than 3000 miRNAs [2,3]. MiRNAs are produced through multiple processes of larger precursor transcripts referred to as primary miRNAs, which are generated by RNA polymerase II or RNA polymerase III. Primary miRNAs transcribed from genome DNAs are cleaved into precursor miRNAs, which have a short hairpin structure, and subsequently, are exported from the nucleus to the cytoplasm. Lastly, the duplex RNA is processed by degrading one of the strands (the passenger strand) and leaving the other strand as a mature guide miRNA [4,5]. It is also possible to have arm switching, in which the mature guide sequence from a pre-miRNA may shift from one arm to the other in different tissues [6]. In addition to the mature miRNA, isoforms (isomiRs) are also produced that are variants of the mature miRNA. Numerous studies have demonstrated that these isomiRs have functional importance [7,8].

MiRNAs not only reside intracellularly, but also can be released from cells in extracellular vesicles, such as exosomes (vesicles with a characteristic size of ~30–150 nm in diameter), and in apoptotic bodies [9–11]. Moreover, extracellular miRNAs in complex with proteins have been described [12]. These circulating miRNAs can be taken up by recipient cells, and in this way, play a role in intercellular communication [13]. It is estimated that approximately 10% of secreted miRNAs are encapsulated in extracellular vesicles, whereas 90% are secreted in a vesicle–free state as complexes with proteins [14]. Plasma from healthy blood donors and media from cell cultures were shown to contain miRNAs associated with the protein argonaute 2 (Ago2), and Ago2-miRNA complexes were stable for at least two months at room temperature. It is not known whether these Ago2-miRNA complexes are byproducts of dying or dead cells or if they are actively released from living cells [12,15]. Recently, neuropilin-1 was identified as a receptor for Ago2-miRNA complexes, suggesting a selective uptake of this protein-miRNA by target cells [16]. Other proteins that have been reported to be associated with extracellular miRNAs are high-density lipoproteins (HDL) and nucleophosmin 1 (NPM1) [17–19]. HDL-miRNAs exhibit a distinct expression pattern in relation to different pathological conditions and may thus have biomarker potentials [17,18]. Extracellular miRNAs in complex with NPM1 were detected in a serum-free medium of HepG2 (human hepatocellular), A549 (human lung carcinoma), T98 (human glioblastoma) and BSEA2B (normal human bronchial epithelium) cells [19]. Zernecke and co-workers found that endothelial cell-derived apoptotic bodies generated during atherosclerosis were enriched in miR-126 [20]. Apoptotic bodies have been reported in Merkel cell carcinoma (MCC) [21], but the presence of miRNAs has not been investigated thus far.

Mature miRNAs inhibit gene expression at the posttranscriptional level by binding to complementary sequences in mRNA targets, which prevent their translation or induce their degradation [22]. However, miRNAs can also activate gene expression by binding to target sequences in promoters [23]. MiRNAs can interfere with numerous cellular processes, including cell proliferation, differentiation, development, apoptosis, angiogenesis, metabolism, and immune responses [24–29]. An aberrant expression of miRNAs is involved in pathogenic processes, including cancer [30–35].

Many human miRNAs are expressed in a cell-type, cellular-process, and disease-specific manner. Moreover, miRNAs are relatively stable. This makes miRNAs relevant as biomarkers for physiological and pathogenic processes. For cancer, in particular, interest in identifying circulating miRNAs as prognostic and diagnostic markers is growing. Both mature miRNA and isomiR profiles may be used as biomarkers [36].

2. Merkel Cell Carcinoma

MCC is an aggressive type of cancer as trabecular cell carcinoma of the skin, which was first described by Cyril Toker in 1972 [37]. Later, he showed that the cellular origin of this cancer was Merkel cells; hence, these tumors were renamed MCC. Merkel cells were originally described as Tastzellern or touch cells in the skin by Frederick Sigmund Merkel in 1875 (for a recent review, see [38]), and are located in the basal layer of the skin (in particular, around hair follicles) and mucosa. They serve as mechanoreceptors for gentle touch stimulation, and are associated with afferent sensory nerves to form the Merkel cell-neurite complex. The exact origin of Merkel cells remains controversial. It has been suggest that these cells originate from one of the neurocrest derivatives [39–41], keratinocytes, epidermal fibroblasts, early B cells or hair follicle stem cells [42–45].

MCC is associated with a poor prognosis, as more than one-third of patients die from the disease compared to ~15% for malignant melanoma. Approximately half of MCC patients with advanced diseases survive for nine months or less [46,47]. The highest worldwide incidence of MCCs is found in Australia (1.6 cases per 100,000 persons), followed by Northern America (0.6/100,000) and Europe (~0.3/100,000). The higher incidence in Australia is attributed to high year-round UV exposure [37,47–50]. The median age at diagnosis is roughly 75 years old, while only 12% of MCC patients are younger than 60 years of age [51]. MCC mostly presents on sun-exposed areas, such as the head and neck and the extremities, and can also occur on the buttocks, oral mucosa, the penis and vulva [52–54].

UV light exposure is a major factor for MCC, but immune deficiencies, fair skin, age (immune senescence), association with other cancer, and chronic inflammation can also be contributing factors [47,49,55]. In 2008, a novel virus was identified in eight out of 10 MCC samples [56]. This virus was named Merkel cell polyomavirus (MCPyV), and has subsequently been detected in 80% of all examined MCC samples [57,58]. The oncogenic potentials of this virus are predominantly attributed to two of its viral proteins: large T-antigen (LTAg) and small t-antigen (STAg). Similar to the LTAg and

STAg of other polyomaviruses, MCPyV LTAg and STAg can transform cells in vitro and induce tumors in animal models [59–66]. Serological studies of healthy adults showed that ~40–85% of the individuals have antibodies against MCPyV. An age-dependence seroprevalence was measured, increasing from ~10–20% (age 1–5 years) to ~80% (>70 years) [67–73]. MCPyV seems to be a normal inhabitant of dermal fibroblasts in the skin, and infectious virus particles are chronically shed, thereby suggesting that direct physical contact may be one mode of transmission [74,75]. Whereas the virus is found in an episomal state in non-malignant cells, a characteristic for all virus-positive MCC tumors is that the viral genome is integrated into a clonal pattern, and that a non-sense mutation is present in the LTAg-encoding gene encoding a C-terminal truncated protein. Whether this mutation occurs before or after integration, or whether both scenarios can occur is not known [76–78]. The truncated LTAg has lost its ability to support viral replication, but has retained its oncogenic potentials [61,79–81].

The MCPyV genome encodes a single miRNA precursor expressed from the late strand, which can produce two mature miRNAs referred to as MCV-miR-M1-5p and MCV-miR-M1-3p [82,83]. MCV-miR-M1-5p seems to be more abundant than MCV-miR-M1-3p in MCPyV-infected neuroectodermal tumor PFSK-1 cells [84], and in HEK293T cells transfected with an expression plasmid encompassing the pre-miR-M1 sequence [85]. The seed sequence is 5'-CUGGAAG-3' or 5'-GGAAGAA-3' for MCV-miR1-5p and 5'-UGCUGGA3'- for MCV-miR-M1-3p [82–84]. The MCV-miR-M1-5p and MCV-miR-3p sequences are perfectly complementary to the coding sequences of LTAg, hence suggesting that they can repress translation of the LTAg mRNA. Indeed, using a dual-luciferase reporter assay, MCV-miR-M1 was shown to attenuate the expression of LTAg [82,86], whereas Theiss and collaborators showed that this miRNA down-regulates expression of LTAg, limits viral replication, and is necessary to establish a long-term persistent infection of MCPyV-infected neuroectodermal tumor PFSK-1 cells [84]. Based on the seed region, predicted human target genes include genes encoding proteins involved in transcription, cell communication, immune response, apoptosis, autophagy and proteasomal degradation, but it remains to be established if they are genuine targets [82,83,86]. Using HEK293 cells that stably express MCV-miR-M1-5p or MCV-miR-M1-3p, SP100 mRNA was verified as a bona fide target for MCV-miR-M1-5p, but not MCV-miR-M1-3p [86]. This protein is implicated in the innate immune response against dsDNA viruses, including MCPyV [87]. The authors also found that CXCL8 transcript levels were significantly different expressed in stably expressing MCV-miR-M1 cells compared to control cells, but despite the putative MCV-miR-M1-3p seed sequence, this change was indirect and mediated by SP100 [86]. The results of this work suggest that MCPyV uses its miRNA to evade the immune system in order to establish infection, but it also illustrates that the presence of a putative miRNA seed sequence in a target mRNA does not imply that this transcript is targeted by the miRNA. The role of MCV-miR-M1 in MCC tumorigenesis is less clear. Examining the expression of viral miRNA MCV-miR-M1-5p in MCC samples showed that up to 29–80% of the specimens expressed detectable levels of MCV-miR-M1-5p [82,83]. However, the levels of this MCPyV-encoded miRNA in MCCs are low, and are estimated to be less than 0.005% of total miRNA levels [82]. This was confirmed by studies in the MCPyV-positive MCC cell lines WaGa and MKL-1, in which MCV-miR-M1 made up 0.001% of all mature miRNAs [84], and 0.0067%, 0.007%, and 0.0025% of MKL-1a, MKL-1b, MKL-1c cells, respectively [85]. Because its absence or low or undetectable levels, MCV-miR-M1-5p's biological relevance in cancer and its value as biomarkers are doubtful. The expression of MCV-miR-M1-3p in virus-positive MCCs has not been examined, but as mentioned above, this miRNA is even less abundant than the 5p strand in MCPyV-infected cells.

3. Merkel Cell Carcinoma and MicroRNAs

3.1. Intracellular MicroRNAs

Ning and colleagues determined the microRNAome by next-generation sequencing of three MCCs, one melanoma, one squamous cell carcinoma (SCC), one basal cell carcinoma (BCC) and one normal skin sample to identify miRNAs specific to MCC [88]. They found that eight miRNAs were upregulated in MCC, while three were downregulated compared to non-MCC cutaneous tumors and normal skin (see Table 1). This differential expression of these miRNAs was confirmed by quantitative reverse transcriptase PCR (qRT-PCR) on a total RNA isolated from 20 MCC samples and from the MCPyV-positive MS-1 MCC cell line. In situ hybridization also confirmed high expression of miR-182 in a MCC sample, but low in surrounding tissue and normal skin. The viral status in the MCC samples was not given. However, the authors also evaluated the expression of four of these MCC- and MS-1-enriched expressed miRNAs (miR-182, miR-183, miR-190b, and miR-340) in the MCPyV-negative cell line MCC13 and found that they demonstrated low expression in these cells. Thus, miR-182, miR-183, miR-190b and miR-340 may be used as biomarkers for MCPyV-positive MCC.

Table 1. Differential expressed miRNAs in MCC or MCC-derived cell lines. See the text for details.

Sample (*n*)	Method	miR↑ [1]	miR↓ [2]	Reference
MCC (3)	NGS [3]	miR-7 miR-9 miR-182 miR-183 miR-190b miR-340 miR-502-3p miR-873	miR-125b miR-374c miR-3170	[88]
MS-1	NGS	miR-182 miR-183 miR-190b miR-340		[88]
MCPyV-positive (15) vs. MCPyV-negative MCC (13)	Microarray	miR-30a miR-34 miR-142-3p miR-1539	miR-181d	[89]
MCC (14), MKL-1, MKL-2, MS-1 [4] versus BCC, normal skin, MCC13, MCC26, UiOS [5]	NGS	miR-375 [6]		[90]
MCPyV-positive vs. MCPyV-negative MCC	Microarray	miR-30a-3p miR-30a-5p miR-34a miR-375 miR-769-5p	miR-203	[91]
Primary vs. metastatic MCC	Microarray	miR-150		[91]

[1] Higher levels compared to MCPyV-negative MCCs; [2] lower levels compared to MCPyV-negative MCCs; [3] next-generation sequencing; [4] MCPyV-positive MCC cell lines; [5] MCPyV-negative MCC cell lines; [6] this miR was elevated in MCC tumors and cell lines, independently of the virus state.

Although tumor-promoting and tumor-inhibiting properties have been attributed to these miRNAs, their biological relevance in MCC remains to be investigated. MiR-182 stimulates metastasis and proliferation, but exerts opposite effects depending on the cancer type [92–95]. The over-expression of miR-183 inhibits cell migration and invasion in vitro (e.g., [96–98]), while other studies demonstrated that miR-183 stimulates cell proliferation and migration [99,100], and is a prognostic biomarker for breast cancer [101]. MiR-190b was shown to inhibit cell proliferation and induce apoptosis in osteosarcoma U2OS cells [102]. MiR-340 has tumor suppressing properties by inhibiting proliferation, invasion and metastasis, and stimulating apoptosis [103–107]. However, miR-340 has also been shown to promote tumor growth in gastric cancer [108]. The miRNAs that had a higher expression in MCPyV-negative MCCs, compared to virus-positive tumors, include miR-125b, miR-374c and miR-3170 [88]. MiR-125b can suppress and promote cancer progression, and further down-regulate γδ T cell activation and cytotoxicity [109], as well as cells involved in anti-tumor surveillance [110]. MiR-374c was identified as a novel miRNA in cervical tumors, and miR-3170 was first identified in the miRNAome of melanoma, which has also been found in breast cancer tumors [111,112]. A possible role of these two miRNAs in cancer has not been elucidated, nor has the plausible involvement in MCC been solved.

A comparison of the microRNAome of MCPyV-positive and MCPyV-negative MCCs revealed that approximately 2.5- to 5-fold higher levels of miR-30a, miR-34a, miR-142-3p and miR-1539 in virus-positive MCCs and 3.5-fold higher levels of miR-181d in virus-negative MCCs [89]. MiR-30a has a dual role in cancer and can act as an oncogene or an onco-suppressor in different cancers, and several of its target genes have been identified (reviewed in [113]). Although miR-34a is a tumor suppressor [114], Veija et al. speculated that an over-expression of miR-34a could decrease p53 expression, thereby interfering with apoptosis, angiogenesis and DNA repair [89]. MiR-142-3p can inhibit cell proliferation and invasion, but high levels have also been correlated with cancer progression [115]. MiR-181d can act as a tumor suppressor [116], but the down-regulation of miR-181d resulted in a decreased proliferation and migration of pancreatic cancer cells [117]. To the best of our knowledge, the role of miR-1539 has not been investigated. The prognostic value of miR-30a, miR-142-3p, miR-1539 and miR-181d is jeopardized, because qRT-PCR validation demonstrated that only miR-34a was significantly under-expressed in virus-negative MCCs compared to virus-positive MCC samples [89]. Whether any of these miRNAs contribute to MCC tumorigenesis remains to be established.

Deep sequencing of RNA purified from normal skin ($n = 5$), BCC ($n = 5$), MCC ($n = 14$), and MCPyV-negative (MCC13, MCC26, UiOS) and MCPyV-positive (MKL-1, MKL-2; MS-1) MCC cells showed that miR-375 is specific for MCC [90]. The miR-375 concentrations were 60-fold higher in the MCC group than in the non-MCC (normal skin and BCC) group. The enrichment of miR-375 seems to be independent of the viral state, because elevated miR-375 levels were found in both virus-negative and virus-positive tumors and tumor cell lines. Of the five skin samples that were examined, three were MCPyV positive, one was virus negative and one was not tested. Of the five BCC samples, four were virus negative and one was not tested. Similarly, since no increased miR-375 levels were found in the virus-positive non-MCC samples, the presence of the virus seems not to affect the expression of miR-375. Although not discussed by the authors, there was a tendency for higher expression levels of miR-9 and miR-188 in MCC samples. MiR-188 suppresses proliferation in different cancers [118–121]. MiR-9 can stimulate or inhibit cell proliferation and metastasis depending on the type of cancer, whereas high expression levels in most cancers are associated with poor survival of the patients, except for ovarian cancer patients, in which an inverse correlation was found [122,123]. MiR-375 has been described as a tumor suppressor known to impede cell proliferation, to prevent cancer cell migration, and to inhibit autophagy, thereby generating an antitumor effect in liver cancer [124–129]. Therefore, it seems surprising that this miRNA is over-expressed in MCC. Nonetheless, an over-expression of miR-375 has also been implied in prostate carcinogenesis and disease progression, while an up-regulation of miR-375 is associated with a poor prognosis in pediatric acute myeloid leukemia [130,131], thus indicating a dual role for miR-375 in cancer. Moreover, miR-375 was shown to inhibit autophagy in hepatocellular carcinoma [132], but whether this role of miR-375 is of importance in MCC is unknown.

A comparison of the intracellular miRNA expression profiles in 10 MCPyV-negative and 16 MCPyV-positive MCCs by a miRNA microarray-based method identified 36 over-expressed and 20 under-expressed miRNAs in virus-positive MCCs compared to virus-negative MCCs [91]. Among these, a significant over-expression of miR-30a-3p, miR-30a-5p, miR-34a, miR-375 and miR-769-5p, and a significant under-expression of miR-203, were confirmed by qRT-PCR. A putative role of miR-30a, miR-34a and miR-375 in oncogenesis was described above. MiR-769 expression was strongly increased in human melanoma cells and clinical tissues compared with their corresponding controls. The over-expression of miR-769 promoted cell proliferation in the human melanoma cell line A375 [133]. MiR-769 may exert these functions by targeting glycogen synthase kinase 3B, while a similar mechanism may be operational in MCC oncogenesis. It is not known whether MCPyV LTAg and/or STAg stimulate the expression of miR-30a-3p, miR-30a-5p, miR-34a, miR-375 and miR-769-5p. The possible involvement of miR-203 in MCC oncogenesis was examined by over-expressing miR-203 in three MCPyV-negative MCC cell lines [91]. This resulted in reduced cell growth, more cells in G1 and less in the G2 phase, but no apparent effect on apoptosis compared to cells transfected with miRNA mimic control. Moreover, survivin expression was reduced. The over-expression of miR-203 in the MCPyV-positive WaGa MCC cell line had no significant effect on cell proliferation, cell cycle progression and survivin expression levels. These results suggest that miR-203 only regulates survivin expression in virus-negative MCCs, but not in MCPyV-positive MCCs, in which LTAg seems to repress survivin expression by sequestering pRb [134]. The same group also examined differentially expressed miRNAs in primary and metastatic MCC tumors [91]. They found that 92 miRNAs were over-expressed in metastasis compared to primary tumors. The four most up-regulated miRNAs were miR-150, miR-142-3p, miR-483-5p and miR-630, but qRT-PCR validation revealed that only miR-150 was significantly overexpressed.

Xie et al. found that miR-375 was specifically over-expressed in MCPyV-positive MCCs, while Renswick et al. reported that miR-375 was specific for MCC, independent of the viral state in the tumors [90,91]. The discrepancy in these results may be explained by the differences in MCC samples that were examined, or because different methods (next-generation sequencing versus miRNA microarray) were used.

The role of MCPyV on miRNA expression in non-small cell lung cancer (NSCLC) was investigated by Lasithiotaki and co-workers [135]. The expression of miR-21, miR-145, miR-146a, miR-155, miR-302c, miR-367 and miR-376c was examined by qRT-PCR in MCPyV-positive and MCPyV-negative NSCLC. MiR-21 and miR-376c were up-regulated, whereas miR-145 was down-regulated in virus-positive NSCLC ($n = 8$) compared to virus-negative NSCLC ($n = 16$). MiR-21 and miR-376c expression levels were also higher in virus-positive NSCLC versus adjacent healthy tissue samples ($n = 10$; 5 MCPyV-positive and 5 MCPyV-negative), while miR-145 levels in MCPyV-negative NSCLC was higher than in control samples. To the best of our knowledge, none of the miRNAs investigated by Lasithiotaki et al. have been described in MCC, except miR-146a which was enriched ~8-fold in exosomes derived from MCPyV-negative MCC13 and MCC26 cell lines compared to virus-positive MKL-1 and MKL-2 cell lines (A.K., D.H.C., B.S., U.M., University of Tromsø, Norway, 2018).

3.2. Extracellular MicroRNAs and Merkel Cell Carcinoma

The presence of extracellular miRNA-protein complexes secreted by MCC cell lines or in MCC patients has not been examined thus far. Likewise, the occurrence of miRNAs in apoptotic bodies has not been investigated although apoptotic bodies have been reported in MCC [21]. We have applied next-generation sequencing to examine the microRNAome in exosomes purified from the MCPyV-negative MCC13 and MCC26 and the MCPyV-positive MKL-1 and MKL-2 MCC cell lines. On average, there were 20.4 million reads per sample (three independent exosome samples of each cell line), with the number of miRNAs per sample varying approximately between 200 and 400. Of the previously identified intracellular miRNAs identified in MCC samples of MCC cell lines (Table 1), our preliminary results confirmed the presence of miR-30a, miR-125b, mi-183, miR-190b and miR-375 in exosomes. MCV-miR-M1 was not detected in any of our samples (A.K., D.H.C., B.S., U.M., University of Tromsø, Norway, 2018).

4. MicroRNAs as Biomarkers and Therapeutic Targets in Merkel Cell Carcinoma

MiRNAs are key components of cells in both normal and pathogenic states. The miRNA expression pattern of normal cells versus malignant cells differs, and cancer-cell-specific miRNAs are being used as biomarkers in different cancers [136–141]. Exosomal miRNAs have become attractive cancer biomarkers because exosomes are easily obtainable from body fluids, such as blood and urine without the requirement of a biopsy sample of the tumor. Exosomal miRNAs in plasma or urine are used as biomarkers for different malignancies, including melanoma, breast, colon, prostate, renal and gastric cancer [142–148]. Only a few studies have examined the miRNAome of MCC and miRNAs are not yet used as biomarkers. One of the pitfalls of using MCC-derived miRNAs as biomarkers is that the miRNA expression pattern of normal Merkel cells has not been determined because these cells are rare. Hence, it is not known whether the MCC-derived miRNAs are specific for the malignant cells or also expressed by non-malignant Merkel cells. Another problem is the lack of common miRNAs among the MCC expressed miRNAs identified so far by independent studies. MiR-30a, miR-34 and miR-375 were reported by Renwick et al. and Xie et al., but not by others (Table 1) [90,91]. We found miR-30 and miR-375 in exosomes derived from MCC cell lines, but it is not known whether these miRNAs are also present in non-transformed Merkel cells. MiR-30a is expressed in different cell types, including normal dermal fibroblasts, keratinocytes and endothelial cells [149–151], whereas miR-375 is present in normal epithelial, pituitary and pancreatic β-cells [152–154], jeopardizing the value of these miRNAs as specific biomarkers for MCC. Additional tumor specimens must be investigated in order to isolate MCC-specific miRNAs and their potential use as biomarkers should be verified.

MiRNAs may also be used to determine the viral state in the MCC. Levels of miR-30a and miR-34 were increased in MCPyV-positive MCCs compared to MCPyV-negative [89,91]; hence, these miRNAs may be applied to distinguish between virus-positive and virus-negative cancers. MiR-375 was found to be a specific miRNA for virus-positive MCCs [90,91], but this could not be confirmed by others who found this miRNA in both MCPyV-positive and MCPyV-negative MCC [89] and in exosomes of both virus-positive and virus-negative MCC cell lines (our unpublished results). Whether miR-375 can be used as a hallmark for MCPyV-positive MCCs needs further investigation.

As for diagnostic purposes, real-time PCR methods with primers against exosomal miRNAs specific for MCPyV-positive MCCs could replace the commonly used PCR with MCPyV sequence-specific on DNA extracted from a tumor sample. The advantage of real-time PCR on a circulating miRNA is that body fluids, such as blood or urine, are readily accessible sources and more convenient for the MCC patient to obtain than using a biopsy, and that the number of miRNA molecules can be estimated.

Another criterion for a useful biomarker is that it can predict the outcome of the disease. Studies by Xie et al. found that higher levels of miR-150 were associated with a worse prognosis of MCC [91]. Quantifying MCC-specific miRNA levels may also provide information on the disease progression and the efficiency of treatment in the case of miRNA-target therapy. There is a need for new and improved therapy of MCC patients. As of today, MCC treatment includes surgery, radiotherapy and chemotherapy. In a few cases, a spontaneous regression of primary and metastatic MCC has been reported (see e.g., [155–158]). Recently, immunotherapy based on blocking the PD1-PDL-1 pathway by either anti-PD1 antibodies (pembrolizumab, nivolumab) or anti-PDL-1 antibodies (avelumab) has demonstrated favorable responses, with a six-month progression-free survival in 40–85%, and even complete resolution of the tumors in some patients [159–167]. Avelumab became the first Food and Drug Administration-approved drug for the treatment of MCC (https://www.fda.gov/newsevents/newsroom/pressannouncements/ucm548278.htm, 10 May 2018). Treatment with the anti-CTLA-4 antibody ipilimumab has also shown beneficial effects in metastatic MCC [168,169]. However, not all patients have a positive response, so the development of additional therapies is necessary. Drugs against MCC-specific miRNAs or their target transcripts can supplement immunotherapy. Clinical trials with miRNAs against some pathological conditions except MCC have been initiated [170–172]. One of the miRNA-based clinical trials includes miR-34, which is expressed in MCC [89,91].

Finally, miRNAs may also help in solving the enigma of the origin of Merkel cells. MiRNA signatures may be an alternative to immunhistochemical staining. This requires the identification of cell-specific miRNAs of neural crest cells, keratinocytes, epidermal fibroblasts, early B cells and hair follicle stem cells, cells that have been suggested to be the origin of Merkel cells (see Section 2).

In conclusion, multi-center microRNAome studies on a large number of MCC samples or biofluids of patients are required to identify valuable miRNA biomarkers. These data should be linked to parameters, such as the clinical features of the patient, the stage of the tumor (primary or metastatic), viral states and LTAg and STAg expression, age and gender of the patient. MiRNA profiling may be used in determining the prognosis and progression of the disease, and monitoring the response to therapy (Figure 1). Exosomes have been found in a number of biological fluids including plasma, urine, breast milk, semen, cerebrospinal fluid and saliva [173]. Exosomal miRNAs, in addition to intracellular miRNAs, may therefore be easily accessible biomarkers.

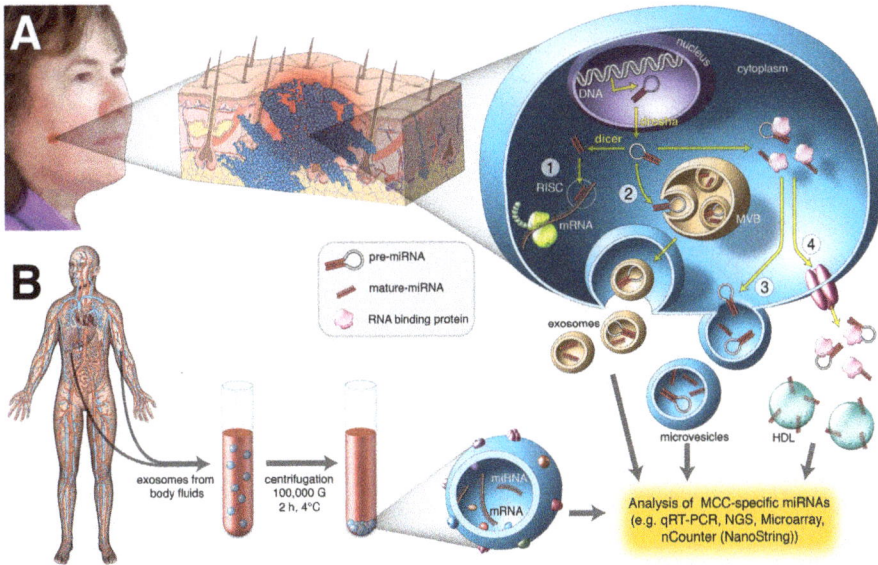

Figure 1. Detection of MCC-specific miRNAs from tumor biopsies or from body fluids. (**A**) The presence of intracellular or/and extracellular MCC-specific miRNAs is examined by qRT-PCR using specific primers. Intracellular miRNAs are amplified from a total RNA isolated from MCC tumor tissue, while extracellular miRNAs are amplified from a RNA extracted from purified exosomes or from the extracellular environment. The biogenesis of a miRNA is shown. A pre-miRNA is transported from the nucleus to the cytoplasm, and when processed to mature miRNA, it binds target mRNA (step 1). Pre-miRNAs and miRNAs can also be enclosed in vesicles and excreted in exosomes (step 2) or other extracellular vesicles (step 3). Pre-miRNAs and miRNAs can also release from the cell in complex with RNA-binding proteins, such as Argonaut 2 or nucleophosmin-1, or in complex with high-density lipoproteins (step 4); (**B**) circulating exosomes are purified from body fluids (e.g., blood, urine, lymphatic fluid, saliva) and a total RNA is extracted. MCC-specific miRNAs are subsequently detected by qRT-PCR applying specific primers, next-generation sequencing (NGS), microarray or nCounter.

5. Future Challenges

MiRNAs can be used as reliable biomarkers in several cancers [137–141], and miRNA-based cancer therapy is being developed and tested [170,171]. As outlined above, little research has been done on MCC-specific miRNAs. It is reasonable to wonder whether there is any clinical value for miRNA in MCC and if so, what could it be?

- MCC-specific miRNAs as biomarkers still have a long way to go. Consensus intracellular MCC miRNAs have not yet been identified, and circulating miRNAs have not been investigated.

We are currently studying the exosome miRNAome of MCC cell lines with the aim of identifying MCC-specific extracellular miRNAs. Cell–cell communication is important in the tumor microenvironment and one way of communication is exosomes [174]. Thus, analyzing the miRNAome of exosomes may provide clues on how tumor cells promote survival, growth and metastasis by modulating the tumor microenvironment [175].

- The intracellular microRNAome or circulating miRNA may be used to discriminate between MCPyV-negative and MCPyV-positive MCCs. So far, unambiguous miRNAs that allow distinguishing between virus-negative and virus-positive tumors have not yet been described.
- Can MCC-specific miRNAs be used as therapeutic targets? The biological importance of miRNAs in MCC oncogenesis is incompletely understood. In fact, most of the currently reported miRNAs in MCC have dual functions (oncogenic or tumor suppressive roles) in other cell systems, so that targeting their expression may be a double-edged sword. The exact contributing role of miRNAs in MCC is required to design efficient and specific therapies.
- Affordable and easy laboratory tests based on these MCC-specific miRNA biomarkers should be developed to improve the diagnosis, prognosis, and progression of this cancer.

Author Contributions: A.K., D.H.C, B.S. and U.M. wrote the paper.

Funding: Studies in our laboratory were funded by the Olav and Erna Aakre Foundation for Cancer (Tromsø, Norway) and Raagholt Stiftelsen (Oslo, Norway). The publication charges for this article have been funded by a grant from the publication fund of UiT the Artic University of Norway.

Acknowledgments: We greatly acknowledge Roy Lyså for preparing Figure 1.

Conflicts of Interest: The authors declare no conflicts of interest.

References

1. He, L.; Hannon, G.J. MicroRNAs: Small RNAs with a big role in gene regulation. *Nat. Rev. Genet.* **2004**, *5*, 522–531. [CrossRef] [PubMed]
2. Kozomara, A.; Griffiths-Jones, S. miRBase: Annotating high confidence microRNAs using deep sequencing data. *Nucleic Acids Res.* **2014**, *42*, D68–D73. [CrossRef] [PubMed]
3. Londin, E.; Loher, P.; Telonis, A.G.; Quann, K.; Clark, P.; Jing, Y.; Hatzimichael, E.; Kirino, Y.; Honda, S.; Lally, M.; et al. Analysis of 13 cell types reveals evidence for the expression of numerous novel primate- and tissue-specific microRNAs. *Proc. Natl. Acad. Sci. USA* **2015**, *112*, E1106–E1115. [CrossRef] [PubMed]
4. Kim, V.N.; Han, J.; Siomi, M.C. Biogenesis of small RNAs in animals. *Nat. Rev. Mol. Cell Biol.* **2009**, *10*, 126–139. [CrossRef] [PubMed]
5. Ha, M.; Kim, V.N. Regulation of microRNA biogenesis. *Nat. Rev. Mol. Cell Biol.* **2014**, *15*, 509–524. [CrossRef] [PubMed]
6. Marco, A.; Ninova, M.; Griffiths-Jones, S. Multiple products from microRNA transcripts. *Biochem. Soc. Trans.* **2013**, *41*, 850–854. [CrossRef] [PubMed]
7. Neilsen, C.T.; Goodall, G.J.; Bracken, C.P. IsomiRs—The overlooked repertoire in the dynamic microRNAome. *Trends Genet.* **2012**, *28*, 544–549. [CrossRef] [PubMed]
8. Tan, G.C.; Dibb, N. IsomiRs have functional importance. *Malays. J. Pathol.* **2015**, *37*, 73–81. [PubMed]
9. Lunavat, T.R.; Cheng, L.; Kim, D.K.; Bhadury, J.; Jang, S.C.; Lasser, C.; Sharples, R.A.; Lopez, M.D.; Nilsson, J.; Gho, Y.S.; et al. Small RNA deep sequencing discriminates subsets of extracellular vesicles released by melanoma cells—Evidence of unique microRNA cargos. *RNA Biol.* **2015**, *12*, 810–823. [CrossRef] [PubMed]
10. Zhang, J.; Li, S.; Li, L.; Li, M.; Guo, C.; Yao, J.; Mi, S. Exosome and exosomal microRNA: Trafficking, sorting, and function. *Genom. Proteom. Bioinform.* **2015**, *13*, 17–24. [CrossRef] [PubMed]
11. Turchinovich, A.; Tonevitsky, A.G.; Burwinkel, B. Extracellular miRNA: A Collision of Two Paradigms. *Trends Biochem. Sci.* **2016**, *41*, 883–892. [CrossRef] [PubMed]
12. Turchinovich, A.; Weiz, L.; Langheinz, A.; Burwinkel, B. Characterization of extracellular circulating microRNA. *Nucleic Acids Res.* **2011**, *39*, 7223–7233. [CrossRef] [PubMed]
13. Hannafon, B.N.; Ding, W.Q. Intercellular communication by exosome-derived microRNAs in cancer. *Int. J. Mol. Sci.* **2013**, *14*, 14240–14269. [CrossRef] [PubMed]

14. Kai, K.; Dittmar, R.L.; Sen, S. Secretory microRNAs as biomarkers of cancer. *Semin. Cell Dev. Biol.* **2018**, *78*, 22–36. [CrossRef] [PubMed]

15. Arroyo, J.D.; Chevillet, J.R.; Kroh, E.M.; Ruf, I.K.; Pritchard, C.C.; Gibson, D.F.; Mitchell, P.S.; Bennett, C.F.; Pogosova-Agadjanyan, E.L.; Stirewalt, D.L.; et al. Argonaute2 complexes carry a population of circulating microRNAs independent of vesicles in human plasma. *Proc. Natl. Acad. Sci. USA* **2011**, *108*, 5003–5008. [CrossRef] [PubMed]

16. Prud'homme, G.J.; Glinka, Y.; Lichner, Z.; Yousef, G.M. Neuropilin-1 is a receptor for extracellular miRNA and AGO2/miRNA complexes and mediates the internalization of miRNAs that modulate cell function. *Oncotarget* **2016**, *7*, 68057–68071. [CrossRef] [PubMed]

17. Vickers, K.C.; Palmisano, B.T.; Shoucri, B.M.; Shamburek, R.D.; Remaley, A.T. MicroRNAs are transported in plasma and delivered to recipient cells by high-density lipoproteins. *Nat. Cell Biol.* **2011**, *13*, 423–433. [CrossRef] [PubMed]

18. Niculescu, L.S.; Simionescu, N.; Sanda, G.M.; Carnuta, M.G.; Stancu, C.S.; Popescu, A.C.; Popescu, M.R.; Vlad, A.; Dimulescu, D.R.; Simionescu, M.; et al. MiR-486 and miR-92a Identified in Circulating HDL Discriminate between Stable and Vulnerable Coronary Artery Disease Patients. *PLoS ONE* **2015**, *10*, e0140958. [CrossRef] [PubMed]

19. Wang, K.; Zhang, S.; Weber, J.; Baxter, D.; Galas, D.J. Export of microRNAs and microRNA-protective protein by mammalian cells. *Nucleic Acids Res.* **2010**, *38*, 7248–7259. [CrossRef] [PubMed]

20. Zernecke, A.; Bidzhekov, K.; Noels, H.; Shagdarsuren, E.; Gan, L.; Denecke, B.; Hristov, M.; Koppel, T.; Jahantigh, M.N.; Lutgens, E.; et al. Delivery of microRNA-126 by apoptotic bodies induces CXCL12-dependent vascular protection. *Sci. Signal.* **2009**, *2*, ra81. [CrossRef] [PubMed]

21. Mori, Y.; Hashimoto, K.; Tanaka, K.; Cui, C.Y.; Mehregan, D.R.; Stiff, M.A. A study of apoptosis in Merkel cell carcinoma: An immunohistochemical, ultrastructural, DNA ladder, and TUNEL labeling study. *Am. J. Dermatopathol.* **2001**, *23*, 16–23. [CrossRef] [PubMed]

22. Iwakawa, H.O.; Tomari, Y. The Functions of MicroRNAs: MRNA Decay and Translational Repression. *Trends Cell Biol.* **2015**, *25*, 651–665. [CrossRef] [PubMed]

23. Place, R.F.; Li, L.C.; Pookot, D.; Noonan, E.J.; Dahiya, R. MicroRNA-373 induces expression of genes with complementary promoter sequences. *Proc. Natl. Acad. Sci. USA* **2008**, *105*, 1608–1613. [CrossRef] [PubMed]

24. Wilfred, B.R.; Wang, W.X.; Nelson, P.T. Energizing miRNA research: A review of the role of miRNAs in lipid metabolism, with a prediction that miR-103/107 regulates human metabolic pathways. *Proc. Natl. Acad. Sci. USA* **2008**, *105*, 1608–1613. [CrossRef] [PubMed]

25. Stefani, G.; Slack, F.J. Small non-coding RNAs in animal development. *Nat. Rev. Mol. Cell Biol.* **2008**, *9*, 219–230. [CrossRef] [PubMed]

26. Shi, Y.; Jin, Y. MicroRNA in cell differentiation and development. *Sci. China C Life Sci.* **2009**, *52*, 205–511. [CrossRef] [PubMed]

27. O'Connell, R.M.; Rao, D.S.; Chaudhuri, A.A.; Baltimore, D. Physiological and pathological roles for microRNAs in the immune system. *Nat. Rev. Immunol.* **2010**, *10*, 111–122. [CrossRef] [PubMed]

28. Subramanian, S.; Steer, C.J. MicroRNAs as gatekeepers of apoptosis. *J. Cell. Physiol.* **2010**, *223*, 289–298. [CrossRef] [PubMed]

29. Landskroner-Eiger, S.; Moneke, I.; Sessa, W.C. miRNAs as modulators of angiogenesis. *Cold Spring Harb. Perspect. Med.* **2013**, *3*, a006643. [CrossRef] [PubMed]

30. Zhao, Y.; Srivastava, D. A developmental view of microRNA function. *Trends Biochem. Sci.* **2007**, *32*, 189–197. [CrossRef] [PubMed]

31. Visone, R.; Croce, C.M. MiRNAs and cancer. *Am. J. Pathol.* **2009**, *174*, 1131–1138. [CrossRef] [PubMed]

32. Ameres, S.L.; Zamore, P.D. Diversifying microRNA sequence and function. *Nat. Rev. Mol. Cell Biol.* **2013**, *14*, 475–488. [CrossRef] [PubMed]

33. Vidigal, J.A.; Ventura, A. The biological functions of miRNAs: Lessons from in vivo studies. *Trends Cell Biol.* **2015**, *25*, 137–147. [CrossRef] [PubMed]

34. Bracken, C.P.; Scott, H.S.; Goodall, G.J. A network-biology perspective of microRNA function and dysfunction in cancer. *Nat. Rev. Genet.* **2016**, *17*, 719–732. [CrossRef] [PubMed]

35. Lou, W.; Liu, J.; Gao, Y.; Zhong, G.; Chen, D.; Shen, J.; Bao, C.; Xu, L.; Pan, J.; Cheng, J.; et al. MicroRNAs in cancer metastasis and angiogenesis. *Oncotarget* **2017**, *8*, 115787–115802. [CrossRef] [PubMed]

36. Telonis, A.G.; Magee, R.; Loher, P.; Chervoneva, I.; Londin, E.; Rigoutsos, I. Knowledge about the presence or absence of miRNA isoforms (isomiRs) can successfully discriminate amongst 32 TCGA cancer types. *Nucleic Acids Res.* **2017**, *45*, 2973–2985. [CrossRef] [PubMed]

37. Schrama, D.; Ugurel, S.; Becker, J.C. Merkel cell carcinoma: Recent insights and new treatment options. *Curr. Opin. Oncol.* **2012**, *24*, 141–149. [CrossRef] [PubMed]

38. Erovic, I.; Erovic, B.M. Merkel cell carcinoma: The past, the present, and the future. *J. Skin Cancer* **2013**, *2013*, 929364. [CrossRef] [PubMed]

39. Godlewski, J.; Kowalczyk, A.; Kozielec, Z.; Pidsudko, Z.; Kmiec, A.; Siedlecka-Kroplewska, K. Plasticity of neuropeptidergic neoplasm cells in the primary and metastatic Merkel cell carcinoma. *Folia Histochem. Cytobiol.* **2013**, *51*, 168–173. [CrossRef] [PubMed]

40. Visscher, D.; Cooper, P.H.; Zarbo, R.J.; Crissman, J.D. Cutaneous neuroendocrine (Merkel cell) carcinoma: An immunophenotypic, clinicopathologic, and flow cytometric study. *Mod. Pathol.* **1989**, *2*, 331–338. [PubMed]

41. Love, J.E.; Thompson, K.; Kilgore, M.R.; Westerhoff, M.; Murphy, C.E.; Papanicolau-Sengos, A.; McCormick, K.A.; Shankaran, V.; Vandeven, N.; Miller, F.; et al. CD200 Expression in Neuroendocrine Neoplasms. *Am. J. Clin. Pathol.* **2017**, *148*, 236–242. [CrossRef] [PubMed]

42. Jankowski, M.; Kopinski, P.; Schwartz, R.; Czajkowski, R. Merkel cell carcinoma: Is this a true carcinoma? *Exp. Dermatol.* **2014**, *23*, 792–794. [CrossRef] [PubMed]

43. Sauer, C.M.; Haugg, A.M.; Chteinberg, E.; Rennspiess, D.; Winnepenninckx, V.; Speel, E.J.; Becker, J.C.; Kurz, A.K.; Zur Hausen, A. Reviewing the current evidence supporting early B-cells as the cellular origin of Merkel cell carcinoma. *Crit. Rev. Oncol. Hematol.* **2017**, *116*, 99–105. [CrossRef] [PubMed]

44. Tilling, T.; Moll, I. Which are the cells of origin in merkel cell carcinoma? *J. Skin Cancer* **2012**, *2012*, 680410. [CrossRef] [PubMed]

45. Sunshine, J.C.; Jahchan, N.S.; Sage, J.; Choi, J. Are there multiple cells of origin of Merkel cell carcinoma? *Oncogene* **2018**, *37*, 1409–1416. [CrossRef] [PubMed]

46. Tai, P. Merkel cell cancer: Update on biology and treatment. *Curr. Opin. Oncol.* **2008**, *20*, 196–200. [CrossRef] [PubMed]

47. Becker, J.C.; Stang, A.; DeCaprio, J.A.; Cerroni, L.; Lebbe, C.; Veness, M.; Nghiem, P. Merkel cell carcinoma. *Nat. Rev. Dis. Primers* **2017**, *3*, 17077. [CrossRef] [PubMed]

48. Agelli, M.; Clegg, L.X.; Becker, J.C.; Rollison, D.E. The etiology and epidemiology of merkel cell carcinoma. *Curr. Probl. Cancer* **2010**, *34*, 14–37. [CrossRef] [PubMed]

49. Kaae, J.; Hansen, A.V.; Biggar, R.J.; Boyd, H.A.; Moore, P.S.; Wohlfahrt, J.; Melbye, M. Merkel cell carcinoma: Incidence, mortality, and risk of other cancers. *J. Natl. Cancer Inst.* **2010**, *102*, 793–801. [CrossRef] [PubMed]

50. Stang, A.; Becker, J.C.; Nghiem, P.; Ferlay, J. The association between geographic location and incidence of Merkel cell carcinoma in comparison to melanoma: An international assessment. *Eur. J. Cancer* **2018**, *94*, 47–60. [CrossRef] [PubMed]

51. Harms, K.L.; Healy, M.A.; Nghiem, P.; Sober, A.J.; Johnson, T.M.; Bichakjian, C.K.; Wong, S.L. Analysis of Prognostic Factors from 9387 Merkel Cell Carcinoma Cases Forms the Basis for the New 8th Edition AJCC Staging System. *Ann. Surg. Oncol.* **2016**, *23*, 3564–3571. [CrossRef] [PubMed]

52. Tomic, S.; Warner, T.F.; Messing, E.; Wilding, G. Penile Merkel cell carcinoma. *Urology* **1995**, *45*, 1062–1065. [CrossRef]

53. Roy, S.; Das, I.; Nandi, A.; Roy, R. Primary Merkel cell carcinoma of the oral mucosa in a young adult male: Report of a rare case. *Indian J. Pathol. Microbiol.* **2015**, *58*, 214–216. [CrossRef] [PubMed]

54. Nguyen, A.H.; Tahseen, A.I.; Vaudreuil, A.M.; Caponetti, G.C.; Huerter, C.J. Clinical features and treatment of vulvar Merkel cell carcinoma: A systematic review. *Gynecol. Oncol. Res. Pract.* **2017**, *4*, 2. [CrossRef] [PubMed]

55. Oram, C.W.; Bartus, C.L.; Purcell, S.M. Merkel cell carcinoma: A review. *Cutis* **2016**, *97*, 290–295. [PubMed]

56. Feng, H.; Shuda, M.; Chang, Y.; Moore, P.S. Clonal integration of a polyomavirus in human Merkel cell carcinoma. *Science* **2008**, *319*, 1096–1100. [CrossRef] [PubMed]

57. Coursaget, P.; Samimi, M.; Nicol, J.T.; Gardair, C.; Touze, A. Human Merkel cell polyomavirus: Virological background and clinical implications. *APMIS* **2013**, *121*, 755–769. [CrossRef] [PubMed]

58. Liu, W.; MacDonald, M.; You, J. Merkel cell polyomavirus infection and Merkel cell carcinoma. *Curr. Opin. Virol.* **2016**, *20*, 20–27. [CrossRef] [PubMed]

59. Baez, C.F.; Brandao Varella, R.; Villani, S.; Delbue, S. Human Polyomaviruses: The Battle of Large and Small Tumor Antigens. *Virology* **2017**, *8*. [CrossRef] [PubMed]

60. Wendzicki, J.A.; Moore, P.S.; Chang, Y. Large T and small T antigens of Merkel cell polyomavirus. *Curr. Opin. Virol.* **2015**, *11*, 38–43. [CrossRef] [PubMed]

61. Demetriou, S.K.; Ona-Vu, K.; Sullivan, E.M.; Dong, T.K.; Hsu, S.W.; Oh, D.H. Defective DNA repair and cell cycle arrest in cells expressing Merkel cell polyomavirus T antigen. *Int. J. Cancer* **2012**, *131*, 1818–1827. [CrossRef] [PubMed]

62. Cheng, J.; Rozenblatt-Rosen, O.; Paulson, K.G.; Nghiem, P.; DeCaprio, J.A. Merkel cell polyomavirus large T antigen has growth-promoting and inhibitory activities. *J. Virol.* **2013**, *87*, 6118–6126. [CrossRef] [PubMed]

63. Verhaegen, M.E.; Mangelberger, D.; Harms, P.W.; Vozheiko, T.D.; Weick, J.W.; Wilbert, D.M.; Saunders, T.L.; Ermilov, A.N.; Bichakjian, C.K.; Johnson, T.M.; et al. Merkel cell polyomavirus small T antigen is oncogenic in transgenic mice. *J. Investig. Dermatol.* **2015**, *135*, 1415–1424. [CrossRef] [PubMed]

64. Spurgeon, M.E.; Cheng, J.; Bronson, R.T.; Lambert, P.F.; DeCaprio, J.A. Tumorigenic activity of merkel cell polyomavirus T antigens expressed in the stratified epithelium of mice. *Cancer Res.* **2015**, *75*, 1068–1079. [CrossRef] [PubMed]

65. Shuda, M.; Guastafierro, A.; Geng, X.; Shuda, Y.; Ostrowski, S.M.; Lukianov, S.; Jenkins, F.J.; Honda, K.; Maricich, S.M.; Moore, P.S.; et al. Merkel Cell Polyomavirus Small T Antigen Induces Cancer and Embryonic Merkel Cell Proliferation in a Transgenic Mouse Model. *PLoS ONE* **2015**, *10*, e0142329. [CrossRef] [PubMed]

66. Verhaegen, M.E.; Mangelberger, D.; Harms, P.W.; Eberl, M.; Wilbert, D.M.; Meireles, J.; Bichakjian, C.K.; Saunders, T.L.; Wong, S.Y.; Dlugosz, A.A. Merkel Cell Polyomavirus Small T Antigen Initiates Merkel Cell Carcinoma-like Tumor Development in Mice. *Cancer Res.* **2017**, *77*, 3151–3157. [CrossRef] [PubMed]

67. Kean, J.M.; Rao, S.; Wang, M.; Garcea, R.L. Seroepidemiology of human polyomaviruses. *PLoS Pathog.* **2009**, *5*, e1000363. [CrossRef] [PubMed]

68. Tolstov, Y.L.; Pastrana, D.V.; Feng, H.; Becker, J.C.; Jenkins, F.J.; Moschos, S.; Chang, Y.; Buck, C.B.; Moore, P.S. Human Merkel cell polyomavirus infection II. MCV is a common human infection that can be detected by conformational capsid epitope immunoassays. *Int. J. Cancer* **2009**, *125*, 1250–1256. [CrossRef] [PubMed]

69. Pastrana, D.V.; Tolstov, Y.L.; Becker, J.C.; Moore, P.S.; Chang, Y.; Buck, C.B. Quantitation of human seroresponsiveness to Merkel cell polyomavirus. *PLoS Pathog.* **2009**, *5*, e1000578. [CrossRef] [PubMed]

70. Viscidi, R.P.; Rollison, D.E.; Sondak, V.K.; Silver, B.; Messina, J.L.; Giuliano, A.R.; Fulp, W.; Ajidahun, A.; Rivanera, D. Age-specific seroprevalence of Merkel cell polyomavirus, BK virus, and JC virus. *Clin. Vaccine Immunol.* **2011**, *18*, 1737–1743. [CrossRef] [PubMed]

71. Nicol, J.T.; Robinot, R.; Carpentier, A.; Carandina, G.; Mazzoni, E.; Tognon, M.; Touze, A.; Coursaget, P. Age-specific seroprevalences of merkel cell polyomavirus, human polyomaviruses 6, 7, and 9, and trichodysplasia spinulosa-associated polyomavirus. *Clin. Vaccine Immunol.* **2013**, *20*, 363–368. [CrossRef] [PubMed]

72. Zhang, C.; Liu, F.; He, Z.; Deng, Q.; Pan, Y.; Liu, Y.; Zhang, C.; Ning, T.; Guo, C.; Liang, Y.; et al. Seroprevalence of Merkel cell polyomavirus in the general rural population of Anyang, China. *PLoS ONE* **2014**, *9*, e106430. [CrossRef] [PubMed]

73. Antonsson, A.; Neale, R.E.; O'Rourke, P.; Wockner, L.; Michel, A.; Pawlita, M.; Waterboer, T.; Green, A.C. Prevalence and stability of antibodies to thirteen polyomaviruses and association with cutaneous squamous cell carcinoma: A population-based study. *Clin. Virol.* **2018**, *101*, 34–37. [CrossRef] [PubMed]

74. Schowalter, R.M.; Pastrana, D.V.; Pumphrey, K.A.; Moyer, A.L.; Buck, C.B. Merkel cell polyomavirus and two previously unknown polyomaviruses are chronically shed from human skin. *Cell Host Microbe* **2010**, *7*, 509–515. [CrossRef] [PubMed]

75. Martel-Jantin, C.; Pedergnana, V.; Nicol, J.T.; Leblond, V.; Tregouet, D.A.; Tortevoye, P.; Plancoulaine, S.; Coursaget, P.; Touze, A.; Abel, L.; et al. Merkel cell polyomavirus infection occurs during early childhood and is transmitted between siblings. *J. Clin. Virol.* **2013**, *58*, 288–291. [CrossRef] [PubMed]

76. Stakaityte, G.; Wood, J.J.; Knight, L.M.; Abdul-Sada, H.; Adzahar, N.S.; Nwogu, N.; Macdonald, A.; Whitehouse, A. Merkel cell polyomavirus: Molecular insights into the most recently discovered human tumour virus. *Cancers* **2014**, *6*, 1267–1297. [CrossRef] [PubMed]

77. Arora, R.; Chang, Y.; Moore, P.S. MCV and Merkel cell carcinoma: A molecular success story. *Curr. Opin. Virol.* **2012**, *2*, 489–498. [CrossRef] [PubMed]

78. Chang, Y.; Moore, P.S. Merkel cell carcinoma: A virus-induced human cancer. *Annu. Rev. Pathol.* **2012**, *7*, 123–144. [CrossRef] [PubMed]
79. Houben, R.; Adam, C.; Baeurle, A.; Hesbacher, S.; Grimm, J.; Angermeyer, S.; Henzel, K.; Hauser, S.; Elling, R.; Brocker, E.B.; et al. An intact retinoblastoma protein-binding site in Merkel cell polyomavirus large T antigen is required for promoting growth of Merkel cell carcinoma cells. *Int. J. Cancer* **2012**, *130*, 847–856. [CrossRef] [PubMed]
80. Hesbacher, S.; Pfitzer, L.; Wiedorfer, K.; Angermeyer, S.; Borst, A.; Haferkamp, S.; Scholz, C.J.; Wobser, M.; Schrama, D.; Houben, R. RB1 is the crucial target of the Merkel cell polyomavirus Large T antigen in Merkel cell carcinoma cells. *Oncotarget* **2016**, *7*, 32956–32968. [CrossRef] [PubMed]
81. Shuda, M.; Kwun, H.J.; Feng, H.; Chang, Y.; Moore, P.S. Human Merkel cell polyomavirus small T antigen is an oncoprotein targeting the 4E-BP1 translation regulator. *J. Clin. Investig.* **2011**, *121*, 3623–3634. [CrossRef] [PubMed]
82. Seo, G.J.; Chen, C.J.; Sullivan, C.S. Merkel cell polyomavirus encodes a microRNA with the ability to autoregulate viral gene expression. *Virology* **2009**, *383*, 183–187. [CrossRef] [PubMed]
83. Lee, S.; Paulson, K.G.; Murchison, E.P.; Afanasiev, O.K.; Alkan, C.; Leonard, J.H.; Byrd, D.R.; Hannon, G.J.; Nghiem, P. Identification and validation of a novel mature microRNA encoded by the Merkel cell polyomavirus in human Merkel cell carcinomas. *J. Clin. Virol.* **2011**, *52*, 272–275. [CrossRef] [PubMed]
84. Theiss, J.M.; Gunther, T.; Alawi, M.; Neumann, F.; Tessmer, U.; Fischer, N.; Grundhoff, A. A Comprehensive Analysis of Replicating Merkel Cell Polyomavirus Genomes Delineates the Viral Transcription Program and Suggests a Role for mcv-miR-M1 in Episomal Persistence. *PLoS Pathog.* **2015**, *11*, e1004974. [CrossRef] [PubMed]
85. Chen, C.J.; Cox, J.E.; Azarm, K.D.; Wylie, K.N.; Woolard, K.D.; Pesavento, P.A.; Sullivan, C.S. Identification of a polyomavirus microRNA highly expressed in tumors. *Virology* **2015**, *476*, 43–53. [CrossRef] [PubMed]
86. Akhbari, P.; Tobin, D.; Poterlowicz, K.; Roberts, W.; Boyne, J.R. MCV-miR-M1 targets the host-cell immune response resulting in the attenuation of neutrophil chemotaxis. *J. Investig. Dermatol.* **2018**. [CrossRef] [PubMed]
87. Neumann, F.; Czech-Sioli, M.; Dobner, T.; Grundhoff, A.; Schreiner, S.; Fischer, N. Replication of Merkel cell polyomavirus induces reorganization of promyelocytic leukemia nuclear bodies. *J. Gen. Virol.* **2016**, *97*, 2926–2938. [CrossRef] [PubMed]
88. Ning, M.S.; Kim, A.S.; Prasad, N.; Levy, S.E.; Zhang, H.; Andl, T. Characterization of the Merkel Cell Carcinoma miRNome. *J. Skin Cancer* **2014**, *2014*, 289548. [CrossRef] [PubMed]
89. Veija, T.; Sahi, H.; Koljonen, V.; Bohling, T.; Knuutila, S.; Mosakhani, N. miRNA-34a underexpressed in Merkel cell polyomavirus-negative Merkel cell carcinoma. *Virchows Arch.* **2015**, *466*, 289–295. [CrossRef] [PubMed]
90. Renwick, N.; Cekan, P.; Masry, P.A.; McGeary, S.E.; Miller, J.B.; Hafner, M.; Li, Z.; Mihailovic, A.; Morozov, P.; Brown, M.; et al. Multicolor microRNA FISH effectively differentiates tumor types. *J. Clin. Investig.* **2013**, *123*, 2694–2702. [CrossRef] [PubMed]
91. Xie, H.; Lee, L.; Caramuta, S.; Hoog, A.; Browaldh, N.; Bjornhagen, V.; Larsson, C.; Lui, W.O. MicroRNA expression patterns related to merkel cell polyomavirus infection in human merkel cell carcinoma. *J. Investig. Dermatol.* **2014**, *134*, 507–517. [CrossRef] [PubMed]
92. Segura, M.F.; Hanniford, D.; Menendez, S.; Reavie, L.; Zou, X.; Alvarez-Diaz, S.; Zakrzewski, J.; Blochin, E.; Rose, A.; Bogunovic, D.; et al. Aberrant miR-182 expression promotes melanoma metastasis by repressing FOXO3 and microphthalmia-associated transcription factor. *Proc. Natl. Acad. Sci. USA* **2009**, *106*, 1814–1819. [CrossRef] [PubMed]
93. Tang, L.; Chen, F.; Pang, E.J.; Zhang, Z.Q.; Jin, B.W.; Dong, W.F. MicroRNA-182 inhibits proliferation through targeting oncogenic ANUBL1 in gastric cancer. *Oncol. Rep.* **2015**, *33*, 1707–1716. [CrossRef] [PubMed]
94. Feng, Y.A.; Liu, T.E.; Wu, Y. microRNA-182 inhibits the proliferation and migration of glioma cells through the induction of neuritin expression. *Oncol. Lett.* **2015**, *10*, 1197–1203. [CrossRef] [PubMed]
95. Zhang, X.; Ma, G.; Liu, J.; Zhang, Y. MicroRNA-182 promotes proliferation and metastasis by targeting FOXF2 in triple-negative breast cancer. *Oncol. Lett.* **2017**, *14*, 4805–4811. [CrossRef] [PubMed]
96. Lowery, A.J.; Miller, N.; Dwyer, R.M.; Kerin, M.J. Dysregulated miR-183 inhibits migration in breast cancer cells. *BMC Cancer* **2010**, *10*, 502. [CrossRef] [PubMed]
97. Miao, F.; Zhu, J.; Chen, Y.; Tang, N.; Wang, X.; Li, X. MicroRNA-183-5p promotes the proliferation, invasion and metastasis of human pancreatic adenocarcinoma cells. *Oncol. Lett.* **2016**, *11*, 134–140. [CrossRef] [PubMed]

98. Yang, X.; Wang, L.; Wang, Q.; Li, L.; Fu, Y.; Sun, J. MiR-183 inhibits osteosarcoma cell growth and invasion by regulating LRP6-Wnt/beta-catenin signaling pathway. *Biochem. Biophys. Res. Commun.* **2018**, *496*, 1197–1203. [CrossRef] [PubMed]

99. Ren, L.H.; Chen, W.X.; Li, S.; He, X.Y.; Zhang, Z.M.; Li, M.; Cao, R.S.; Hao, B.; Zhang, H.J.; Qiu, H.Q.; et al. MicroRNA-183 promotes proliferation and invasion in oesophageal squamous cell carcinoma by targeting programmed cell death. *Br. J. Cancer* **2014**, *111*, 2003–2013. [CrossRef] [PubMed]

100. Ruan, H.; Liang, X.; Zhao, W.; Ma, L.; Zhao, Y. The effects of microRNA-183 promots cell proliferation and invasion by targeting MMP-9 in endometrial cancer. *Biomed. Pharmacother.* **2017**, *89*, 812–818. [CrossRef] [PubMed]

101. Song, C.; Zhang, L.; Wang, J.; Huang, Z.; Li, X.; Wu, M.; Li, S.; Tang, H.; Xie, X. High expression of microRNA-183/182/96 cluster as a prognostic biomarker for breast cancer. *Sci. Rep.* **2016**, *6*, 24502. [CrossRef] [PubMed]

102. Kang, M.; Xia, P.; Hou, T.; Qi, Z.; Liao, S.; Yang, X. MicroRNA-190b inhibits tumor cell proliferation and induces apoptosis by regulating Bcl-2 in U2OS osteosarcoma cells. *Die Pharm.* **2017**, *72*, 279–282. [CrossRef]

103. Huang, K.; Tang, Y.; He, L.; Dai, Y. MicroRNA-340 inhibits prostate cancer cell proliferation and metastasis by targeting the MDM2-p53 pathway. *Oncol. Rep.* **2016**, *35*, 887–895. [CrossRef] [PubMed]

104. Yuan, J.; Ji, H.; Xiao, F.; Lin, Z.; Zhao, X.; Wang, Z.; Zhao, J.; Lu, J. MicroRNA-340 inhibits the proliferation and invasion of hepatocellular carcinoma cells by targeting JAK1. *Biochem. Biophys. Res. Commun.* **2017**, *483*, 578–584. [CrossRef] [PubMed]

105. Maskey, N.; Li, D.; Xu, H.; Song, H.; Wu, C.; Hua, K.; Song, J.; Fang, L. MicroRNA-340 inhibits invasion and metastasis by downregulating ROCK1 in breast cancer cells. *Oncol. Lett.* **2017**, *14*, 2261–2267. [CrossRef] [PubMed]

106. Arivazhagan, R.; Lee, J.; Bayarsaikhan, D.; Kwak, P.; Son, M.; Byun, K.; Salekdeh, G.H.; Lee, B. MicroRNA-340 inhibits the proliferation and promotes the apoptosis of colon cancer cells by modulating REV3L. *Oncotarget* **2018**, *9*, 5155–5168. [CrossRef] [PubMed]

107. Qu, F.; Wang, X. microRNA-340 induces apoptosis by downregulation of BAG3 in ovarian cancer SKOV3 cells. *Pharmazie* **2017**, *72*, 482–486. [CrossRef] [PubMed]

108. Yin, G.; Zhou, H.; Xue, Y.; Yao, B.; Zhao, W. MicroRNA-340 promotes the tumor growth of human gastric cancer by inhibiting cyclin G2. *Oncol. Rep.* **2016**, *36*, 1111–1118. [CrossRef] [PubMed]

109. Yin, H.; Sun, Y.; Wang, X.; Park, J.; Zhang, Y.; Li, M.; Yin, J.; Liu, Q.; Wei, M. Progress on the relationship between miR-125 family and tumorigenesis. *Exp. Cell Res.* **2015**, *339*, 252–260. [CrossRef] [PubMed]

110. Zou, C.; Zhao, P.; Xiao, Z.; Han, X.; Fu, F.; Fu, L. Gammadelta T cells in cancer immunotherapy. *Oncotarget* **2017**, *8*, 8900–8909. [CrossRef] [PubMed]

111. Stark, M.S.; Tyagi, S.; Nancarrow, D.J.; Boyle, G.M.; Cook, A.L.; Whiteman, D.C.; Parsons, P.G.; Schmidt, C.; Sturm, R.A.; Hayward, N.K. Characterization of the Melanoma miRNAome by Deep Sequencing. *PLoS ONE* **2010**, *5*, e9685. [CrossRef] [PubMed]

112. Persson, H.; Kvist, A.; Rego, N.; Staaf, J.; Vallon-Christersson, J.; Luts, L.; Loman, N.; Jonsson, G.; Naya, H.; Hoglund, M.; et al. Identification of new microRNAs in paired normal and tumor breast tissue suggests a dual role for the ERBB2/Her2 gene. *Cancer Res.* **2011**, *71*, 78–86. [CrossRef] [PubMed]

113. Yang, X.; Chen, Y.; Chen, L. The Versatile Role of microRNA-30a in Human Cancer. *Cell. Physiol. Biochem.* **2017**, *41*, 1616–1632. [CrossRef] [PubMed]

114. Farooqi, A.A.; Tabassum, S.; Ahmad, A. MicroRNA-34a: A Versatile Regulator of Myriads of Targets in Different Cancers. *Int. J. Mol. Sci.* **2017**, *18*, E2089. [CrossRef] [PubMed]

115. Shrestha, A.; Mukhametshina, R.T.; Taghizadeh, S.; Vasquez-Pacheco, E.; Cabrera-Fuentes, H.; Rizvanov, A.; Mari, B.; Carraro, G.; Bellusci, S. MicroRNA-142 is a multifaceted regulator in organogenesis, homeostasis, and disease. *Dev. Dyn.* **2017**, *246*, 285–290. [CrossRef] [PubMed]

116. Wang, X.F.; Shi, Z.M.; Wang, X.R.; Cao, L.; Wang, Y.Y.; Zhang, J.X.; Yin, Y.; Luo, H.; Kang, C.S.; Liu, N.; et al. MiR-181d acts as a tumor suppressor in glioma by targeting K-ras and Bcl-2. *J. Cancer Res. Clin. Oncol.* **2012**, *138*, 573–584. [CrossRef] [PubMed]

117. Zhang, G.; Liu, D.; Long, G.; Shi, L.; Qiu, H.; Hu, G.; Hu, G.; Liu, S. Downregulation of microRNA-181d had suppressive effect on pancreatic cancer development through inverse regulation of KNAIN2. *Tumour Biol.* **2017**, *39*. [CrossRef] [PubMed]

118. Zhang, H.; Qi, S.; Zhang, T.; Wang, A.; Liu, R.; Guo, J.; Wang, Y.; Xu, Y. miR-188-5p inhibits tumour growth and metastasis in prostate cancer by repressing LAPTM4B expression. *Oncotarget* **2015**, *6*, 6092–6104. [CrossRef] [PubMed]

119. Wang, L.; Liu, H. microRNA-188 is downregulated in oral squamous cell carcinoma and inhibits proliferation and invasion by targeting SIX1. *Tumour Biol.* **2016**, *37*, 4105–4113. [CrossRef] [PubMed]

120. Li, N.; Shi, H.; Zhang, L.; Li, X.; Gao, L.; Zhang, G.; Shi, Y.; Guo, S. MiR-188 Inhibits Glioma Cell Proliferation and Cell Cycle Progression through Targeting ss-catenin. *Oncol. Res.* **2017**. [CrossRef]

121. Peng, Y.; Shen, X.; Jiang, H.; Chen, Z.; Wu, J.; Zhu, Y.; Zhou, Y.; Li, J. MiR-188-5p suppresses gastric cancer cell proliferation and invasion via targeting ZFP91. *Oncol. Res.* **2018**. [CrossRef] [PubMed]

122. Yuva-Aydemir, Y.; Simkin, A.; Gascon, E.; Gao, F.B. MicroRNA-9: Functional evolution of a conserved small regulatory RNA. *RNA Biol.* **2011**, *8*, 557–564. [CrossRef] [PubMed]

123. Sun, H.; Shao, Y.; Huang, J.; Sun, S.; Liu, Y.; Zhou, P.; Yang, H. Prognostic value of microRNA-9 in cancers: A systematic review and meta-analysis. *Oncotarget* **2016**, *7*, 67020–67032. [CrossRef] [PubMed]

124. Chang, Y.; Lin, J.; Tsung, A. Manipulation of autophagy by MIR375 generates antitumor effects in liver cancer. *Autophagy* **2012**, *8*, 1833–1834. [CrossRef] [PubMed]

125. Shi, Z.C.; Chu, X.R.; Wu, Y.G.; Wu, J.H.; Lu, C.W.; Lu, R.X.; Ding, M.C.; Mao, N.F. MicroRNA-375 functions as a tumor suppressor in osteosarcoma by targeting PIK3CA. *Tumour Biol.* **2015**, *36*, 8579–8584. [CrossRef] [PubMed]

126. Cui, F.; Wang, S.; Lao, I.; Zhou, C.; Kong, H.; Bayaxi, N.; Li, J.; Chen, Q.; Zhu, T.; Zhu, H. miR-375 inhibits the invasion and metastasis of colorectal cancer via targeting SP1 and regulating EMT-associated genes. *Oncol. Rep.* **2016**, *36*, 487–493. [CrossRef] [PubMed]

127. Xu, L.; Wen, T.; Liu, Z.; Xu, F.; Yang, L.; Liu, J.; Feng, G.; An, G. MicroRNA-375 suppresses human colorectal cancer metastasis by targeting Frizzled 8. *Oncotarget* **2016**, *7*, 40644–40656. [CrossRef] [PubMed]

128. Osako, Y.; Seki, N.; Kita, Y.; Yonemori, K.; Koshizuka, K.; Kurozumi, A.; Omoto, I.; Sasaki, K.; Uchikado, Y.; Kurahara, H.; et al. Regulation of MMP13 by antitumor microRNA-375 markedly inhibits cancer cell migration and invasion in esophageal squamous cell carcinoma. *Int. J. Oncol.* **2016**, *49*, 2255–2264. [CrossRef] [PubMed]

129. Wei, R.; Yang, Q.; Han, B.; Li, Y.; Yao, K.; Yang, X.; Chen, Z.; Yang, S.; Zhou, J.; Li, M.; et al. microRNA-375 inhibits colorectal cancer cells proliferation by downregulating JAK2/STAT3 and MAP3K8/ERK signaling pathways. *Oncotarget* **2017**, *8*, 16633–16641. [CrossRef] [PubMed]

130. Wang, Z.; Hong, Z.; Gao, F.; Feng, W. Upregulation of microRNA-375 is associated with poor prognosis in pediatric acute myeloid leukemia. *Mol. Cell. Biochem.* **2013**, *383*, 59–65. [CrossRef] [PubMed]

131. Costa-Pinheiro, P.; Ramalho-Carvalho, J.; Vieira, F.Q.; Torres-Ferreira, J.; Oliveira, J.; Goncalves, C.S.; Costa, B.M.; Henrique, R.; Jeronimo, C. MicroRNA-375 plays a dual role in prostate carcinogenesis. *Clin. Epigenet.* **2015**, *7*, 42. [CrossRef] [PubMed]

132. Liu, L.; Liao, J.Z.; He, X.X.; Li, P.Y. The role of autophagy in hepatocellular carcinoma: Friend or foe. *Oncotarget* **2017**, *8*, 57707–57722. [CrossRef] [PubMed]

133. Qiu, H.J.; Lu, X.H.; Yang, S.S.; Weng, C.Y.; Zhang, E.K.; Chen, F.C. MiR-769 promoted cell proliferation in human melanoma by suppressing GSK3B expression. *Biomed. Pharmacother.* **2016**, *82*, 117–123. [CrossRef] [PubMed]

134. Arora, R.; Shuda, M.; Guastafierro, A.; Feng, H.; Toptan, T.; Tolstov, Y.; Normolle, D.; Vollmer, L.L.; Vogt, A.; Domling, A.; et al. Survivin is a therapeutic target in Merkel cell carcinoma. *Sci. Transl. Med.* **2012**, *4*, 133ra56. [CrossRef] [PubMed]

135. Lasithiotaki, I.; Tsitoura, E.; Koutsopoulos, A.; Lagoudaki, E.; Koutoulaki, C.; Pitsidianakis, G.; Spandidos, D.A.; Siafakas, N.M.; Sourvinos, G.; Antoniou, K.M. Aberrant expression of miR-21, miR-376c and miR-145 and their target host genes in Merkel cell polyomavirus-positive non-small cell lung cancer. *Oncotarget* **2017**, *8*, 112371–112383. [CrossRef] [PubMed]

136. Hayes, J.; Peruzzi, P.P.; Lawler, S. MicroRNAs in cancer: Biomarkers, functions and therapy. *Trends Mol. Med.* **2014**, *20*, 460–469. [CrossRef] [PubMed]

137. Lan, H.; Lu, H.; Wang, X.; Jin, H. MicroRNAs as potential biomarkers in cancer: Opportunities and challenges. *Biomed. Res. Int.* **2015**, *2015*, 125094. [CrossRef] [PubMed]

138. Larrea, E.; Sole, C.; Manterola, L.; Goicoechea, I.; Armesto, M.; Arestin, M.; Caffarel, M.M.; Araujo, A.M.; Araiz, M.; Fernandez-Mercado, M.; et al. New Concepts in Cancer Biomarkers: Circulating miRNAs in Liquid Biopsies. *Int. J. Mol. Sci.* **2016**, *17*, E627. [CrossRef] [PubMed]

139. Barger, J.F.; Rahman, M.A.; Jackson, D.; Acunzo, M.; Nana-Sinkam, S.P. Extracellular miRNAs as biomarkers in cancer. *Food Chem. Toxicol.* **2016**, *98*, 66–72. [CrossRef] [PubMed]

140. Moretti, F.; D'Antona, P.; Finardi, E.; Barbetta, M.; Dominioni, L.; Poli, A.; Gini, E.; Noonan, D.M.; Imperatori, A.; Rotolo, N.; et al. Systematic review and critique of circulating miRNAs as biomarkers of stage I-II non-small cell lung cancer. *Oncotarget* **2017**, *8*, 94980–94996. [CrossRef] [PubMed]

141. Wang, H.; Peng, R.; Wang, J.; Qin, Z.; Xue, L. Circulating microRNAs as potential cancers biomarkers: The advantage and disavantage. *Clin. Epigenet.* **2018**, *10*, 59. [CrossRef] [PubMed]

142. Huang, X.; Yuan, T.; Liang, M.; Du, M.; Xia, S.; Dittmar, R.; Wang, D.; See, W.; Costello, B.A.; Quevedo, F.; et al. Exosomal miR-1290 and miR-375 as prognostic markers in castration-resistant prostate cancer. *Oncol. Rep.* **2016**, *35*, 887–895. [CrossRef] [PubMed]

143. Pfeffer, S.R.; Grossmann, K.F.; Cassidy, P.B.; Yang, C.H.; Fan, M.; Kopelovich, L.; Leachman, S.A.; Pfeffer, L.M. Detection of Exosomal miRNAs in the Plasma of Melanoma Patients. *J. Clin. Med.* **2015**, *4*, 2012–2027. [CrossRef] [PubMed]

144. Imamura, T.; Komatsu, S.; Ichikawa, D.; Miyamae, M.; Okajima, W.; Ohashi, T.; Kiuchi, J.; Nishibeppu, K.; Kosuga, T.; Konishi, H.; et al. Low plasma levels of miR-101 are associated with tumor progression in gastric cancer. *Oncotarget* **2017**, *8*, 106538–106550. [CrossRef] [PubMed]

145. Zhao, Q.; Deng, S.; Wang, G.; Liu, C.; Meng, L.; Qiao, S.; Shen, L.; Zhang, Y.; Lu, J.; Li, W.; et al. A direct quantification method for measuring plasma MicroRNAs identified potential biomarkers for detecting metastatic breast cancer. *Oncotarget* **2016**, *7*, 21865–21874. [CrossRef] [PubMed]

146. Foj, L.; Ferrer, F.; Serra, M.; Arevalo, A.; Gavagnach, M.; Gimenez, N.; Filella, X. Exosomal and Non-Exosomal Urinary miRNAs in Prostate Cancer Detection and Prognosis. *Prostate* **2017**, *77*, 573–583. [CrossRef] [PubMed]

147. Wang, J.; Yan, F.; Zhao, Q.; Zhan, F.; Wang, R.; Wang, L.; Zhang, Y.; Huang, X. Circulating exosomal miR-125a-3p as a novel biomarker for early-stage colon cancer. *Sci. Rep.* **2017**, *7*, 4150. [CrossRef] [PubMed]

148. Butz, H.; Nofech-Mozes, R.; Ding, Q.; Khella, H.W.Z.; Szabo, P.M.; Jewett, M.; Finelli, A.; Lee, J.; Ordon, M.; Stewart, R.; et al. Exosomal MicroRNAs Are Diagnostic Biomarkers and Can Mediate Cell-Cell Communication in Renal Cell Carcinoma. *Eur. Urol. Focus* **2016**, *2*, 210–218. [CrossRef] [PubMed]

149. Jiang, Q.; Lagos-Quintana, M.; Liu, D.; Shi, Y.; Helker, C.; Herzog, W.; le Noble, F. miR-30a regulates endothelial tip cell formation and arteriolar branching. *Hypertension* **2013**, *62*, 592–598. [CrossRef] [PubMed]

150. Alsaleh, G.; Francois, A.; Philippe, L.; Gong, Y.Z.; Bahram, S.; Cetin, S.; Pfeffer, S.; Gottenberg, J.E.; Wachsmann, D.; Georgel, P.; et al. MiR-30a-3p negatively regulates BAFF synthesis in systemic sclerosis and rheumatoid arthritis fibroblasts. *PLoS ONE* **2014**, *9*, e111266. [CrossRef] [PubMed]

151. Muther, C.; Jobeili, L.; Garion, M.; Heraud, S.; Thepot, A.; Damour, O.; Lamartine, J. An expression screen for aged-dependent microRNAs identifies miR-30a as a key regulator of aging features in human epidermis. *Aging* **2017**, *9*, 2376–2396. [CrossRef] [PubMed]

152. Kapsimali, M.; Kloosterman, W.P.; de Bruijn, E.; Rosa, F.; Plasterk, R.H.; Wilson, S.W. MicroRNAs show a wide diversity of expression profiles in the developing and mature central nervous system. *Genome Biol.* **2007**, *8*, R173. [CrossRef] [PubMed]

153. Lu, T.X.; Lim, E.J.; Wen, T.; Plassard, A.J.; Hogan, S.P.; Martin, L.J.; Aronow, B.J.; Rothenberg, M.E. MiR-375 is downregulated in epithelial cells after IL-13 stimulation and regulates an IL-13-induced epithelial transcriptome. *Mucosal Immunol.* **2012**, *5*, 388–396. [CrossRef] [PubMed]

154. Eliasson, L. The small RNA miR-375—A pancreatic islet abundant miRNA with multiple roles in endocrine beta cell function. *Mol. Cell. Endocrinol.* **2017**, *456*, 95–101. [CrossRef] [PubMed]

155. Branch, S.; Maloney, K.; Purcell, S.M. Spontaneous regression of Merkel cell carcinoma. *Cutis* **2018**, *101*, 301–305. [PubMed]

156. Ahmadi Moghaddam, P.; Cornejo, K.M.; Hutchinson, L.; Tomaszewicz, K.; Dresser, K.; Deng, A.; O'Donnell, P. Complete Spontaneous Regression of Merkel Cell Carcinoma After Biopsy: A Case Report and Review of the Literature. *Am. J. Dermatopathol.* **2016**, *38*, e154–e158. [CrossRef] [PubMed]

157. Cirillo, F. Spontaneous Regression of Primitive Merkel Cell Carcinoma. *Rare Tumors* **2015**, *7*, 5961. [CrossRef] [PubMed]

158. Walsh, N.M. Complete spontaneous regression of Merkel cell carcinoma (1986–2016): A 30 year perspective. *J. Cutan. Pathol.* **2016**, *43*, 1150–1154. [CrossRef] [PubMed]

159. Mantripragada, K.; Birnbaum, A. Response to Anti-PD-1 Therapy in Metastatic Merkel Cell Carcinoma Metastatic to the Heart and Pancreas. *Cureus* **2015**, *7*, e403. [CrossRef] [PubMed]

160. Nghiem, P.T.; Bhatia, S.; Lipson, E.J.; Kudchadkar, R.R.; Miller, N.J.; Annamalai, L.; Berry, S.; Chartash, E.K.; Daud, A.; Fling, S.P.; et al. PD-1 Blockade with Pembrolizumab in Advanced Merkel-Cell Carcinoma. *N. Engl. J. Med.* **2016**, *374*, 2542–2552. [CrossRef] [PubMed]

161. Kaufman, H.L.; Russell, J.; Hamid, O.; Bhatia, S.; Terheyden, P.; D'Angelo, S.P.; Shih, K.C.; Lebbe, C.; Linette, G.P.; Milella, M.; et al. Avelumab in patients with chemotherapy-refractory metastatic Merkel cell carcinoma: A multicentre, single-group, open-label, phase 2 trial. *Lancet Oncol.* **2016**, *17*, 1374–1385. [CrossRef]

162. Walocko, F.M.; Scheier, B.Y.; Harms, P.W.; Fecher, L.A.; Lao, C.D. Metastatic Merkel cell carcinoma response to nivolumab. *J. Immunother. Cancer* **2016**, *4*, 79. [CrossRef] [PubMed]

163. Thiem, A.; Kneitz, H.; Schummer, P.; Herz, S.; Schrama, D.; Houben, R.; Goebeler, M.; Schilling, B.; Gesierich, A. Coincident Metastatic Melanoma and Merkel Cell Carcinoma with Complete Remission on Treatment with Pembrolizumab. *Acta Derm. Venereol.* **2017**, *97*, 1252–1254. [CrossRef] [PubMed]

164. Zhao, C.; Tella, S.H.; Del Rivero, J.; Kommalapati, A.; Ebenuwa, I.; Gulley, J.; Strauss, J.; Brownell, I. Anti-PD-L1 Treatment Induced Central Diabetes Insipidus. *J. Clin. Endocrinol. Metab.* **2018**, *103*, 365–369. [CrossRef] [PubMed]

165. Kaufman, H.L.; Russell, J.S.; Hamid, O.; Bhatia, S.; Terheyden, P.; D'Angelo, S.P.; Shih, K.C.; Lebbe, C.; Milella, M.; Brownell, I.; et al. Updated efficacy of avelumab in patients with previously treated metastatic Merkel cell carcinoma after >/=1 year of follow-up: JAVELIN Merkel 200, a phase 2 clinical trial. *J. Immunother. Cancer* **2018**, *6*, 7. [CrossRef] [PubMed]

166. Eshghi, N.; Lundeen, T.F.; MacKinnon, L.; Avery, R.; Kuo, P.H. 18F-FDG PET/CT for Monitoring Response of Merkel Cell Carcinoma to the Novel Programmed Cell Death Ligand 1 Inhibitor Avelumab. *Clin. Nucl. Med.* **2018**, *43*, e142–e144. [CrossRef] [PubMed]

167. Topalian, S.L.; Bhatia, S.; Hollebecque, A.; Awada, A.; de Boer, J.P.; Kudchadkar, R.R.; Goncalves, A.; Delord, J.-P.; Martens, U.M.; Picazo, J.M.L.; et al. Non-comparative, open-label, multiple cohort, phase 1/2 study to evaluate nivolumab (NIVO) in patients with virus-associated tumors (CheckMate 358): Efficacy and safety in Merkel cell carcinoma (MCC). *Cancer Res.* **2017**, *77*. [CrossRef]

168. Winkler, J.K.; Dimitrakopoulou-Strauss, A.; Sachpekidis, C.; Enk, A.; Hassel, J.C. Ipilimumab has efficacy in metastatic Merkel cell carcinoma: A case series of five patients. *Eur. Acad. Dermatol. Venereol.* **2017**, *31*, e389–e391. [CrossRef] [PubMed]

169. Williams, B.A.; Hardee, M.E.; Hutchins, L.F.; Shalin, S.; Gao, L. A case of Merkel cell carcinoma treatment with anti-CTLA-4 antibody (Ipilimumab). *J. Clin. Case Rep.* **2015**, *5*, 12. [CrossRef]

170. Shah, M.Y.; Ferrajoli, A.; Sood, A.K.; Lopez-Berestein, G.; Calin, G.A. microRNA Therapeutics in Cancer—An Emerging Concept. *EBioMedicine* **2016**, *12*, 34–42. [CrossRef] [PubMed]

171. Rupaimoole, R.; Slack, F.J. MicroRNA therapeutics: Towards a new era for the management of cancer and other diseases. *Nat. Rev. Drug Discov.* **2017**, *16*, 203–222. [CrossRef] [PubMed]

172. Chakraborty, C.; Sharma, A.R.; Sharma, G.; Doss, C.G.P.; Lee, S.S. Therapeutic miRNA and siRNA: Moving from Bench to Clinic as Next Generation Medicine. *Mol. Ther. Nucleic Acids* **2017**, *8*, 132–143. [CrossRef] [PubMed]

173. Verma, M.; Lam, T.K.; Hebert, E.; Divi, R.L. Extracellular vesicles: Potential applications in cancer diagnosis, prognosis, and epidemiology. *BMC Clin. Pathol.* **2015**, *15*, 6. [CrossRef] [PubMed]

174. Maia, J.; Caja, S.; Strano Moraes, M.C.; Couto, N.; Costa-Silva, B. Exosome-Based Cell-Cell Communication in the Tumor Microenvironment. *Front. Cell Dev. Biol.* **2018**, *6*, 18. [CrossRef] [PubMed]

175. Graner, M.W.; Schnell, S.; Olin, M.R. Tumor-derived exosomes, microRNAs, and cancer immune suppression. *Semin. Immunopathol.* **2018**. [CrossRef] [PubMed]

International Journal of
Molecular Sciences

MDPI

Review

MiRNA Dysregulation in Childhood Hematological Cancer

Jaqueline Carvalho de Oliveira [1]**, Gabriela Molinari Roberto** [2]**, Mirella Baroni** [2]**,
Karina Bezerra Salomão** [2]**, Julia Alejandra Pezuk** [3] **and María Sol Brassesco** [4]**,***

[1] Department Genetics, Federal University of Paraná, 80060-000 Paraná, Brazil; jaqbiomed@yahoo.com.br
[2] Department of Pediatrics, Ribeirão Preto School of Medicine, University of São Paulo,
    14049-900 Ribeirão Preto, Brazil; gabi_cdl@hotmail.com (G.M.R.);
    mirella_baroni@yahoo.com.br (M.B.); karina_slm@hotmail.com (K.B.S.)
[3] Programa de Pós-graduação em Farmácia, Anhanguera University of São Paulo, UNIAN/SP,
    05145-200 São Paulo, Brazil; julia.pezuk@hotmail.com
[4] Departamento de Biologia, Faculty of Philosophy, Sciences and Letters at Ribeirão Preto,
    University of São Paulo, 14040-901 Ribeirão Preto, Brazil
* Correspondence: solbrassesco@usp.br; Tel.: +55-16-3315-9144; Fax: +55-16-3315-4886

Received: 20 July 2018; Accepted: 8 September 2018; Published: 10 September 2018

Abstract: For decades, cancer biology focused largely on the protein-encoding genes that have clear roles in tumor development or progression: cell-cycle control, apoptotic evasion, genome instability, drug resistance, or signaling pathways that stimulate growth, angiogenesis, or metastasis. MicroRNAs (miRNAs), however, represent one of the more abundant classes of cell modulators in multicellular organisms and largely contribute to regulating gene expression. Many of the ~2500 miRNAs discovered to date in humans regulate vital biological processes, and their aberrant expression results in pathological and malignant outcomes. In this review, we highlight what has been learned about the roles of miRNAs in some of the most common human pediatric leukemias and lymphomas, along with their value as diagnostic/prognostic factors.

Keywords: miRNA; cancer; children; leukemia; lymphoma; review

---

1. Introduction

Following the first microRNA (miRNA) discovery in 1993 [1], a constantly increasing number of miRNAs have been described and investigated. In 2002, the first miRNA dysregulation associated with human disease revealed deletion of miR-15 and miR-16 as a frequent event in chronic lymphocytic leukemia patients [2]. A year later, another couple of miRNAS, miR-143 and miR-145, were described as downregulated in colon adenocarcinoma [3].

In 2004, Calin and colleagues mapped 186 miRNAs and found that over 50% of miRNA genes are located at cancer-associated genomic regions or in fragile sites, and that those located in deleted regions are generally downregulated in cancer samples [4]. In the same year, let-7 was associated with shortened postoperative survival in lung cancer [5] and it was identified as a specific miRNA profile in B cell chronic lymphocytic leukemia [6].

Thenceforth, many miRNAs have been identified to have an essential role in human carcinogenesis and progression. Several reports have shown that miRNAs are central in cancer pathways by acting as "oncomiRs" or "tumor-suppressive miRNAs" and are often related to apoptosis, cell proliferation, angiogenesis, metastasis, and drug resistance [7,8]. Furthermore, miRNA regulation is dependent on the expression of its multiple mRNA targets, which are not always constitutively expressed; consequently, a unique miRNA may have different effects under diverse conditions [9].

The biological behavior of pediatric tumors is heterogeneous with several aspects distinguishing them from their adult counterparts, including their location, cellular environment, cell of origin, and genetic mutations. From treatment perspectives, they are also heterogeneous, ranging from generally curable low-grade tumors to highly aggressive forms that are usually fatal. Consequently, the identification of molecular markers that can effectively predict prognosis, and might contribute to the development of new therapeutic approaches, is still needed.

Like in adults, dysregulation of miRNAs is a hallmark in childhood cancer. Herein, we will compile current information about the role of miRNAs in the biology of leading hematological cancers in a pediatric setting.

2. Leukemias

Leukemia, a cancer of the bone marrow (BM) that hampers normal hematopoiesis, is the most common childhood malignancy, accounting for about 30% of all pediatric cancer. The two major subtypes seen in children are acute lymphoblastic leukemia (ALL), and acute myeloid leukemia (AML) [10,11], though a small fraction may present chronic myeloid leukemia (CML) and juvenile myelomonocytic leukemia (JMML) [12,13].

2.1. Acute Lymphoblastic Leukemia

ALL represents 80% of all leukemia types in children [10] and in recent decades clinicians have seen a significant improvement in event-free survival (EFS) rates, currently exceeding 80% in developed nations [14]. This advancement was notably facilitated by multiagent chemotherapy regimens and risk-adapted therapy, where the study of laboratory-based outcome variables consents the allocation of treatment [14,15].

In 2007, Mi and colleagues [16] showed that miRNA signatures could accurately discriminate ALL from AML. This study, although samples from adult and pediatric patients were analyzed indiscriminately, was the first suggestion of miRNAs' involvement in childhood leukemia. Thereafter, many research groups also utilized miRNA-expression analyses and proved this strategy to be useful in the refinement of ALL classification schemes. Nowadays, disruption of miRNA expression and function in ALL is the most broadly studied and well-characterized among pediatric leukemias (Figure 1).

Differentially expressed miRNAs in childhood ALL (cALL) were firstly described in 2009 [17]. Examining 40 newly diagnosed pre-B ALL samples, miR-222, miR-339, and miR-142-3p were found overexpressed, along with the downregulation of miR-451 and miR-373 when compared to normal cells [17]. Additionally, a subsequent report [18] examined miRNA profiles in pediatric ALL samples in comparison to normal CD34+ cells, and gave evidence of the upregulation of miR-128a, miR-142, miR-150, miR-181, miR-30e-5p, miR-193, miR-34b, miR-365, miR-582, and miR-708, and the downregulation of miR-100, miR-125b, miR-99a, miR-196b, and miR-let-7e. Later on, several studies reported dysregulated miRNA expression in pediatric ALL samples compared to normal cells. For example, a number of publications described increased expression of miR-21, miR-34, miR-128, miR-142, miR-146a, miR-181b, miR-195, and miR-708 [19–23], and decreased levels of miR-18, miR-181a, miR-99a, miR-100, miR-145, let-7, and miR-196b in ALL cells [19,21,24,25].

The biological heterogeneity and distinct-lineage origins of ALL are well-established [26]. Such heterogeneity is also reflected with respect to miRNA expression profiles, for example, miR-18a is lower in childhood ALL when compared to the adult counterparts [24]. Among the dysregulation of miRNAs evinced in multiples studies, specific miRNA profiles have been described for specific ALL subtypes [18]. The same group [27] later identified unique miRNA expression patterns for each pediatric ALL subtype and measured the expression levels of 397 miRNAs in samples from 81 patients. The authors were able to differentiate many of the major subtypes of ALL, such as T-cell, *MLL*-rearranged, *ETV6/RUNX1*-positive, *E2A/PBX1*-positive, and hyperdiploid. However, conclusive evidence for discriminative miRNA expression was not found in *BCR/ABL* positive and "B-other groups".

Moreover, the downregulation of miR-let-7b (~70-fold) in *MLL*-rearranged ALL; miR-100 in *ETV6/RUNX1*-positive; miR-17-3p, miR-17-5p, miR-29c-3p, miR-92a-3p, miR-214-3p, miR-214-5p, miR-708 in T-ALL; and miR-31, miR-24, miR-708, and miR-128 were associated with *PAX5*-deleted ALL [18,20,28,29]. Contrariwise, higher expression of miR-125b, miR-196, miR-223, and miR-708 were found in patients with *ETV6/RUNX1* translocation, and miR-24 and miR-542 were associated with PAR1 deletion, ALL [27,29,30] increased miR-181b and miR-128a in *MLL*-rearranged [20,27,31], higher miR-100 and miR-21 in B-ALL, and higher miR-196b in T-ALL [19,32–34]. Moreover, a difference on miRNA expression was found when infant and childhood T-ALL were compared [35].

The clinical importance of miRNA profiling was also verified by the description of an association with treatment resistance and EFS. Specific miRNA profiles were described for several commonly used drugs (Figure 2). A study published in 2015 by Hamzeh et al. showed that miRNA-related dysregulated pathways were associated to resistance to asparaginase (L_ASP), daunorubicin (DAUNO), prednisolone (PREDS), and vincristine (VIN) [36]. Several miRNAs have been attributed an association with leukemia treatment resistance, such as miR-34 [37,38], miR-128b, and miR-223 [39]. Zhang et al. [23] described an miRNA signature (miR-18a, miR-532, miR-218, miR-625, miR-193a, miR-638, miR-550, and miR-633) that was able to predict prednisone (PRED) response in childhood ALL patients. Later, in a separate study involving cALL samples, it was shown that the expression profiles of the same group of miRNAs could similarly be used to predict early response to this glucocorticoid [40]. Furthermore, miR-124 was upregulated in prednisone and glucocorticoid resistance [41]. Alternatively, it was shown that prednisolone significantly increased miR-16-1 and miR-15a expression [42]. Additionally, it has been demonstrated that the restoration of miR-128b and miR-221 co-operatively sensitizes MLL/AF4(+) ALL cell lines to glucocorticoids [43], while exogenous expression of miR-335 in ALL cells renders cells to PREDS-mediated apoptosis [44].

Comparatively, in an attempt to elucidate miRNA signatures that indicate sensitivity to other chemotherapeutics used in ALL treatment, 397 miRNA were verified by Schotte and colleagues [27]. From those, 17 were related to resistance to one or more drugs. Among them, miR-99a, miR-100, miR-125b, and miR-126 were associated with VIN and DAUNO resistance, while miR-625 was associated with VIN and PREDS resistance. The expression of miR-125b, together with miR-99a and/or miR-100 overexpression, is also linked to vincristine resistance [45,46], while the overexpression of miR-652-3p increases sensitivity to vincristine and cytarabin (CYT) [47]. Recently, downregulation miR-326 was associated with multidrug resistance [48].

Furthermore, miR-3117-3p polymorphism has been associated with vincristine-induced neurotoxicity [49]. Moreover, miR-5189, miR-595, miR-6083, and some polymorphisms related to miRNA can also be linked to methotrexate (MTX) response, and miR-1206 can be used to predict MTX toxicity [50–52].

On the other hand, the association of distinct miRNA expression patterns in relation to risk stratification in childhood ALL has been scrutinized in the literature. A report by Zhang et al. [23] alluded to the association of miRNA expression with prognostic parameters such Central Nervous system (CNS) relapse, specific risk category, and disease recurrence. More than 20% of patients with CNS relapse showed a threefold increase of miR-7, miR-198, and miR-633, and a decrease of miR-126, miR-345, miR-222, and miR-551a at a one-year follow-up. Some of these findings were later confirmed in a report by Xu and colleagues [40].

The expression of some miRNAs can be used to monitor disease progression, such as with miR-128, miR-146a, miR-155, miR-181a, and miR-195 [21]. In addition, miR-210 has been proposed as a good prognostic factor and a useful predictor of drug sensitivity [53]. Moreover, a systematic investigation by Schotte et al. [27] verified a correlation with the probability of disease-free survival (DFS) and expression levels of 31 distinct miRNAs. Among those, 14 miRNAs were considered independent prognostic factors that allowed the distinction of a group of patients with favorable expression profiles and a five-year DFS of $89.4 \pm 7\%$ from those with less favorable miRNA profiles, with a five-year DFS rate of $60.8 \pm 12\%$.

In parallel, Han et al. [54], using panel of matched samples (diagnosis/remission and diagnosis/relapse), described the altered expression of miR-223, miR-23a, let-7g, miR-181, miR-708, and miR-130b in relapsed samples, and miR-27a, miR-223, miR-23a, miR-181, and miR-128b in samples taken at remission. MiRNAs expression associated with shorter EFS or other clinical markers in cALL were also described [39,53,55–58]. An association of miR-708, miR-223, and miR-27a with relapse-free survival (RFS) was also demonstrated, as well as a prediction for relapse in patients with altered expression of miR-210 and for miR-143/miR-182 [59,60]. Moreover, the low expression of miR-151-5p and miR-451, high expression of miR-1290, or a combination of all three predicted inferior RFS [61].

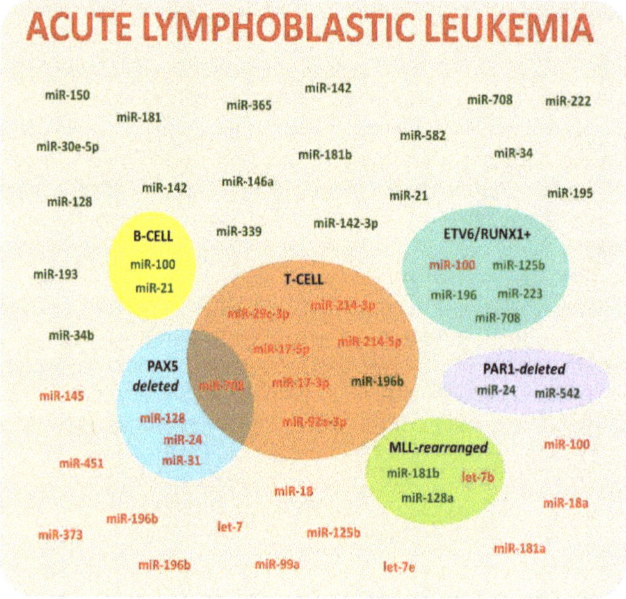

Figure 1. Dysregulated microRNAs (miRNAs) in childhood acute lymphoblastic leukemia. Hyperexpressed and hypoexpressed miRNAs in acute lymphoblastic leukemia within cellular (B-cell or T-cell) or molecular (*ETV6/RUNX1+*, *PAX5*-deleted, *MLL*-rearranged, or *PAR1*-deleted) subgroups are denoted in green and red, respectively.

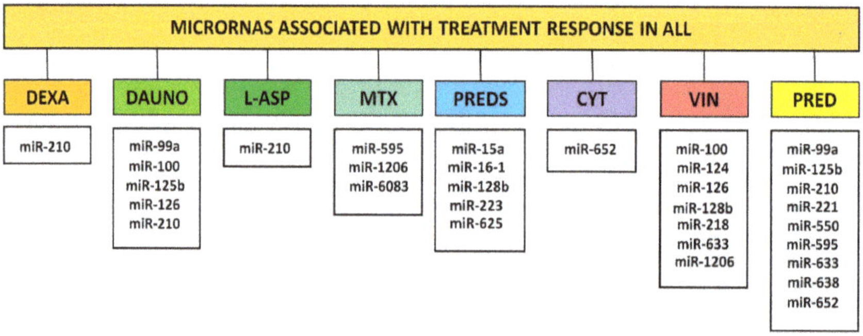

Figure 2. Dysregulated miRNAS that are associated with treatment response. Several miRNAs have been attributed an association with differential responses to dexamethasone (DEXA), daunorrubicin (DAUNO), L-asparaginase (L-ASP), methotrexate (MTX), prednisolone (PREDS), cytarabin (CYT), vincristine (VIN), or prednisone (PRED).

2.2. *Acute Myeloid Leukemia*

Acute myeloid leukemia (AML) is the second most common type of pediatric leukemia, representing 17% of all hematological cancer in children [10] and five-year survival estimates of approximately 65% in developed countries [62]. MiRNA dysregulation has also been described in this pathology, though to a much lower extent when compared to ALL (Figure 3).

The first analysis of miRNA signatures in childhood AML (cAML) was also described by Mi and colleagues [16]. Nonetheless, adult and pediatric patients were analyzed indiscriminately and, among the 93 AML cases analyzed, only 17 were under 18 years old (nine patients under 12 years-old).

In 2009, samples from patients diagnosed with AML showed low levels of miR-34b, while in vitro exogenous expression of this miRNA caused cell-cycle abnormalities, reduced anchorage-independent growth, and altered CREB (cAMP response element-binding protein) target gene expression, suggesting suppressor potential [63]. Furthermore, in 2013, the same group demonstrated the hypermethylation of miR-34b promoter in AML [63].

Hypermethylation of the miR-663 promoter was also observed in another pediatric AML cohort and, consequently, a significantly lower expression of this miRNA was observed compared to normal bone-marrow control samples [64].

Several research groups have widely explored the study of individual miRNAs in AML. Emmrich et al. [65] observed that miR-582 and miR-9 are downregulated in t(8;21) AML; miR-500a and miR-192/194 are downregulated in AML with inv (16); and miR-181a, miR-1331, and miR-126 are downregulated while miR-187 is increased in *MLL*-rearranged AML. In 2017, Obulkasim et al. [66] published a signature prolife of 47 miRNAs to distinguished different AML cytogenetic subtypes. Moreover, miR-155 was proposed as a potential diagnostic biomarker for all AML, whereas miR-196b is specific for subgroups M4–M5 [67,68]. Another study stated that high miR-155 expression is also an adverse prognostic factor in pediatric NK-AML and is associated with worse EFS and overall survival (OS) [69].

In addition, miR-193b-3p was described as downregulated and proposed as an independent indicator for poor prognosis in pediatric AML, independent of patient age or genetics [70]. MiR-146b was described as an independent poor prognostic factor, while high expression of miR-181c and miR-4786 appeared to be favorable factors [71]. High expression of miR-196b in diagnostic marrow samples of pediatric AML was also associated with an unfavorable outcome [72].

Upregulation of miR-100 and miR-375 was also correlated with poor RFS OS [73], and downregulation of miR-29a was associated with advanced clinical features and poor prognosis of pediatric patients [74]. More, recently, an miRNA-based predictor of poststandard induction chemotherapy outcome in cAML was created to identify EFS in children with AML, and it offers the potential for improved patient stratification and management [75].

On the other hand, Danen-van Oorschot and colleagues [76] showed high levels of miR-196a and -b expression in pediatric patients carrying *MLL* fissions, *NPM1* mutations, or *FLT3/ITD*. In contrast, *CEBPA*-mutated cases presented low expression of miR-196a and -b. Alternatively, high expression of miR-155 was also observed in *FLT3/ITD* and *NPM1*-mutated cases, while downregulation of miR-29a was mostly detected in *MLL*-rearranged samples [76]. Moreover, the miR-106b~25 cluster has shown to be upregulated in relapse pediatric AML with MLL rearrangements [77].

Another comprehensive overview of miRNA expression showed that samples with core-binding factor AML and promyelocytic leukemia differed from each other and could be distinguished from *MLL*-rearranged AML subtypes by differentially expressed miRNAs that included miR-126, -146a, -181a/b, -100, and miR-125b [78].

MiR-99a was also found highly expressed in pediatric-onset AML, while significantly underexpressed during complete remission. Additionally, in vitro studies suggested a potential oncogenic role [79]. Moreover, forced expression of miR-9 reduced leukemic growth and induced monocytes differentiation of t(8;21) AML cell lines in vitro and in vivo, being characterized as a tumor-suppressor miRNA that acts in a strict cell context-dependent manner [65].

In parallel, many miRNAs have been related to AML by regulating cell proliferation, including downregulation of miR-122 as an aggressive progression marker [80], and miR-181a as a regulator of G1/S transition [81]. Others, like miR-126 and miR-182, were highly expressed in AML cell lines and inhibition of miR-126 significantly induced cell death through apoptosis [82,83]

2.3. Chronic Myeloid Leukemia

Chronic myeloid leukemia (CML) is a rare childhood hematological malignancy representing around 3% of all leukemias, with an annual incidence of one per million children and young people aged <15 years [84].

The characteristic reciprocal translocation t(9;22)(q34;q11) that leads to the formation of *BCR/ABL* chimeric oncoprotein is present in 90–95% of childhood CML [85]. In the era of therapy with specific tyrosine kinase inhibitors, the two-year survival among children with CML is 81–89%; nonetheless, age-group analysis evidenced that risk of death was three times higher for children younger than five years versus those aged 10–14 years [86].

Independent of age, many miRNAs have been described as active regulators of *ABL1* and *BCR/ABL*. For example, it was demonstrated that *ABL1* is a direct target of miR-203. This miRNA is silenced in CML and its restoration reduces *ABL1* and *BCR/ABL1* expression, decreasing cell growth. Additionally, one of the molecular mechanisms of imatinib is the demethylation of miR-203 in *BCR/ABL*-positive leukemia cells [87,88]. Nevertheless, the specific role of miRNAs in pediatric CML has been little explored. In 2013, miR-99a expression levels were evaluated in eight CML patients (including four samples before therapy and four samples with complete remission) and 12 pediatric controls. Although with a small number of patients, miR-99a expression was significantly increased in samples collected at diagnosis and decreased in samples after treatment [79].

A more recent study, aiming to evaluate profibrotic changes in childhood CML, analyzed 16 pediatric and 16 adult CML samples with and without fibrosis (each $n = 8$), as well as 18 non-neoplastic controls. Fiber accumulation in BM represents an adverse prognostic factor in adult CML, but, in children, this event is unknown. Nonetheless, among many gene-expression profiles investigated, two were miRNAs: miR-10 (previously associated with CML) and miR-146b (previously associated with fibrosis). MiR-10 was not associated with disease subtypes, fibrosis, or age. MiR-146b, on the other hand, showed lower expression levels in most pediatric samples when compared to adult counterparts, but no clear associations were found [89]. Other studies evaluating miRNA pathways specifically in childhood CML were not found.

2.4. Juvenile Myelomonocytic Leukemia

Juvenile myelomonocytic leukemia (JMML) is a rare myeloid progenitor disorder that occurs in young children with an annual incidence of as much as 1.2 per million children, accounting for less than 3% of all childhood hematologic malignancies [90].

Patients with JMML respond poorly to chemotherapy and have poor prognosis. The EFS is between 24–54% after hematopoietic stem-cell transplantation and is less than 10% without transplant [91]. Somatic defects in *RAS, NF1, PTPN11,* or *CBL* are detected in 85% of patients, evidencing RAS/MAPK-pathway (Rat Sarcoma virus /mitogen-activated protein kinase) activation as an important mechanism in JMML pathogenesis [92].

Aiming to evaluate miRNAs' role in JMML, miR-let-7a-1/miR-let-7f-1 and the 3'UTR of *NRAS* or *KRAS* were sequenced in BM cells from 10 JMML patients. *RAS* is a known target of miR-let-7 but, in this report, there was no evidence of any mutations in let-7 or in let-7-binding sites that might lead to its upregulation in JMML [93]. On the other hand, reduced levels of most members of the miR-let-7 miRNA family were evidenced in a novel fetal-like subgroup of JMML patients with LIN28B protein overexpression [94].

Analyses of 20 JMML samples by Ripperger and coworkers through comparative genomic hybridization found two patients with an almost identical partial gain of chromosome 8, suggesting

8p11.21q11.21 as a critical region. This band includes 31 protein-coding genes and two noncoding RNAs among which miR-486 is a known regulator of phosphatase and tensin homolog (*PTEN*) and the transcription factor forkhead box O1 (*FOXO1*) [95].

In 2013, downregulation of miR-34b was described in JMML patients (*n* = 17) [63], but this association was not confirmed by Liu et al., analyzing a bigger JMML cohort (*n* = 47) [96]. Nonetheless, these authors described high expression levels of miR-183 (13.8 vs. 4.2, *p* < 0.001) with significant linear correlation with monocyte percentage and in samples with *PTPN11* mutations [96]. MiR-223 and miR-15a were also found upregulated in JMML BM harboring *PTPN11* mutations (11 from 19 analyzed patients), but not those without *PTPN11* defects [97].

More recently, distinctive miRNA signatures associated with the *PTPN11*, *KRAS*, and *NRAS* molecular subtypes of JMML were also described. From a panel of miRNAs, miR-630, miR-3195, miR-575, miR-4508, miR-224-5p, miR-320e, miR-494, miR-548ai, miR-222-3p, miR-23a-3p, and miR-338-3p were found upregulated, while miR-150-5p, let-7g-5p, miR-1260a, let-7a-5p, miR-4454, miR-148a-3p, miR-146b-5p, miR-342-3p, let-7f-5p, miR-26a-5p, let-7d-5p, miR-30b-5p, miR-29b-3p, and miR-29a-3p were described as downregulated. Of note, miR-150-5p was found to target *STAT5b* (Signal transducer and activator of transcription 5b), and its induced overexpression in mononuclear cells from JMML patients decreased proliferation rates [98].

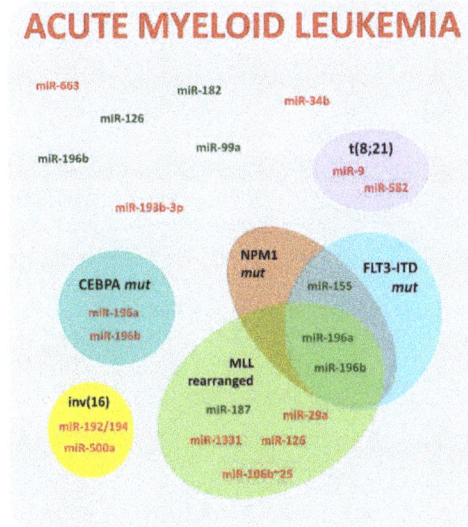

Figure 3. Dysregulated miRNAS in childhood acute myeloid leukemia. Hyperexpressed and hypoexpressed miRNAs in acute lymphoblastic myeloid and cellular/molecular subgroups are denoted in green and red, respectively.

3. Lymphomas

Lymphomas stricto sensu comprise any neoplasm of the lymphatic tissue. In the pediatric setting, lymphomas represent the third most common malignancy.

The World Health Organization (WHO) groups lymphomas by cell type and defining phenotypic, molecular, or cytogenetic characteristics [99]. Basically, there are two main categories of lymphoma, Hodgkin (HL) and non-Hodgkin lymphoma (NHL). HL most commonly affects adolescents and accounts for 4–7% of overall childhood cancer [100], while NHL is more frequently diagnosed in children younger than 15 years of age and represents 7–10% of pediatric malignancies [101,102]. NHL has a wide range of histological appearances and clinical features at presentation and, despite classification refinement, some groups remain heterogeneous. Nonetheless, Burkitt lymphoma (BL), diffuse large B-cell lymphoma (DLBCL), primary mediastinal large B-cell lymphoma (PMLBCL), anaplastic large cell lymphoma (ALCL), and lymphoblastic lymphoma (LL) comprise childhood NHLs [103].

Over the last decade, the pattern of miRNA expression of different types of pediatric lymphomas has been extensively studied. A substantial number of miRNAs have been described as dysregulated and contributing to better classifying this type of tumor, though, for many, their roles in tumor development are still unclear (Figure 4). Moreover, the lack of information about miRNAs in some forms, such as PMLBCL and classical HL, is evident.

3.1. Burkitt Lymphoma

BL represents about 30% of all pediatric NHL and is considered a highly aggressive tumor [102]. The study of miRNA in BL is usually focused on the establishment of miRNA profiles and the understanding of tumor transformation and progression. On this regard, miRNA expression could discriminate BL from other lymphomas, pediatric from adult samples, and Epstein–Barr virus (EBV) BL-positive from negative cases [104–108].

In 2004, miR-155 was the first described as deregulated in lymphoma. This miRNA is encoded by the human *BIC* gene and it was found overexpressed in pediatric BL. Metzler et al. [102] suggested that this RNA could be acting in co-operation with c-Myc on B-cell transformation. MiR-155 expression induces polyclonal expansion in B-cells, favoring the occurrence of secondary mutations and leading to full transformation [109]. This miRNA has as a direct target *SHIP1* (Src homology-2 domain-containing inositol 5-phosphatase 1), whose suppression in hematopoietic cells leads to mieloproliferative diseases [110]. In addition, miR-155 showed low expression in pediatric BL, inversely associated with the downregulation of the nuclear interactor of ARF (ADP-ribosylation factor) and Mdm2 (murine double minute 2) (NIAM, the protein-coding transcript splice variant of *TBRG1* locus), a protein with a tumor-suppressor function [111].

It was demonstrated that miR-155 was expressed only in BL EBV-positive cases, which account for 70% of all pediatric BL [112,113]. Secreted vesicles (exosomes) from EBV-positive Raji cells could deliver miR-155 to other recipient cell lines, such as retinal-pigment epithelial cells (ARPE-19), and miR-155 increased transcriptional and translational levels of VEGF-A in ARPE-19 cells [114].

Other studies have also demonstrated a close association between EBV infection and miRNA dysregulation in BL. High levels of miR-155 and miR-146a were found in response to the viral latent membrane protein-1 (LMP1) through NF-κB (nuclear factor kappa B) modulation, although the precise mechanism is still unclear [115,116]. LMP1 seems to stabilize BIC mRNA via p38/MAPK, and the LMP1–BIC axis contributes to EBV-induced lymphomagenesis [117]. Additionally, LMP1 induces miR-34a expression, leading to EBV-transformed cell growth [118]. However, a tumor-suppressor effect of LMP1 has also been described through the upregulation of miR-29b, which represses *TCL1* (T-cell leukemia/lymphoma protein 1) and leads to tumor cell-proliferation reduction [119].

Epstein–Barr nuclear antigen 1 (EBNA1) also has a role on BL development by miR-127 induction. This miRNA impairs B-cell differentiation by decreasing *BLIMP-1* (PR domain zinc finger protein 1) and *XBP-1* (X-box binding protein 1) expression, which leads to BCL-6 overexpression and IRF-4 (interferon regulatory factor 4) downregulation [120]. The presence of the EBV virus determines a profile of miRNAs in pediatric and adults BLs, therefore 28 miRNAs were differentially expressed in positive EBV cases, including EBV-encoded and host miRNAs [107].

During EBV infection, the virus-control miRNAs expression of host cells and expressed two clusters of miRNAs. A member of the miR-BART cluster, EBV-BART-6-3p target interleukin-6 receptor (IL-6R) and impair the immune system [121,122]. Synergistically with EBV-BASRT-6-3p, host cellular miR-142 and miR-197 targeted and reduced expression of IL-6R in Ramos BL cell lines [123,124]. Therefore, miRNA detection could be a specific and sensitive tool to recognize EBV vestiges once EBV-negative samples classified by immunohistochemistry demonstrated the presence of EBV-miRNAs, suggesting that EBV might contribute to lymphomagenesis [125].

Among the dysregulated miRNAs in BL, there are also those regulated by the NF-κB pathway and those regulated by c-Myc, transcription factors that regulate cell proliferation, growth, and apoptosis. Among these, miR-23a, miR-26a, miR-29b, miR-30d, miR-146a, miR-146b-5p, miR-155, and miR-221 were found statistically significantly downregulated in BL compared to other lymphomas [126]. In Raji cells, miR-520a was associated with the regulation of the AKT1 (v-akt murine thymoma viral oncogene homolog 1) and NF-κB signaling pathways, and mimics of this miRNA inhibited growth and proliferation, and promoted apoptosis [127].

Alterations on c-Myc expression or function of are one of the most frequent abnormalities in human malignancy; in BL, the recurrent t(14;18) chromosomal translocation juxtaposes this oncogene to the regulatory elements of the immunoglobulin resulting in the constitutive expression of c-Myc [128]. This activation promotes the expression of cluster miR-17–92 [129], which has a causative role in lymphomagenesis, regulating proapoptotic proteins and cell-cycle regulators [130]. In the p53-mutated BL cell line Raji, c-Myc is an alternative target of Inauhzin (INZ) via miRNA pathways, including miR-24 and miR-34a. INZ is a small molecule that activates p53 and inhibits tumor growth [131]. In fact, De Falco et al. (2015) showed four miRNAs (miR-29a, miR-29b, miR-513a-5p, and miR-628-3p) differentially expressing between MYC translocation-positive and negative BL. These miRNAs targets are involved in gene expression, proliferation, and DNA modification. In MYC translocation-negative, overexpression of DNA methyltransferase (DNMT) was associated with hypoexpression of the miR-29 family [132]. Association between high expression of DNMT1 and decrease in the miR-29 family was observed in a pediatric cohort (n = 71), suggesting a methylation control of has-miR-29 [133].

Analysis of known Myc-targeted miRNAs demonstrated significant association between BL with Myc translocation in a cohort composed by 61% of pediatric BL [104]. Those miRNAs included an upregulated cluster (miR-17-92 and its paralogs miR-18b, miR20b, miR-106a), and a set of downregulated miRNAs (miR-23a, miR-29c, miR-29b, miR150, miR146a). Upregulated miRNAs in BL were expressed at significantly lower levels in normal B cells, T cells and stromal cells, but were noted in BL cell lines (Daudi and Raji). BL cases with high expression of miR-17-92 cluster members showed significantly repression of these target genes in BL [104]. The investigation of the miR-17–92 cluster (miR-17, miR-19a, miR-19b, miR-20, and miR-92 a) expression in pediatric BL showed correlation between the upregulation of miR-17 and miR-20a with a lack of proapoptotic BIM (Bcl-2-like protein 11) expression. MiR-17 was a predictor of shortened OS, and inhibition of this miRNA in Daudi cells induced BIM expression [133].

MiR-26a and miR-28, on the other hand, were found underexpressed in BL. Their ectopic expression reduces proliferation, impairs cell-cycle progression, and increases apoptosis by targeting different proteins [134,135]. Another important c-Myc target, miR-150, which targets *MYB* and *survivin*, was found downregulated in BL. Overexpression of this miRNA in BL cells reduced proliferation rates and increased apoptosis [136,137].

Moreover, miR-181b, which is located in the intron of the *FAMLF* (familial acute myelogenous leukemia related factor) gene, showed an inverse correlation with FAMLF expression, and an interaction with its 5′UTR. Downregulation of *FAMLF* by miR-181b inhibited cell viability and arrested cell cycle in Raji BL cells [138,139].

Differential miRNA profiles have also been described in endemic Burkitt Lymphoma (eBL), an aggressive germinal center CG cell cancer that represents a subdivision of BL with high incidence in pediatric patients in equatorial Africa. eBL is associated with EBV and *Plasmodium falciparum* malaria coinfection, and shows c-Myc overexpression [140]. An integrative analysis compared normal germinal center (GC) B cells with eBL and evidenced 49 miRNAs with differential gene expression. Of these, 27 miRNAs were downregulated (including let-7 family members) and 22 upregulated (among them miR-17–92 cluster) in eBL samples. Enrichment of pathways showed the interaction of these miRNAs with marked tumor suppression (*PTEN*, *AXIN1*, *ATM*, *NLK*), and important proto-oncogenes and tumor-promoting genes as *MYC* [141].

A comparison between eBL jaw- and abdominal-tumor biopsies showed no discernible clustering based on tumor-site designation. MiR-10a-5p was the only miRNA with differential expression, and it was lower in jaw eBL compared to abdominal, and it presented reduced expression in nonsurvivor patients. MiR-10a-5p could target 473 genes, and enrichment of pathways showed its importance in cancer, focal adhesion, EPV infection, and apoptosis pathways [142]. Therefore, in pediatric BL, EBV infection and c-Myc translocation promote lymphomagenesis through the deregulation of several miRNAs.

3.2. Diffuse Large B-Cell Lymphoma

DLBCL represents 10% to 20% of childhood lymphomas and it is more frequently found in children older than 10 years of age [128]. This tumor is divided in two distinct subtypes according to the cell of origin: the activated B cell-like (ABC) and the germinal center B cell-like (GCB) [143], though most pediatric DLBCL patients are diagnosed with the GCB form [144].

The clinical presentation of pediatric DLBCL and BL are very similar. Currently, they are recognized as two different entities; however, in the pediatric group, there is a significant overlap of features resulting in a group of unclassifiable lymphomas. Nonetheless, through microarray technology, different research groups were able to define a collection of distinct miRNAs that constitutes a DLBCL signature [106,145]. This analysis also enables us to differentiate ABC from GCB-subtypes.

Ten miRNAs (miR-146b, miR-146a, miR-21, miR-155, miR-500, miR-222, miR-221, miR-363, miR-574, and miR-574*) were found to be more upregulated in ABC than in the GCB lymphoma type, suggesting that the high levels of these miRNAs are not due to tumor malignancy but associated with the cell of origin [146,147]. MiR-155 was one of the first miRNAs found upregulated in ABC-DLBCL [147,148]. Its aberrant expression seems to be a consequence of an autocrine stimulation by TNFα (tumor necrosis factor alfa) rather than chromosomal translocations like in BL tumors [149]. Initially, no correlation with prognosis was found when all DLBCL tumors were considered [145,150]; however, when only the ABC group was examined, high miR-155 expression was associated with better survival rates. In this case, the five-year survival probability changed from 15% for patients with low miR-155 to 53% for patients with high miR-155. Moreover, higher expression of miR-222 was also associated with inferior overall and progression-free survival [150].

Other studies later indicated miR-155 and miR-146a as potential diagnostic and prognostic indicators in DLBCL. Patients with low expression of these miRNAs were associated with high complete remission, high overall response rate, and better five-year OS when patients were treated with the R-CHOP protocol (rituximab, cyclophosphamide, doxorubicin, VIN, and PREDS) [151]. Furthermore, in DLBCL patients, higher expression of miR-28, miR-214, miR-339*, and miR-5586 was associated with better outcome, while upregulation of miR-324 was associated with poor prognosis [152].

Although ABC-DLBCL miRNA signatures have been better studied, the expression of few miRNAs was also associated with GCB-DLBCL, such as the amplification of the 17–92 cluster [153] and high levels of miR-106a and miR-181b [154].

Additionally, the role of some of these miRNAs in DLBCL development has been elucidated over the last few years. MiR-125a and miR-25b, for example, are overexpressed in DLBCL and target *TNFαIP3*, an NF-κB negative regulator. They participate in a positive self-regulatory loop where miR-125 is also regulated by NF-κB, what is probably an important mechanism to keep the constitutive activation of the NF-κB pathway in DLBCL pathogenesis [155]. In addition, miR-34a repression was described to cause high-grade transformation of B-cell lymphoma by altering *FOXP1* (Forkhead Box P1) expression [156]. Interestingly, mice treated with miR-34a mimics results in a 95% reduction in DLBCL tumor growth due to its strong proapoptotic properties, suggesting an alternative therapeutic strategy [157].

3.3. Primary Mediastinal Large B-Cell Lymphoma

PMLBCL was first described in the 1980s and is considered a distinct clinicopathologic entity of DLBCL [158]. PMLBCL is characterized by a rapidly growing mediastinal mass that arises from mature thymic medulla B-cells, frequently accompanied by local invasiveness and occasionally with distant metastasis. Uncommon, but not rare, this clinicopathological entity occurs more often in young adult females [159] and constitutes 2–3% of all N and 6–10% of all diffuse large-cell lymphomas [160,161].

Recently, a large population-based study was able to estimate the incidence of PMLBCL. Based on slightly more than 400 patients in the United States, the annual incidence rate was estimated at 0.4 per million. Females had significantly higher incidence than males (ratio 3:1) and the peak of occurrence was recognized at 30–39 years. The five-year survival rate of 85% and prognosis were also reduced with advancing age [162].

In patients aged <18 years (22/451 cases), PMLBCL incidence was 4.9% [162]. Other reports, in which young patients showed inferior outcomes, compared adult counterparts or children with other B-NHL histological subtypes, with EFS rates ranging between 70% and 80% [163–167].

In the literature, there are few studies on miRNA expression profiles in PMLBCL. The first report was given by Kluiver et al. [168], who showed positivity for *BIC* and miR-155 in one cell line and eight PMLBCL samples derived from a tissue bank. Later, Iqbal et al. [104] described an miRNA signature that allowed the distinction between PMLBCL from other DLBCL, including upregulation of miR-193b and miR-365, and underexpression of miR-629, miR-423-5p, and miR-15a. Higher levels of miR-92a were also described as a classifier of PMBLBCL [169]. Moreover, a recent study by Malpeli et al. [170] showed that the polycistron miR-17–92 cluster, miR-29 family, miR-150, and miR-497 had the highest power of discrimination between B-cell NHL types, though only eight PMLBCL samples were included. Nonetheless, none of these studies gave any specific details about pediatric samples, and the mean age of patients was always reported above 27 years old.

3.4. Anaplastic Large-Cell Lymphoma

Anaplastic large-cell lymphoma (ALCL) is an intermediate grade NHL and accounts for approximately 10% of pediatric NHL [171]. Most pediatric ALCL present the chromosomal translocation t(2;5) (p23;q35). In 80% of cases, that translocation results in the expression of a fusion gene called NPM–ALK that encodes a potent oncogenic tyrosine kinase [172].

Several miRNAs have been described as promoting this neoplasia and they seem to express and act differently in ALK^+ and ALK^- tumors and cell lines [173,174]. The suggested signature for ALK^+ cells includes seven miRNA, five of them being upregulated (miR-512*, miR-886, miR-886*, miR-708, and miR-135b) and two downregulated (miR-146a and miR-155). High expression of miR-886 and miR-886* seems to be related with higher AKT expression, since treatment with AKT inhibitors leads to a reduction on these miRNAs levels. It has been shown that miR-886 might act deregulating apoptosis by targeting the proapoptotic gene *BAX* [173]. Furthermore, miR-16 is downregulated in AKT^+, resulting in VEGF expression, tumor growth, and angiogenesis [175].

Besides the above-mentioned miRNAs, the 17–92 cluster has also been found overexpressed in AKT^+ ALCL [176]. These miRNAs are transcriptionally regulated by STAT3, a major substrate for ALK, and promote survival and growth of this tumor. Among the known targets of this cluster, BIM and TGFβRII have been described. An autoregulatory loop between STAT3 and miR-17–92 was also characterized, suggesting an involvement of this cluster in the pathogenesis of this tumor [177].

Conversely, miR-155 showed low expression in ALCL ALK^+ tumors and cell lines, and its inhibition is mediated by methylation. SR278 transfection (pediatric ALCL ALK-positive cells) with pre-miR-155 reduced expression levels of miR-155 targets (C/EBPβ, SOCS1) by binding sites in their 3′-UTR. The action of miR-155 in the immune system was demonstrated through reducing IL-8 and IL-22 transcript levels [178]. C/EBPβ downregulation evidenced the role of this transcription factor in miRNA regulation, mainly miR-181a*, miR-181, and miR-203. MiR-181a showed low expression in ALK$^+$ ALCL cases; this miRNA coordinates T-cell differentiation and modulates TCR antigen expression, being involved in innate and adaptive immune response [174].

MiR-29a was found remarkably reduced in ALK$^+$ when compared to ALK-ALCL, where it regulates MCL-1, contributing with apoptosis blockage [179]. Moreover, ALK knockdown results in increased miR-96 levels, while miR-96 overexpression leads to a reduction in ALK protein levels and decreases cell viability and growth, reinforcing the hypothesis that ALK sustains its own expression by exerting a reciprocal negative feedback loop that hinders the expression of miRNAs [180]. ALK-positive cells showed low levels of miR-146a, miR-29c, miR-29b, miR-29a, miR-22, miR-101, miR-150, and miR-125b, while miR-20b was upregulated [176].

The translocation and activity of NPM-ALK are responsible for miR-150 and miR-125b silencing in cell lines, mediated by DNMT1-dependent activity [181,182]. Inhibition of DNMT1 binding to the MIR125B1 promoter decreased BAK1 expression, an miR-125b target. Mir-125b repression and increase of BAK1 is correlated with early relapse in human ALK$^+$ ALCL biopsies [182]. Conversely, miR-101 is found downregulated in both types of ALCL, but, because it targets the mammalian/mechanistic target of the rapamycin (mTOR) pathway, its forced expression only affects ALK$^+$ cell growth [176].

3.5. Lymphoblastic Lymphoma

LL is a rare neoplasm of immature cells committed to the B (B-LBL)- or T-cell lineage (T-LBL) that accounts for approximately 2% of all lymphomas. The annual incidence in children (<15 years) is 3.6 per 100,000, which is then reduced to 0.8 in people older than 25 years old [183]. Studies about the role of miRNAs on this form of lymphoma are scarce and are summarized as follows.

3.5.1. B-Cell Lymphoblastic Lymphoma

B-LBL typically affects children younger than six years, but is also encountered in older children and in adult populations [184]. B-LBL tumor cells are virtually always positive for B-cell markers CD19, CD79a, and CD22, and may be associated with the presence of leukemia rearrangements such as those involving *ETV6*, *MLL*, or *ABL1* [183]. Thus, even though lymph nodes and extranodal sites, such as skin, bone, and soft tissue, are frequently involved, this rare NHL is considered a lymphomatous variant of ALL and is often treated with leukemia-like regimens [183].

As a result, over the last years, high priority has been given to the identification of biological/prognostic features of T-LBL to allow either risk stratification or treatment planning.

3.5.2. T-Cell Lymphoblastic Lymphoma

T-LBL represents 30% of pediatric NHL [185]. The downregulation of miR-193b in T-LBL was first associated with the activation of the GLI/hedgehog pathway promoting cell survival and proliferation by enhancing SMO (smoothened) expression [186]. Later, this miRNA with miR-196b were found involved in the regulation of the PDGF (platelet-derived growth factor) signaling pathway. In addition, miR-221 was specifically found upregulated in T-LBL directly targeting CDKN1B, a cell-cycle regulator [187]. MiR-22, miR-125a, and miR-125b were also identified as upregulated in T-LBL, and seem to have a role on the maintenance of hematopoietic cells contributing to their proliferation and self-renewal abilities [187].

In a cohort with 52% of T-LBL pediatric samples, upregulation of miR-17 and miR-19 and positive MYC protein was associated with unfavorable prognosis. MYC is known to regulate the miR-17–92 cluster. Cox proportional hazard models showed that miR-17, miR-19, and MYC overexpression were independent poor prognostic factors [188]. MiR-241 is upregulated in T-LBL tissue and a direct target of a long noncoding RNA *MEG3* (maternally expressed 3). Overexpression of MEG3 inhibits tumor growth in vitro and in vivo [189].

Downregulation of miR-374b was associated with worse overall survival and increased risk in T-LBL samples. MiR-374b inhibited proliferation and promoted apoptosis in a pediatric T-LBL cell line (SUP-T1) by repressing AKT1 and Wnt-16 [190]. Moreover, upregulation of miR-221-3p and miR-222-3p, and downregulation of miR-203a and miR-205-5p, miR-200a-3p, and miR-375 have shown to play important roles in T-LBLs by dysregulating in the CDKN1C/E2F1/TP53 axis [191].

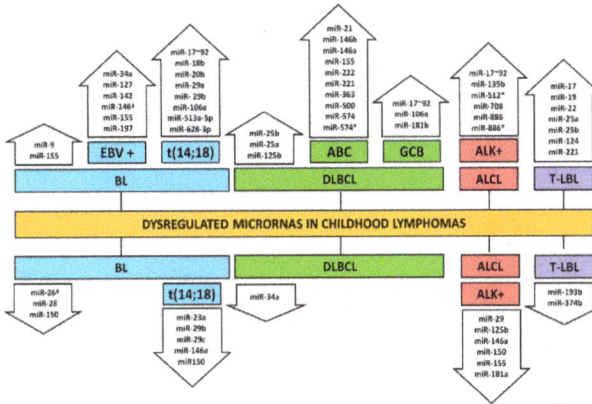

Figure 4. Dysregulated miRNAS in childhood lymphomas. Upside and downside arrows designate hyperexpressed and hypoexpressed miRNAs, respectively; Burkitt lymphoma (BL) (including Epstein–Bar virus and t(14;18) positive cases), diffuse large B-cell lymphoma (DLBCL) (including activated B-cell-like and germinal center B-cell-like subtypes), anaplastic large-cell lymphoma (ALCL) (including the ALK⁺ subtype), and lymphoblastic T-cell lymphoma (T-LBL).

3.6. Hodgkin's Lymphoma

Hodgkin lymphoma (HL) is characterized by multinucleated giant cells (Hodgkin/Reed-Sternberg cells, H/RS) or large mononuclear cell variants (lymphocytic and histiocytic cells) (representing 1% of the tumor) in a background of inflammatory cells that include lymphocytes, histiocytes, neutrophils, eosinophils, plasma cells, and fibroblasts [192]. The annual incidence of HL is 2–3 cases per 100,000 in Europe and the USA, with a bimodal peak, with young adults aged 15–34 being the most affected, followed by those aged 60 and older [193]. HL accounts for 5% to 6% of all childhood cancer and is one of the most curable forms [194]. The five-year EFS in childhood and adolescence exceeds 90% for patients with early-stage and 70% to 80% for those with advanced-stage disease [195].

Over the last decade, efforts have been made in order to identify miRNAs as biomarkers for the refinement of diagnosis and therapy of HL; even so, information is still limited. Some miRNAs have been described as dysregulated in adult samples (mean age 29 years old) by different groups [196–198]. MiR-25, miR-30a/d, miR-26b, miR-182, miR-186, miR-140*, and miR-125a [199], or miR-34a-5p, miR-146a-5p, miR-93-5p, miR-20a-5p, miR-339-3p, miR-324-3p, miR-372, miR-127-3p, miR-155-5p, miR-320a, and miR-370 [200], for instance, have been described as upregulated in tumor samples. Concomitantly, miR-23a, miR-122, miR-93, and miR-144 [199], miR-582-3p, miR-525-3p, miR-448, miR-512-3p, miR-642a-5p, miR-876-5p, miR-532-3p, miR-654-5p, miR-128, miR-145-5p, miR-15b-5p, miR-328, and miR-660-5p were designated as downregulated [200]. Other studies that are based on a limited number of samples have no information about age [201], used data miming [202], or are centered on different cell lines that are all of adult origin [203–206]. Thus, so far, there is no information about dysregulated miRNAs in the pediatric setting.

4. MiRNAs in Clinics

4.1. Circulating MiRNAs as Biomarkers

The pursuit for noninvasive tools for the diagnosis and management of cancer has long encouraged the interest of researchers into the field of circulating nucleic acids. Compelling evidence has shown that genetic and epigenetic cancer markers are also measureable in the plasma and serum of cancer patients and may be useful as a tool for early detection, diagnosis, and follow-up [207,208].

Recently, extracellular circulating miRNAs were detected in secreted membrane vesicles (exosomes), blood serum, and other body fluids. This discovery suggests that miRNAs play a role in intracellular communication in both a paracrine and endocrine manner [209]. Dysregulated expression of miRNAs is implicated in tumorigenesis; therefore, functional characterization of these miRNAs in cancer has received more attention in identifying promising diagnostic and/or prognostic biomarkers. On this regard, MiRNAs are ideal candidates due to their unique expression patterns associated with disease-stage stability and their stability in plasma, easy detection, and recovery [210].

After the first description of circulating miRNAs in lymphoma patients, a significant increase in the number of studies appeared [207], but the amount of research of pediatric cases is smaller than their adult counterparts.

In ALL, a few circulating miRNAs were recently described, and among them miR-146a. Significant higher median levels of miR-100, miR-196a, and miR-146a were reported in blood samples of affected children compared to controls, but the diagnostic efficacy for miR-146a analysis presented superior sensitivity and specificity [211]. These same authors recently reported the significant increased expression of circulating miR-125b-1 and low levels of miR-203 in serum samples from untreated newly diagnosed children with ALL (*n* = 43), as detected by quantitative RT-PCR analysis [212]. They also showed higher levels of miR-125b-1 in T-ALL samples as compared to other ALL phenotypes [212]. More recently, circulating miR-652-3p was found downregulated in serum from ALL patients and levels reported as restored when patients attained in complete remission [47].

In AML patients, Fayyad-Kazan et al. [213] analyzed serum samples from a large cohort of newly diagnosed patients and compared them to normal samples from adult donors. After a two-phase selection and validation process, let-7d, miR-150, miR-339, and miR-342 were found downregulated, while let-7b and miR-523 were upregulated AML compared to control sera [213]. Other results have revealed the presence of two miRNAs, miR-150 and miR-342, significantly downregulated in the plasma of AML patients at diagnosis when compared to healthy controls [214]. Moreover, high serum miR-335 levels were associated with shorter RFS and OS. Furthermore, serum miR-335 and cytogenetic risk were identified as independent prognostic factors for both RFS and OS, suggesting miR-335 as a promising biomarker for pediatric AML [214,215]. Conversely, Zhao and colleagues [216] showed that miR-144 was markedly reduced in both the peripheral blood and bone marrow of AML patients. A similar pattern is commonly observed, miR-34a underexpression in AML patients with intermediate/poor risk cytogenetic and the M5 subtype [217]. More recently, low levels of miR-370 and miR-195 were described in sera of pediatric AML patients and associated with classification M7 subtype, unfavorable karyotype, and shorter RFS and OS [218,219].

In lymphoma, Lawrie et al. [207] showed miR-155, miR-210, and miR-21 in high levels in serum from DLBCL patients compared with healthy controls' sera. Moreover, high miR-21 expression was associated with RFS [207], which was later confirmed by Chen et al. [220] in an independent cohort; thus far, it is the only circulating miRNA in DLBC that has shown consistent results and is now considered a biomarker for diagnosis [221].

More recently, the high levels of miR-155 and miR-22 in plasma from DLBCL patients were associated with shorter overall survival [222,223], while high levels of circulating miR-125b and miR-130a further demonstrated that they were involved in the recurrence, progression, and R-CHOP resistance [224]. Additionally, Khare et al. [199] described increased plasma levels of miR-124 and miR-532-5p, and decreased levels of miR-425, miR-141, miR-145, miR-197, miR-345, miR-424, miR-128, and miR-122 in plasma samples from patients with DLBCL through small-RNA sequencing.

For other lymphoma types, information is restricted to adult cases. MiR-221 has been described as elevated in plasma samples from T-LBL patients, with higher levels associated with a poorer long-term outcome [225]. Overexpression of miR-21 and miR-23a, on the other hand, has been associated with staging, WBC, upregulated serum lactate dehydrogenase (LDH) level, and tumor size ≥6 cm in

BL, while miR-125b expression had an association with staging and upregulated serum LDH [226]. Additionally, a more recent study identified miR-25, miR-30a/d, miR-26b, miR-182, miR-186, miR-140*, and miR-125a to be upregulated, while miR-23a, miR-122, miR-93, and miR-144 were downregulated in HL [199].

4.2. Prognostic Use of MiRNAs

Despite substantial advancement in research and medicine, cancer remains a major public-health problem in our society. Thus, the utility of miRNA expression analysis as diagnostic and prognostic molecular markers is strongly supported. For example, analysis of 217 miRNAs from 334 samples including multiple human cancers provided expression signatures more accurate for cancer-subtype classification than expression-profiling of all known mRNAs does [227]. Additionally, as seen in previous sections, many miRNA dysregulations have also been associated with treatment response (Figure 2). Thereafter, many research groups have shown aberrant-expression profiles of miRNAs in a broad variety of human malignant cancers, especially hematological cancer (Tables S1 and S2). Furthermore, miRNA analysis has some benefits because these molecules are highly resistant to degradation and their expression levels can be obtained in a few hours with small biological samples [9].

4.3. Therapeutic Use of MiRNAs

The progress of miRNAs analysis as molecular markers creates hope for personalized cancer treatments by miRNA modulation. Unfortunately, miRNAs are still not druggable, and clinics are far from reality due to a variety of challenges. Nonetheless, local delivery through encapsulation in lipidic or polymer nanoparticles, or ultrasound-mediated microbubble formulations and hyaluronic acid (HA)/protamine sulfate (PS) interpolyelectrolyte complexes, has shown promising results in mice models [228–233]. Viral vectors for delivering miRNAs into cells have also been widely used in preclinical studies; nonetheless, their safety remains controversial mainly because of lack of safety (i.e., lentivirus), off-target effects, or immune responses [234].

So far, only two strategies have entered clinical trials, though none of them involved hematologic cancer. Miravirsen (Satnaris Pharma®), a locked nucleic acid-modified DNA antisense oligonucleotide that targets the liver-specific miR-122 has demonstrated antiviral activity against hepatitis C and, at phase 2, no dose-limiting adverse events [233,235]. More recently, Mirna Therapeutics Inc. began treating patients with advanced solid tumors with MRX34®, a liposomal injection carrying encapsulated miR-34 showing acceptable safety and evidence of antitumor activity in a subset of patients, despite some liposome-related toxicities [236].

5. Concluding Remarks

Targeted therapies have distinctly transformed the treatment of cancer over the past decade. The utility of miRNA-expression analysis as diagnostic and prognostic molecular markers is strongly supported and, in the near future, it may impact the treatment of hematological cancer.

Supplementary Materials: Supplementary materials can be found at http://www.mdpi.com/1422-0067/19/9/2688/s1.

Author Contributions: J.C.d.O., G.M.R., M.B., K.B.S., J.A.P. and M.S.B. wrote and organized the manuscript. M.S.B. created the figures, and edited and finalized the manuscript. All authors read and approved the final manuscript.

Funding: This research received no external funding.

Conflicts of Interest: The authors declare no conflict of interest.

Abbreviations

ABC	Activated B-cell-like
ALCL	Anaplastic Large-Cell Lymphoma
ALL	Acute Lymphoblastic Leukemia

AML	Acute Myeloid Leukemia
BL	Burkitt Lymphoma
B-LBL	B-cell Lymphoblastic Lymphoma
cALL	Childhood ALL
CML	Chronic Myeloid Leukemia
CNS	Central nervous System
CYT	Cytarabin
DAUNO	Daunorrubicin
DEXA	Dexamethasone
DFS	Disease-Free Survival
DLBCL	Diffuse Large B-cell lymphoma
eBL	Endemic Burkitt Lymphoma
EBNA1	Epstein–Barr nuclear antigen 1
EBV	Epstein–Barr virus
EFS	Event-Free Survival
GCB	Germinal center B cell-like
HL	Hodgkin Lymphoma
INZ	Inauhzin
JMML	Juvenile Myelomonocytic Leukemia
L-ASP	L-asparaginase
LDH	Lactate dehydrogenase
LL	Lymphoblastic Lymphoma
miR	Micro RNA
MiRNAs	Micro RNAs
MTX	Methotrexate
NHL	Non-Hodgkin Lymphoma
OS	Overall Survival
PMLBCL	Primary Mediastinal Large B Cell Lymphoma
PRED	Prednisone
PREDS	Prednisolone
RFS	Relapse-Free Survival
RT-PCR	Reverse Transcription Polymerase Chain Reaction
T-LBL	T-cell Lymphoblastic Lymphoma
VIN	Vincristine
WHO	World Health Organization

References

1. Lee, R.C.; Feinbaum, R.L.; Ambros, V. The *C. elegans* heterochronic gene lin-4 encodes small RNAs with antisense complementarity to lin-14. *Cell* **1993**, *75*, 843–854. [CrossRef]
2. Calin, G.A.; Dumitru, C.D.; Shimizu, M.; Bichi, R.; Zupo, S.; Noch, E.; Aldler, H.; Rattan, S.; Keating, M.; Rai, K.; et al. Nonlinear partial differential equations and applications: Frequent deletions and down-regulation of micro-RNA genes *miR15* and *miR16* at 13q14 in chronic lymphocytic leukemia. *Proc. Natl. Acad. Sci. USA* **2002**, *99*, 15524–15529. [CrossRef] [PubMed]
3. Michael, M.Z.; O' Connor, S.M.; van Holst Pellekaan, N.G.; Young, G.P.; James, R.J. Reduced accumulation of specific microRNAs in colorectal neoplasia. *Mol. Cancer Res.* **2003**, *1*, 882–891. [PubMed]
4. Calin, G.A.; Sevignani, C.; Dumitru, C.D.; Hyslop, T.; Noch, E.; Yendamuri, S.; Shimizu, M.; Rattan, S.; Bullrich, F.; Negrini, M.; et al. Human microRNA genes are frequently located at fragile sites and genomic regions involved in cancers. *Proc. Natl. Acad. Sci. USA* **2004**, *101*, 2999–3004. [CrossRef] [PubMed]
5. Takamizawa, J.; Konishi, H.; Yanagisawa, K.; Tomida, S.; Osada, H.; Endoh, H.; Harano, T.; Yatabe, Y.; Nagino, M.; Nimura, Y.; et al. Reduced Expression of the let-7 MicroRNAs in Human Lung Cancers in Association with Shortened Postoperative Survival Advances in Brief Reduced Expression of the let-7 MicroRNAs in Human Lung Cancers in Association with Shortened Postoperative Survival. *Cancer Res.* **2004**, *64*, 3753–3756. [CrossRef] [PubMed]

6. Calin, G.A.; Liu, C.-G.; Sevignani, C.; Ferracin, M.; Felli, N.; Dumitru, C.D.; Shimizu, M.; Cimmino, A.; Zupo, S.; Dono, M.; et al. MicroRNA profiling reveals distinct signatures in B cell chronic lymphocytic leukemias. *Proc. Natl. Acad. Sci. USA* **2004**, *101*, 11755–11760. [CrossRef] [PubMed]

7. MacDonagh, L.; Gray, S.G.; Finn, S.P.; Cuffe, S.; O'Byrne, K.J.; Barr, M.P. The emerging role of microRNAs in resistance to lung cancer treatments. *Cancer Treat. Rev.* **2015**, *41*, 160–169. [CrossRef] [PubMed]

8. Sun, X.; Charbonneau, C.; Wei, L.; Chen, Q.; Terek, R.M. miR-181a Targets RGS16 to Promote Chondrosarcoma Growth, Angiogenesis, and Metastasis. *Mol. Cancer Res.* **2015**, *13*, 1347–1357. [CrossRef] [PubMed]

9. Kong, Y.W.; Ferland-McCollough, D.; Jackson, T.J.; Bushell, M. microRNAs in cancer management. *Lancet Oncol.* **2012**, *13*, e249–e258. [CrossRef]

10. Metayer, C.; Milne, E.; Clavel, J.; Infante-Rivard, C.; Petridou, E.; Taylor, M.; Schüz, J.; Spector, L.G.; Dockerty, J.D.; Magnani, C.; et al. The Childhood Leukemia International Consortium. *Cancer Epidemiol.* **2013**, *37*, 336–347. [CrossRef] [PubMed]

11. Amitay, E.L.; Keinan-Boker, L. Breastfeeding and Childhood Leukemia Incidence: A Meta-analysis and Systematic Review. *JAMA Pediatr.* **2015**, *169*, e151025. [CrossRef] [PubMed]

12. Madhusoodhan, P.P.; Carroll, W.L.; Bhatla, T. Progress and Prospects in Pediatric Leukemia. *Curr. Probl. Pediatr. Adolesc. Health Care* **2016**, *46*, 229–241. [CrossRef] [PubMed]

13. Seth, R.; Singh, A. Leukemias in Children. *Indian J. Pediatr.* **2015**, *82*, 817–824. [CrossRef] [PubMed]

14. Pui, C.-H.; Evans, W.E. A 50-year journey to cure childhood acute lymphoblastic leukemia. *Semin. Hematol.* **2013**, *50*, 185–196. [CrossRef] [PubMed]

15. Iacobucci, I.; Papayannidis, C.; Lonetti, A.; Ferrari, A.; Baccarani, M.; Martinelli, G. Cytogenetic and molecular predictors of outcome in acute lymphocytic leukemia: Recent developments. *Curr. Hematol. Malig. Rep.* **2012**, *7*, 133–143. [CrossRef] [PubMed]

16. Mi, S.; Lu, J.; Sun, M.; Li, Z.; Zhang, H.; Neilly, M.B.; Wang, Y.; Qian, Z.; Jin, J.; Zhang, Y.; et al. MicroRNA expression signatures accurately discriminate acute lymphoblastic leukemia from acute myeloid leukemia. *Proc. Natl. Acad. Sci. USA* **2007**, *104*, 19971–19976. [CrossRef] [PubMed]

17. Ju, X.; Li, D.; Shi, Q.; Hou, H.; Sun, N.; Shen, B. Differential microRNA expression in childhood B-cell precursor acute lymphoblastic leukemia. *Pediatr. Hematol. Oncol.* **2009**, *26*, 1–10. [CrossRef] [PubMed]

18. Schotte, D.; Chau, J.C.K.; Sylvester, G.; Liu, G.; Chen, C.; van der Velden, V.H.J.; Broekhuis, M.J.C.; Peters, T.C.J.M.; Pieters, R.; den Boer, M.L. Identification of new microRNA genes and aberrant microRNA profiles in childhood acute lymphoblastic leukemia. *Leukemia* **2009**, *23*, 313–322. [CrossRef] [PubMed]

19. De Oliveira, J.C.; Scrideli, C.A.; Brassesco, M.S.; Morales, A.G.; Pezuk, J.A.; Queiroz, R.D.P.; Yunes, J.A.; Brandalise, S.R.; Tone, L.G. Differential MiRNA expression in childhood acute lymphoblastic leukemia and association with clinical and biological features. *Leuk. Res.* **2012**, *36*, 293–298. [CrossRef] [PubMed]

20. De Oliveira, J.C.; Scrideli, C.A.; Brassesco, M.S.; Yunes, J.A.; Brandalise, S.R.; Tone, L.G. MiR-708-5p is differentially expressed in childhood acute lymphoblastic leukemia but not strongly associated to clinical features. *Pediatr. Blood Cancer* **2015**, *62*, 177–178. [CrossRef] [PubMed]

21. Duyu, M.; Durmaz, B.; Gunduz, C.; Vergin, C.; Yilmaz Karapinar, D.; Aksoylar, S.; Kavakli, K.; Cetingul, N.; Irken, G.; Yaman, Y.; et al. Prospective evaluation of whole genome microRNA expression profiling in childhood acute lymphoblastic leukemia. *Biomed. Res. Int.* **2014**, *2014*, 967585. [CrossRef] [PubMed]

22. Panagal, M.; Senthil, R.S.K.; Sivakurunathan, P.; Biruntha, M.; Karthigeyan, M.; Gopinathe, V.; Sivakumare, P.; Sekar, D. MicroRNA21 and the various types of myeloid leukemia. *Cancer Gene Ther.* **2018**, *25*, 161–166. [CrossRef] [PubMed]

23. Zhang, H.; Luo, X.-Q.; Zhang, P.; Huang, L.-B.; Zheng, Y.-S.; Wu, J.; Zhou, H.; Qu, L.-H.; Xu, L.; Chen, Y.-Q. MicroRNA Patterns Associated with Clinical Prognostic Parameters and CNS Relapse Prediction in Pediatric Acute Leukemia. *PLoS ONE* **2009**, *4*, e7826. [CrossRef] [PubMed]

24. Mosakhani, N.; Missiry, M.E.; Vakkila, E.; Knuutila, S.; Vakkila, J. Low Expression of miR-18a as a Characteristic of Pediatric Acute Lymphoblastic Leukemia. *J. Pediatr. Hematol. Oncol.* **2017**, *39*, 585–588. [CrossRef] [PubMed]

25. Nabhan, M.; Louka, M.L.; Khairy, E.; Tash, F.; Ali-Labib, R.; El-Habashy, S. MicroRNA-181a and its target Smad 7 as potential biomarkers for tracking child acute lymphoblastic leukemia. *Gene* **2017**, *628*, 253–258. [CrossRef] [PubMed]

26. Fulci, V.; Colombo, T.; Chiaretti, S.; Messina, M.; Citarella, F.; Tavolaro, S.; Guarini, A.; Foà, R.; Macino, G. Characterization of B- and T-lineage acute lymphoblastic leukemia by integrated analysis of MicroRNA and mRNA expression profiles. *Genes. Chromosom. Cancer* **2009**, *48*, 1069–1082. [CrossRef] [PubMed]

27. Schotte, D.; De Menezes, R.X.; Moqadam, F.A.; Khankahdani, L.M.; Lange-Turenhout, E.; Chen, C.; Pieters, R.; Den Boer, M.L. MicroRNA characterize genetic diversity and drug resistance in pediatric acute lymphoblastic leukemia. *Haematologica* **2011**, *96*, 703–711. [CrossRef] [PubMed]

28. He, Z.; Liao, Z.; Chen, S.; Li, B.; Yu, Z.; Luo, G.; Yang, L.; Zeng, C.; Li, Y. Downregulated miR-17, miR-29c, miR-92a and miR-214 may be related to *BCL11B* overexpression in T cell acute lymphoblastic leukemia. *Asia Pac. J. Clin. Oncol.* **2018**. [CrossRef] [PubMed]

29. Krzanowski, J.; Madzio, J.; Pastorczak, A.; Tracz, A.; Braun, M.; Tabarkiewicz, J.; Pluta, A.; Młynarski, W.; Zawlik, I. Selected miRNA levels are associated with IKZF1 microdeletions in pediatric acute lymphoblastic leukemia. *Oncol. Lett.* **2017**, *14*, 3853–3861. [CrossRef] [PubMed]

30. Gefen, N.; Binder, V.; Zaliova, M.; Linka, Y.; Morrow, M.; Novosel, A.; Edry, L.; Hertzberg, L.; Shomron, N.; Williams, O.; et al. Hsa-mir-125b-2 is highly expressed in childhood ETV6/RUNX1 (TEL/AML1) leukemias and confers survival advantage to growth inhibitory signals independent of p53. *Leukemia* **2010**, *24*, 89–96. [CrossRef] [PubMed]

31. Labib, H.A.; Elantouny, N.G.; Ibrahim, N.F.; Alnagar, A.A. Upregulation of microRNA-21 is a poor prognostic marker in patients with childhood B cell acute lymphoblastic leukemia. *Hematology* **2017**, *22*, 392–397. [CrossRef] [PubMed]

32. Mavrakis, K.J.; Van Der Meulen, J.; Wolfe, A.L.; Liu, X.; Mets, E.; Taghon, T.; Khan, A.A.; Setty, M.; Setti, M.; Rondou, P.; et al. A cooperative microRNA-tumor suppressor gene network in acute T-cell lymphoblastic leukemia (T-ALL). *Nat. Genet.* **2011**, *43*, 673–678. [CrossRef] [PubMed]

33. Mansour, M.R.; Sanda, T.; Lawton, L.N.; Li, X.; Kreslavsky, T.; Novina, C.D.; Brand, M.; Gutierrez, A.; Kelliher, M.A.; Jamieson, C.H.M.; et al. The TAL1 complex targets the FBXW7 tumor suppressor by activating miR-223 in human T cell acute lymphoblastic leukemia. *J. Exp. Med.* **2013**, *210*, 1545–1557. [CrossRef] [PubMed]

34. Kumar, V.; Palermo, R.; Talora, C.; Campese, A.F.; Checquolo, S.; Bellavia, D.; Tottone, L.; Testa, G.; Miele, E.; Indraccolo, S.; et al. Notch and NF-kB signaling pathways regulate miR-223/FBXW7 axis in T-cell acute lymphoblastic.leukemia. *Leukemia* **2014**, *28*, 2324–2335. [CrossRef] [PubMed]

35. Doerrenberg, M.; Kloetgen, A.; Hezaveh, K.; Wössmann, W.; Bleckmann, K.; Stanulla, M.; Schrappe, M.; McHardy, A.C.; Borkhardt, A.; Hoell, J.I. T-cell acute lymphoblastic leukemia in infants has distinct genetic and epigenetic features compared to childhood cases. *Genes Chromosom. Cancer* **2017**, *56*, 159–167. [CrossRef] [PubMed]

36. Mesrian Tanha, H.; Mojtabavi Naeini, M.; Rahgozar, S.; Moafi, A.; Honardoost, M.A. Integrative computational in-depth analysis of dysregulated miRNA-mRNA interactions in drug-resistant pediatric acute lymphoblastic leukemia cells: An attempt to obtain new potential gene-miRNA pathways involved in response to treatment. *Tumour Biol.* **2016**, *37*, 7861–7872. [CrossRef] [PubMed]

37. Bereza, W.; Szczepanek, J.; Laskowska, J.; Tretyn, A. New Candidate Genes for Lack of Sensitivity to Therapy in Pediatric Leukemias. *Curr. Cancer Drug Targets* **2017**, *17*, 333–343. [CrossRef] [PubMed]

38. Cao, L.; Wang, N.; Pan, J.; Hu, S.; Zhao, W.; He, H.; Wang, Y.; Gu, G.; Chai, Y. Clinical significance of microRNA-34b expression in pediatric acute leukemia. *Mol. Med. Rep.* **2016**, *13*, 2777–2784. [CrossRef] [PubMed]

39. Nemes, K.; Csóka, M.; Nagy, N.; Márk, Á.; Váradi, Z.; Dankó, T.; Kovács, G.; Kopper, L.; Sebestyén, A. Expression of Certain Leukemia/Lymphoma Related microRNAs and its Correlation with Prognosis in Childhood Acute Lymphoblastic Leukemia. *Pathol. Oncol. Res.* **2015**, *21*, 597–604. [CrossRef] [PubMed]

40. Xu, C.; Zheng, Y.; Lian, D.; Ye, S.; Yang, J.; Zeng, Z. Analysis of microRNA expression profile identifies novel biomarkers for non-small cell lung cancer. *Tumori* **2015**, *101*, 104–110. [CrossRef] [PubMed]

41. Liang, Y.-N.; Tang, Y.-L.; Ke, Z.-Y.; Chen, Y.-Q.; Luo, X.-Q.; Zhang, H.; Huang, L.-B. MiR-124 contributes to glucocorticoid resistance in acute lymphoblastic leukemia by promoting proliferation, inhibiting apoptosis and targeting the glucocorticoid receptor. *J. Steroid Biochem. Mol. Biol.* **2017**, *172*, 62–68. [CrossRef] [PubMed]

42. Azimi, A.; Hagh, M.; Yousefi, B.; Rahnama, M.; Khorrami, A.; Heydarabad, M.; Najafpour, M.; Hallajzadeh, J.; Ghahremani, A. The Effect of Prednisolone on miR 15a and miR16-1 Expression Levels and Apoptosis in Acute Lymphoblastic Leukemia Cell Line: CCRF-CEM. *Drug Res.* **2016**, *66*, 432–435. [CrossRef] [PubMed]

43. Kotani, A.; Ha, D.; Hsieh, J.; Rao, P.K.; Schotte, D.; den Boer, M.L.; Armstrong, S.A.; Lodish, H.F. miR-128b is a potent glucocorticoid sensitizer in MLL-AF4 acute lymphocytic leukemia cells and exerts cooperative effects with miR-221. *Blood* **2009**, *114*, 4169–4178. [CrossRef] [PubMed]

44. Yan, J.; Jiang, N.; Huang, G.; Tay, J.L.-S.; Lin, B.; Bi, C.; Koh, G.S.; Li, Z.; Tan, J.; Chung, T.-H.; et al. Deregulated MIR335 that targets MAPK1 is implicated in poor outcome of paediatric acute lymphoblastic leukaemia. *Br. J. Haematol.* **2013**, *163*, 93–103. [CrossRef] [PubMed]

45. Umerez, M.; Garcia-Obregon, S.; Martin-Guerrero, I.; Astigarraga, I.; Gutierrez-Camino, A.; Garcia-Orad, A. Role of miRNAs in treatment response and toxicity of childhood acute lymphoblastic leukemia. *Pharmacogenomics* **2018**, *19*, 361–373. [CrossRef] [PubMed]

46. Akbari Moqadam, F.; Lange-Turenhout, E.A.M.; Ariës, I.M.; Pieters, R.; den Boer, M.L. MiR-125b, miR-100 and miR-99a co-regulate vincristine resistance in childhood acute lymphoblastic leukemia. *Leuk. Res.* **2013**, *37*, 1315–1321. [CrossRef] [PubMed]

47. Jiang, Q.; Lu, X.; Huang, P.; Gao, C.; Zhao, X.; Xing, T.; Li, G.; Bao, S.; Zheng, H. Expression of miR-652-3p and Effect on Apoptosis and Drug Sensitivity in Pediatric Acute Lymphoblastic Leukemia. *Biomed. Res. Int.* **2018**, *2018*, 1–10. [CrossRef] [PubMed]

48. Ghodousi, E.S.; Rahgozar, S. MicroRNA-326 and microRNA-200c: Two novel biomarkers for diagnosis and prognosis of pediatric acute lymphoblastic leukemia. *J. Cell. Biochem.* **2018**, *119*, 6024–6032. [CrossRef] [PubMed]

49. Gutierrez-Camino, Á.; Umerez, M.; Martin-Guerrero, I.; García de Andoin, N.; Santos, B.; Sastre, A.; Echebarria-Barona, A.; Astigarraga, I.; Navajas, A.; Garcia-Orad, A. Mir-pharmacogenetics of Vincristine and peripheral neurotoxicity in childhood B-cell acute lymphoblastic leukemia. *Pharmacogenom. J.* **2017**. [CrossRef] [PubMed]

50. Gutierrez-Camino, A.; Oosterom, N.; den Hoed, M.A.H.; Lopez-Lopez, E.; Martin-Guerrero, I.; Pluijm, S.M.F.; Pieters, R.; de Jonge, R.; Tissing, W.J.E.; Heil, S.G.; et al. The miR-1206 microRNA variant is associated with methotrexate-induced oral mucositis in pediatric acute lymphoblastic leukemia. *Pharmacogenet. Genom.* **2017**, *27*, 303–306. [CrossRef] [PubMed]

51. Wang, S.-M.; Zeng, W.-X.; Wu, W.-S.; Sun, L.-L.; Yan, D. Association between *MTHFR* microRNA binding site polymorphisms and methotrexate concentrations in Chinese pediatric patients with acute lymphoblastic leukemia. *J. Gene Med.* **2017**, *19*, e2990. [CrossRef] [PubMed]

52. Iparraguirre, L.; Gutierrez-Camino, A.; Umerez, M.; Martin-Guerrero, I.; Astigarraga, I.; Navajas, A.; Sastre, A.; Garcia de Andoin, N.; Garcia-Orad, A. MiR-pharmacogenetics of methotrexate in childhood B-cell acute lymphoblastic leukemia. *Pharmacogenet. Genom.* **2016**, *26*, 517–525. [CrossRef] [PubMed]

53. Mei, Y.; Gao, C.; Wang, K.; Cui, L.; Li, W.; Zhao, X.; Liu, F.; Wu, M.; Deng, G.; Ding, W.; et al. Effect of microRNA-210 on prognosis and response to chemotherapeutic drugs in pediatric acute lymphoblastic leukemia. *Cancer Sci.* **2014**, *105*, 463–472. [CrossRef] [PubMed]

54. Han, B.-W.; Feng, D.-D.; Li, Z.-G.; Luo, X.-Q.; Zhang, H.; Li, X.-J.; Zhang, X.-J.; Zheng, L.-L.; Zeng, C.-W.; Lin, K.-Y.; et al. A set of miRNAs that involve in the pathways of drug resistance and leukemic stem-cell differentiation is associated with the risk of relapse and glucocorticoid response in childhood ALL. *Hum. Mol. Genet.* **2011**, *20*, 4903–4915. [CrossRef] [PubMed]

55. Ohyashiki, J.H.; Umezu, T.; Kobayashi, C.; Hamamura, R.S.; Tanaka, M.; Kuroda, M.; Ohyashiki, K. Impact on cell to plasma ratio of miR-92a in patients with acute leukemia: In vivo assessment of cell to plasma ratio of miR-92a. *BMC Res. Notes* **2010**, *3*, 347. [CrossRef] [PubMed]

56. Wang, Y.; Li, Z.; He, C.; Wang, D.; Yuan, X.; Chen, J.; Jin, J. MicroRNAs expression signatures are associated with lineage and survival in acute leukemias. *Blood Cells. Mol. Dis.* **2010**, *44*, 191–197. [CrossRef] [PubMed]

57. Rodriguez-Otero, P.; Román-Gómez, J.; Vilas-Zornoza, A.; José-Eneriz, E.S.; Martín-Palanco, V.; Rifón, J.; Torres, A.; Calasanz, M.J.; Agirre, X.; Prosper, F. Deregulation of FGFR1 and CDK6 oncogenic pathways in acute lymphoblastic leukaemia harbouring epigenetic modifications of the MIR9 family. *Br. J. Haematol.* **2011**, *155*, 73–83. [CrossRef] [PubMed]

58. Kaddar, T.; Chien, W.W.; Bertrand, Y.; Pages, M.P.; Rouault, J.P.; Salles, G.; Ffrench, M.; Magaud, J.P. Prognostic value of miR-16 expression in childhood acute lymphoblastic leukemia relationships to normal and malignant lymphocyte proliferation. *Leuk. Res.* **2009**, *33*, 1217–1223. [CrossRef] [PubMed]

59. Mei, Y.; Li, Z.; Zhang, Y.; Zhang, W.; Hu, H.; Zhang, P.; Wu, M.; Huang, D. Low miR-210 and CASP8AP2 expression is associated with a poor outcome in pediatric acute lymphoblastic leukemia. *Oncol. Lett.* **2017**, *14*, 8072–8077. [CrossRef] [PubMed]

60. Piatopoulou, D.; Avgeris, M.; Drakaki, I.; Marmarinos, A.; Xagorari, M.; Baka, M.; Pourtsidis, A.; Kossiva, L.; Gourgiotis, D.; Scorilas, A. Clinical utility of miR-143/miR-182 levels in prognosis and risk stratification specificity of BFM-treated childhood acute lymphoblastic leukemia. *Ann. Hematol.* **2018**, *97*, 1169–1182. [CrossRef] [PubMed]

61. Avigad, S.; Verly, I.R.; Lebel, A.; Kordi, O.; Shichrur, K.; Ohali, A.; Hameiri-Grossman, M.; Kaspers, G.J.; Cloos, J.; Fronkova, E.; et al. miR expression profiling at diagnosis predicts relapse in pediatric precursor B-cell acute lymphoblastic leukemia. *Genes Chromosom. Cancer* **2016**, *55*, 328–339. [CrossRef] [PubMed]

62. Pulte, D.; Gondos, A.; Brenner, H. Trends in 5- and 10-year Survival After Diagnosis with Childhood Hematologic Malignancies in the United States, 1990–2004. *JNCI J. Natl. Cancer Inst.* **2008**, *100*, 1301–1309. [CrossRef] [PubMed]

63. Pigazzi, M.; Manara, E.; Bresolin, S.; Tregnago, C.; Beghin, A.; Baron, E.; Giarin, E.; Cho, E.-C.; Masetti, R.; Rao, D.S.; et al. MicroRNA-34b promoter hypermethylation induces CREB overexpression and contributes to myeloid transformation. *Haematologica* **2013**, *98*, 602–610. [CrossRef] [PubMed]

64. Yan-Fang, T.; Jian, N.; Jun, L.; Na, W.; Pei-Fang, X.; Wen-Li, Z.; Dong, W.; Li, P.; Jian, W.; Xing, F.; et al. The promoter of miR-663 is hypermethylated in Chinese pediatric acute myeloid leukemia (AML). *BMC Med. Genet.* **2013**, *14*, 74. [CrossRef] [PubMed]

65. Emmrich, S.; Katsman-Kuipers, J.E.; Henke, K.; Khatib, M.E.; Jammal, R.; Engeland, F.; Dasci, F.; Zwaan, C.M.; den Boer, M.L.; Verboon, L.; et al. miR-9 is a tumor suppressor in pediatric AML with t(8;21). *Leukemia* **2014**, *28*, 1022–1032. [CrossRef] [PubMed]

66. Obulkasim, A.; Katsman-Kuipers, J.E.; Verboon, L.; Sanders, M.; Touw, I.; Jongen-Lavrencic, M.; Pieters, R.; Klusmann, J.-H.; Zwaan, C.M.; Heuvel-Eibrink, M.M.; et al. Classification of pediatric acute myeloid leukemia based on miRNA expression profiles. *Oncotarget* **2017**, *8*, 33078–33085. [CrossRef] [PubMed]

67. Yan, W.; Xu, L.; Sun, Z.; Lin, Y.; Zhang, W.; Chen, J.; Hu, S.; Shen, B. MicroRNA biomarker identification for pediatric acute myeloid leukemia based on a novel bioinformatics model. *Oncotarget* **2015**, *6*, 26424–26436. [CrossRef] [PubMed]

68. Xu, L.-H.; Guo, Y.; Cen, J.-N.; Yan, W.-Y.; He, H.-L.; Niu, Y.-N.; Lin, Y.-X.; Chen, C.-S.; Hu, S.-Y. Overexpressed miR-155 is associated with initial presentation and poor outcome in Chinese pediatric acute myeloid leukemia. *Eur. Rev. Med. Pharmacol. Sci.* **2015**, *19*, 4841–4850. [PubMed]

69. Ramamurthy, R.; Hughes, M.; Morris, V.; Bolouri, H.; Gerbing, R.B.; Wang, Y.-C.; Loken, M.R.; Raimondi, S.C.; Hirsch, B.A.; Gamis, A.S.; et al. miR-155 expression and correlation with clinical outcome in pediatric AML: A report from Children's Oncology Group. *Pediatr. Blood Cancer* **2016**, *63*, 2096–2103. [CrossRef] [PubMed]

70. Bhayadia, R.; Krowiorz, K.; Haetscher, N.; Jammal, R.; Emmrich, S.; Obulkasim, A.; Fiedler, J.; Schwarzer, A.; Rouhi, A.; Heuser, M.; et al. Endogenous Tumor Suppressor microRNA-193b: Therapeutic and Prognostic Value in Acute Myeloid Leukemia. *J. Clin. Oncol.* **2018**, *36*, 1007–1016. [CrossRef] [PubMed]

71. Zhu, R.; Zhao, W.; Fan, F.; Tang, L.; Liu, J.; Luo, T.; Deng, J.; Hu, Y. A 3-miRNA signature predicts prognosis of pediatric and adolescent cytogenetically normal acute myeloid leukemia. *Oncotarget* **2017**, *8*, 38902–38913. [CrossRef] [PubMed]

72. Xu, L.; Guo, Y.; Yan, W.; Cen, J.; Niu, Y.; Yan, Q.; He, H.; Chen, C.-S.; Hu, S. High level of miR-196b at newly diagnosed pediatric acute myeloid leukemia predicts a poor outcome. *EXCLI J.* **2017**, *16*, 197–209. [CrossRef] [PubMed]

73. Wang, Z.; Hong, Z.; Gao, F.; Feng, W. Upregulation of microRNA-375 is associated with poor prognosis in pediatric acute myeloid leukemia. *Mol. Cell. Biochem.* **2013**, *383*, 59–65. [CrossRef] [PubMed]

74. Zhu, C.; Wang, Y.; Kuai, W.; Sun, X.; Chen, H.; Hong, Z. Prognostic value of miR-29a expression in pediatric acute myeloid leukemia. *Clin. Biochem.* **2013**, *46*, 49–53. [CrossRef] [PubMed]

75. Lim, E.L.; Trinh, D.L.; Ries, R.E.; Wang, J.; Gerbing, R.B.; Ma, Y.; Topham, J.; Hughes, M.; Pleasance, E.; Mungall, A.J.; et al. MicroRNA Expression-Based Model Indicates Event-Free Survival in Pediatric Acute Myeloid Leukemia. *J. Clin. Oncol.* **2017**, *35*, 3964–3977. [CrossRef] [PubMed]

76. Danen-van Oorschot, A.A.; Kuipers, J.E.; Arentsen-Peters, S.; Schotte, D.; de Haas, V.; Trka, J.; Baruchel, A.; Reinhardt, D.; Pieters, R.; Michel Zwaan, C.; et al. Differentially expressed miRNAs in cytogenetic and

molecular subtypes of pediatric acute myeloid leukemia. *Pediatr. Blood Cancer* **2012**, *58*, 715–721. [CrossRef] [PubMed]

77. Verboon, L.J.; Obulkasim, A.; de Rooij, J.D.E.; Katsman-Kuipers, J.E.; Sonneveld, E.; Baruchel, A.; Trka, J.; Reinhardt, D.; Pieters, R.; Cloos, J.; et al. MicroRNA-106b~25 cluster is upregulated in relapsed;-rearranged pediatric acute myeloid leukemia. *Oncotarget* **2016**, *7*, 48412–48422. [CrossRef] [PubMed]

78. Daschkey, S.; Röttgers, S.; Giri, A.; Bradtke, J.; Teigler-Schlegel, A.; Meister, G.; Borkhardt, A.; Landgraf, P. MicroRNAs Distinguish Cytogenetic Subgroups in Pediatric AML and Contribute to Complex Regulatory Networks in AML-Relevant Pathways. *PLoS ONE* **2013**, *8*, e56334. [CrossRef] [PubMed]

79. Zhang, L.; Li, X.; Ke, Z.; Huang, L.; Liang, Y.; Wu, J.; Zhang, X.; Chen, Y.; Zhang, H.; Luo, X. MiR-99a may serve as a potential oncogene in pediatric myeloid leukemia. *Cancer Cell Int.* **2013**, *13*, 110. [CrossRef] [PubMed]

80. Yang, J.; Yuan, Y.; Yang, X.; Hong, Z.; Yang, L. Decreased expression of microRNA-122 is associated with an unfavorable prognosis in childhood acute myeloid leukemia and function analysis indicates a therapeutic potential. *Pathol. Res. Pract.* **2017**, *213*, 1166–1172. [CrossRef] [PubMed]

81. Liu, X.; Liao, W.; Peng, H.; Luo, X.; Luo, Z.; Jiang, H.; Xu, L. miR-181a promotes G1/S transition and cell proliferation in pediatric acute myeloid leukemia by targeting ATM. *J. Cancer Res. Clin. Oncol.* **2016**, *142*, 77–87. [CrossRef] [PubMed]

82. Ding, Q.; Wang, Q.; Ren, Y.; Zhu, H.Q.; Huang, Z. MicroRNA-126 attenuates cell apoptosis by targeting TRAF7 in acute myeloid leukemia cells. *Biochem. Cell Biol.* **2018**. [CrossRef] [PubMed]

83. Sharifi, M.; Fasihi-Ramandi, M.; Sheikhi, A.; Moridnia, A.; Saneipour, M. Apoptosis induction in acute promyelocytic leukemia cells through upregulation of CEBPα by miR-182 blockage. *Mol. Biol. Res. Commun.* **2018**, *7*, 25–33. [CrossRef] [PubMed]

84. Chen, Y.; Wang, H.; Kantarjian, H.; Cortes, J. Trends in chronic myeloid leukemia incidence and survival in the United States from 1975 to 2009. *Leuk. Lymphoma* **2013**, *54*, 1411–1417. [CrossRef] [PubMed]

85. De la Fuente, J.; Baruchel, A.; Biondi, A.; de Bont, E.; Dresse, M.-F.; Suttorp, M.; Millot, F.; International BFM Group (iBFM). Study Group Chronic Myeloid Leukaemia Committee Managing children with chronic myeloid leukaemia (CML). *Br. J. Haematol.* **2014**, *167*, 33–47. [CrossRef] [PubMed]

86. Karalexi, M.A.; Baka, M.; Ryzhov, A.; Zborovskaya, A.; Dimitrova, N.; Zivkovic, S.; Eser, S.; Antunes, L.; Sekerija, M.; Zagar, T.; et al. Survival trends in childhood chronic myeloid leukaemia in Southern-Eastern Europe and the United States of America. *Eur. J. Cancer* **2016**, *67*, 183–190. [CrossRef] [PubMed]

87. Shibuta, T.; Honda, E.; Shiotsu, H.; Tanaka, Y.; Vellasamy, S.; Shiratsuchi, M.; Umemura, T. Imatinib induces demethylation of miR-203 gene: An epigenetic mechanism of anti-tumor effect of imatinib. *Leuk. Res.* **2013**, *37*, 1278–1286. [CrossRef] [PubMed]

88. Faber, J.; Gregory, R.I.; Armstrong, S.A. Linking miRNA regulation to BCR-ABL expression: The next dimension. *Cancer Cell* **2008**, *13*, 467–469. [CrossRef] [PubMed]

89. Hussein, K.; Stucki-Koch, A.; Göhring, G.; Kreipe, H.; Suttorp, M. Increased megakaryocytic proliferation, pro-platelet deposition and expression of fibrosis-associated factors in children with chronic myeloid leukaemia with bone marrow fibrosis. *Leukemia* **2017**, *31*, 1540–1546. [CrossRef] [PubMed]

90. Hasle, H.; Kerndrup, G.; Jacobsen, B.B. Childhood myelodysplastic syndrome in Denmark: Incidence and predisposing conditions. *Leukemia* **1995**, *9*, 1569–1572. [PubMed]

91. Yoshimi, A.; Kojima, S.; Hirano, N. Juvenile Myelomonocytic Leukemia. *Pediatr. Drugs* **2010**, *12*, 11–21. [CrossRef] [PubMed]

92. Stieglitz, E.; Taylor-Weiner, A.N.; Chang, T.Y.; Gelston, L.C.; Wang, Y.-D.; Mazor, T.; Esquivel, E.; Yu, A.; Seepo, S.; Olsen, S.R.; et al. The genomic landscape of juvenile myelomonocytic leukemia. *Nat. Genet.* **2015**, *47*, 1326–1333. [CrossRef] [PubMed]

93. Steinemann, D.; Tauscher, M.; Praulich, I.; Niemeyer, C.M.; Flotho, C.; Schlegelberger, B. Mutations in the let-7 binding site—A mechanism of RAS activation in juvenile myelomonocytic leukemia? *Haematologica* **2010**, *95*, 1616. [CrossRef] [PubMed]

94. Helsmoortel, H.H.; Bresolin, S.; Lammens, T.; Cavé, H.; Noellke, P.; Caye, A.; Ghazavi, F.; de Vries, A.; Hasle, H.; Labarque, V.; et al. *LIN28B* overexpression defines a novel fetal-like subgroup of juvenile myelomonocytic leukemia. *Blood* **2016**, *127*, 1163–1172. [CrossRef] [PubMed]

95. Ripperger, T.; Tauscher, M.; Praulich, I.; Pabst, B.; Teigler-Schlegel, A.; Yeoh, A.; Göhring, G.; Schlegelberger, B.; Flotho, C.; Niemeyer, C.M.; et al. Constitutional trisomy 8p11.21-q11.21 mosaicism:

A germline alteration predisposing to myeloid leukaemia. *Br. J. Haematol.* **2011**, *155*, 209–217. [CrossRef] [PubMed]

96. Liu, Y.L.; Lensing, S.Y.; Yan, Y.; Cooper, T.M.; Loh, M.L.; Emanuel, P.D. Deficiency of CREB and over expression of miR-183 in juvenile myelomonocytic leukemia. *Leukemia* **2013**, *27*, 1585–1588. [CrossRef] [PubMed]

97. Mulero-Navarro, S.; Sevilla, A.; Roman, A.C.; Lee, D.-F.; D'Souza, S.L.; Pardo, S.; Riess, I.; Su, J.; Cohen, N.; Schaniel, C.; et al. Myeloid Dysregulation in a Human Induced Pluripotent Stem Cell Model of PTPN11 -Associated Juvenile Myelomonocytic Leukemia. *Cell Rep.* **2015**, *13*, 504–515. [CrossRef] [PubMed]

98. Leoncini, P.P.; Bertaina, A.; Papaioannou, D.; Flotho, C.; Masetti, R.; Bresolin, S.; Menna, G.; Santoro, N.; Zecca, M.; Basso, G.; et al. MicroRNA fingerprints in juvenile myelomonocytic leukemia (JMML) identified miR-150-5p as a tumor suppressor and potential target for treatment. *Oncotarget* **2016**, *7*, 55395–55408. [CrossRef] [PubMed]

99. Campo, E.; Swerdlow, S.H.; Harris, N.L.; Pileri, S.; Stein, H.; Jaffe, E.S. The 2008 WHO classification of lymphoid neoplasms and beyond: Evolving concepts and practical applications. *Blood* **2011**, *117*, 5019–5032. [CrossRef] [PubMed]

100. Dinand, V.; Arya, L.S. Epidemiology of childhood Hodgkins disease: Is it different in developing countries? *Indian Pediatr.* **2006**, *43*, 141–147. [PubMed]

101. Gualco, G.; Klumb, C.E.; Barber, G.N.; Weiss, L.M.; Bacchi, C.E. Pediatric lymphomas in Brazil. *Clinics* **2010**, *65*, 1267–1277. [CrossRef] [PubMed]

102. Metzler, M.; Wilda, M.; Busch, K.; Viehmann, S.; Borkhardt, A. High expression of precursor microRNA-155/BIC RNA in children with Burkitt lymphoma. *Genes Chromosom. Cancer* **2004**, *39*, 167–169. [CrossRef] [PubMed]

103. Sandlund, J.T. Non-Hodgkin Lymphoma in Children. *Curr. Hematol. Malig. Rep.* **2015**, *10*, 237–243. [CrossRef] [PubMed]

104. Iqbal, J.; Shen, Y.; Huang, X.; Liu, Y.; Wake, L.; Liu, C.; Deffenbacher, K.; Lachel, C.M.; Wang, C.; Rohr, J.; et al. Global microRNA expression profiling uncovers molecular markers for classification and prognosis in aggressive B-cell lymphoma. *Blood* **2015**, *125*, 1137–1145. [CrossRef] [PubMed]

105. Ambrosio, M.R.; Mundo, L.; Gazaneo, S.; Picciolini, M.; Vara, P.S.; Sayed, S.; Ginori, A.; Bello, G.L.; Del Porro, L.; Navari, M.; et al. MicroRNAs sequencing unveils distinct molecular subgroups of plasmablastic lymphoma. *Oncotarget* **2017**, *8*, 107356–107373. [CrossRef] [PubMed]

106. Deffenbacher, K.E.; Iqbal, J.; Sanger, W.; Shen, Y.; Lachel, C.; Liu, Z.; Liu, Y.; Lim, M.S.; Perkins, S.L.; Fu, K.; et al. LYMPHOID NEOPLASIA Molecular distinctions between pediatric and adult mature B-cell non-Hodgkin lymphomas identified through genomic profiling. *Blood* **2012**, *119*, 3757–3766. [CrossRef] [PubMed]

107. Navari, M.; Etebari, M.; De Falco, G.; Ambrosio, M.R.; Gibellini, D.; Leoncini, L.; Piccaluga, P.P. The presence of Epstein-Barr virus significantly impacts the transcriptional profile in immunodeficiency-associated Burkitt lymphoma. *Front. Microbiol.* **2015**, *6*, 556. [CrossRef] [PubMed]

108. Hezaveh, K.; Kloetgen, A.; Bernhart, S.H.; Mahapatra, K.D.; Lenze, D.; Richter, J.; Haake, A.; Bergmann, A.K.; Brors, B.; Burkhardt, B.; et al. ICGC MMML-Seq Project Alterations of microRNA and microRNA-regulated messenger RNA expression in germinal center B-cell lymphomas determined by integrative sequencing analysis. *Haematologica* **2016**, *101*, 1380–1389. [CrossRef] [PubMed]

109. Costinean, S.; Zanesi, N.; Pekarsky, Y.; Tili, E.; Volinia, S.; Heerema, N.; Croce, C.M. Pre-B cell proliferation and lymphoblastic leukemia/high-grade lymphoma in E(mu)-miR155 transgenic mice. *Proc. Natl. Acad. Sci. USA* **2006**, *103*, 7024–7029. [CrossRef] [PubMed]

110. O'Connell, R.M.; Chaudhuri, A.A.; Rao, D.S.; Baltimore, D. Inositol phosphatase SHIP1 is a primary target of miR-155. *Proc. Natl. Acad. Sci. USA* **2009**, *106*, 7113–7118. [CrossRef] [PubMed]

111. Slezak-Prochazka, I.; Kluiver, J.; de Jong, D.; Smigielska-Czepiel, K.; Kortman, G.; Winkle, M.; Rutgers, B.; Koerts, J.; Visser, L.; Diepstra, A.; et al. Inhibition of the miR-155 target NIAM phenocopies the growth promoting effect of miR-155 in B-cell lymphoma. *Oncotarget* **2016**, *7*, 2391–2400. [CrossRef] [PubMed]

112. Kluiver, J.; Haralambieva, E.; de Jong, D.; Blokzijl, T.; Jacobs, S.; Kroesen, B.-J.; Poppema, S.; van den Berg, A. Lack of BIC and microRNA miR-155 expression in primary cases of Burkitt lymphoma. *Gene. Chromosom. Cancer* **2006**, *45*, 147–153. [CrossRef] [PubMed]

113. Chabay, P.A.; Preciado, M.V. EBV primary infection in childhood and its relation to B-cell lymphoma development: A mini-review from a developing region. *Int. J. Cancer* **2013**, *133*, 1286–1292. [CrossRef] [PubMed]

114. Yoon, C.; Kim, J.; Park, G.; Kim, S.; Kim, D.; Hur, D.Y.; Kim, B.; Kim, Y.S. Delivery of miR-155 to retinal pigment epithelial cells mediated by Burkitt's lymphoma exosomes. *Tumor Biol.* **2016**, *37*, 313–321. [CrossRef] [PubMed]

115. Motsch, N.; Pfuhl, T.; Mrazek, J.; Barth, S.; Grässer, F.A. Epstein-Barr virus-encoded latent membrane protein 1 (LMP1) induces the expression of the cellular microRNA miR-146a. *RNA Biol.* **2007**, *4*, 131–137. [CrossRef] [PubMed]

116. Cameron, J.E.; Yin, Q.; Fewell, C.; Lacey, M.; McBride, J.; Wang, X.; Lin, Z.; Schaefer, B.C.; Flemington, E.K. Epstein-Barr virus latent membrane protein 1 induces cellular MicroRNA miR-146a, a modulator of lymphocyte signaling pathways. *J. Virol.* **2008**, *82*, 1946–1958. [CrossRef] [PubMed]

117. Rahadiani, N.; Takakuwa, T.; Tresnasari, K.; Morii, E.; Aozasa, K. Latent membrane protein-1 of Epstein-Barr virus induces the expression of B-cell integration cluster, a precursor form of microRNA-155, in B lymphoma cell lines. *Biochem. Biophys. Res. Commun.* **2008**, *377*, 579–583. [CrossRef] [PubMed]

118. Forte, E.; Salinas, R.E.; Chang, C.; Zhou, T.; Linnstaedt, S.D.; Gottwein, E.; Jacobs, C.; Jima, D.; Li, Q.-J.; Dave, S.S.; et al. The Epstein-Barr virus (EBV)-induced tumor suppressor microRNA MiR-34a is growth promoting in EBV-infected B cells. *J. Virol.* **2012**, *86*, 6889–6898. [CrossRef] [PubMed]

119. Anastasiadou, E.; Boccellato, F.; Vincenti, S.; Rosato, P.; Bozzoni, I.; Frati, L.; Faggioni, A.; Presutti, C.; Trivedi, P. Epstein-Barr virus encoded LMP1 downregulates TCL1 oncogene through miR-29b. *Oncogene* **2010**, *29*, 1316–1328. [CrossRef] [PubMed]

120. Onnis, A.; Navari, M.; Antonicelli, G.; Morettini, F.; Mannucci, S.; De Falco, G.; Vigorito, E.; Leoncini, L. Epstein-Barr nuclear antigen 1 induces expression of the cellular microRNA hsa-miR-127 and impairing B-cell differentiation in EBV-infected memory B cells. New insights into the pathogenesis of Burkitt lymphoma. *Blood Cancer J.* **2012**, *2*, e84. [CrossRef] [PubMed]

121. Kanda, T.; Miyata, M.; Kano, M.; Kondo, S.; Yoshizaki, T.; Iizasa, H. Clustered microRNAs of the Epstein-Barr virus cooperatively downregulate an epithelial cell-specific metastasis suppressor. *J. Virol.* **2015**, *89*, 2684–2697. [CrossRef] [PubMed]

122. Ambrosio, M.R.; Navari, M.; Di Lisio, L.; Leon, E.A.; Onnis, A.; Gazaneo, S.; Mundo, L.; Ulivieri, C.; Gomez, G.; Lazzi, S.; et al. The Epstein Barr-encoded BART-6-3p microRNA affects regulation of cell growth and immuno response in Burkitt lymphoma. *Infect. Agent. Cancer* **2014**, *9*, 12. [CrossRef] [PubMed]

123. Zhou, L.; Bu, Y.; Liang, Y.; Zhang, F.; Zhang, H.; Li, S. Epstein-Barr Virus (EBV)-BamHI-A Rightward Transcript (BART)-6 and Cellular MicroRNA-142 Synergistically Compromise Immune Defense of Host Cells in EBV-Positive Burkitt Lymphoma. *Med. Sci. Monit.* **2016**, *22*, 4114–4120. [CrossRef] [PubMed]

124. Zhang, Y.-M.; Yu, Y.; Zhao, H.-P. EBV-BART-6-3p and cellular microRNA-197 compromise the immune defense of host cells in EBV-positive Burkitt lymphoma. *Mol. Med. Rep.* **2017**, *15*, 1877–1883. [CrossRef] [PubMed]

125. Mundo, L.; Ambrosio, M.R.; Picciolini, M.; Lo Bello, G.; Gazaneo, S.; Del Porro, L.; Lazzi, S.; Navari, M.; Onyango, N.; Granai, M.; et al. Unveiling Another Missing Piece in EBV-Driven Lymphomagenesis: EBV-Encoded MicroRNAs Expression in EBER-Negative Burkitt Lymphoma Cases. *Front. Microbiol.* **2017**, *8*, 229. [CrossRef] [PubMed]

126. Lenze, D.; Leoncini, L.; Hummel, M.; Volinia, S.; Liu, C.G.; Amato, T.; De Falco, G.; Githanga, J.; Horn, H.; Nyagol, J.; et al. The different epidemiologic subtypes of Burkitt lymphoma share a homogenous micro RNA profile distinct from diffuse large B-cell lymphoma. *Leukemia* **2011**, *25*, 1869–1876. [CrossRef] [PubMed]

127. Wang, X.; Wang, P.; Zhu, Y.; Zhang, Z.; Zhang, J.; Wang, H. MicroRNA-520a attenuates proliferation of Raji cells through inhibition of AKT1/NF-κB and PERK/eIF2α signaling pathway. *Oncol. Rep.* **2016**, *36*, 1702–1708. [CrossRef] [PubMed]

128. Allen, C.E.; Kelly, K.M.; Bollard, C.M. Pediatric lymphomas and histiocytic disorders of childhood. *Pediatr. Clin. North Am.* **2015**, *62*, 139–165. [CrossRef] [PubMed]

129. O'Donnell, K.A.; Wentzel, E.A.; Zeller, K.I.; Dang, C.V.; Mendell, J.T. c-Myc-regulated microRNAs modulate E2F1 expression. *Nature* **2005**, *435*, 839–843. [CrossRef] [PubMed]

130. Jin, H.Y.; Oda, H.; Lai, M.; Skalsky, R.L.; Bethel, K.; Shepherd, J.; Kang, S.G.; Liu, W.-H.; Sabouri-Ghomi, M.; Cullen, B.R.; et al. MicroRNA-17~92 plays a causative role in lymphomagenesis by coordinating multiple oncogenic pathways. *EMBO J.* **2013**, *32*, 2377–2391. [CrossRef] [PubMed]

131. Jung, J.H.; Liao, J.-M.; Zhang, Q.; Zeng, S.; Nguyen, D.; Hao, Q.; Zhou, X.; Cao, B.; Kim, S.-H.; Lu, H. Inauhzin(c) inactivates c-Myc independently of p53. *Cancer Biol. Ther.* **2015**, *16*, 412–419. [CrossRef] [PubMed]

132. De Falco, G.; Ambrosio, M.R.; Fuligni, F.; Onnis, A.; Bellan, C.; Rocca, B.J.; Navari, M.; Etebari, M.; Mundo, L.; Gazaneo, S.; et al. Burkitt lymphoma beyond MYC translocation: N-MYC and DNA methyltransferases dysregulation. *BMC Cancer* **2015**, *15*, 668. [CrossRef] [PubMed]

133. Robaina, M.C.; Mazzoccoli, L.; Arruda, V.O.; Reis, F.R.D.S.; Apa, A.G.; de Rezende, L.M.M.; Klumb, C.E. Deregulation of DNMT1, DNMT3B and miR-29s in Burkitt lymphoma suggests novel contribution for disease pathogenesis. *Exp. Mol. Pathol.* **2015**, *98*, 200–207. [CrossRef] [PubMed]

134. Sander, S.; Bullinger, L.; Klapproth, K.; Fiedler, K.; Kestler, H.A.; Barth, T.F.E.; Möller, P.; Stilgenbauer, S.; Pollack, J.R.; Wirth, T. MYC stimulates EZH2 expression by repression of its negative regulator miR-26a. *Blood* **2008**, *112*, 4202–4212. [CrossRef] [PubMed]

135. Ristau, J.; Staffa, J.; Schrotz-King, P.; Gigic, B.; Makar, K.W.; Hoffmeister, M.; Brenner, H.; Ulrich, A.; Schneider, M.; Ulrich, C.M.; et al. Suitability of Circulating miRNAs as Potential Prognostic Markers in Colorectal Cancer. *Cancer Epidemiol. Biomark. Prev.* **2014**, *23*, 2632–2637. [CrossRef] [PubMed]

136. Chen, S.; Wang, Z.; Dai, X.; Pan, J.; Ge, J.; Han, X.; Wu, Z.; Zhou, X.; Zhao, T. Re-expression of microRNA-150 induces EBV-positive Burkitt lymphoma differentiation by modulating c-Myb in vitro. *Cancer Sci.* **2013**, *104*, 826–834. [CrossRef] [PubMed]

137. Wang, M.; Yang, W.; Li, M.; Li, Y. Low expression of miR-150 in pediatric intestinal Burkitt lymphoma. *Exp. Mol. Pathol.* **2014**, *96*, 261–266. [CrossRef] [PubMed]

138. Chen, W.L.; Luo, D.F.; Gao, C.; Ding, Y.; Wang, S.Y. The consensus sequence of FAMLF alternative splice variants is overexpressed in undifferentiated hematopoietic cells. *Braz. J. Med. Biol. Res.* **2015**, *48*, 603–609. [CrossRef] [PubMed]

139. Li, J.G.; Ding, Y.; Huang, Y.M.; Chen, W.L.; Pan, L.L.; Li, Y.; Chen, X.L.; Chen, Y.; Wang, S.Y.; Wu, X.N. FAMLF is a target of miR-181b in Burkitt lymphoma. *Braz. J. Med. Biol. Res.* **2017**, *50*. [CrossRef] [PubMed]

140. Hecht, J.L.; Aster, J.C. Molecular Biology of Burkitt's Lymphoma. *J. Clin. Oncol.* **2000**, *18*, 3707–3721. [CrossRef] [PubMed]

141. Oduor, C.I.; Kaymaz, Y.; Chelimo, K.; Otieno, J.A.; Ong'echa, J.M.; Moormann, A.M.; Bailey, J.A. Integrative microRNA and mRNA deep-sequencing expression profiling in endemic Burkitt lymphoma. *BMC Cancer* **2017**, *17*, 761. [CrossRef] [PubMed]

142. Oduor, C.I.; Movassagh, M.; Kaymaz, Y.; Chelimo, K.; Otieno, J.; Ong'echa, J.M.; Moormann, A.M.; Bailey, J.A. Human and Epstein-Barr Virus miRNA Profiling as Predictive Biomarkers for Endemic Burkitt Lymphoma. *Front. Microbiol.* **2017**, *8*, 501. [CrossRef] [PubMed]

143. Alizadeh, A.A.; Eisen, M.B.; Davis, R.E.; Ma, C.; Lossos, I.S.; Rosenwald, A.; Boldrick, J.C.; Sabet, H.; Tran, T.; Yu, X.; et al. Distinct types of diffuse large B-cell lymphoma identified by gene expression profiling. *Nature* **2000**, *403*, 503–511. [CrossRef] [PubMed]

144. Oschlies, I.; Klapper, W.; Zimmermann, M.; Krams, M.; Wacker, H.-H.; Burkhardt, B.; Harder, L.; Siebert, R.; Reiter, A.; Parwaresch, R. Diffuse large B-cell lymphoma in pediatric patients belongs predominantly to the germinal-center type B-cell lymphomas: A clinicopathologic analysis of cases included in the German BFM (Berlin-Frankfurt-Munster) Multicenter Trial. *Blood* **2006**, *107*, 4047–4052. [CrossRef] [PubMed]

145. Roehle, A.; Hoefig, K.P.; Repsilber, D.; Thorns, C.; Ziepert, M.; Wesche, K.O.; Thiere, M.; Loeffler, M.; Klapper, W.; Pfreundschuh, M.; et al. MicroRNA signatures characterize diffuse large B-cell lymphomas and follicular lymphomas. *Br. J. Haematol.* **2008**, *142*, 732–744. [CrossRef] [PubMed]

146. Lawrie, C.H.; Soneji, S.; Marafioti, T.; Cooper, C.D.O.; Palazzo, S.; Paterson, J.C.; Cattan, H.; Enver, T.; Mager, R.; Boultwood, J.; et al. Microrna expression distinguishes between germinal center B cell-like and activated B cell-like subtypes of diffuse large B cell lymphoma. *Int. J. Cancer* **2007**, *121*, 1156–1161. [CrossRef] [PubMed]

147. Eis, P.S.; Tam, W.; Sun, L.; Chadburn, A.; Li, Z.; Gomez, M.F.; Lund, E.; Dahlberg, J.E. Accumulation of miR-155 and BIC RNA in human B cell lymphomas. *Proc. Natl. Acad. Sci. USA* **2005**, *102*, 3627–3632. [CrossRef] [PubMed]

148. Rai, D.; Karanti, S.; Jung, I.; Dahia, P.L.M.; Aguiar, R.C.T. Coordinated expression of microRNA-155 and predicted target genes in diffuse large B-cell lymphoma. *Cancer Genet. Cytogenet.* **2008**, *181*, 8–15. [CrossRef] [PubMed]

149. Pedersen, I.M.; Otero, D.; Kao, E.; Miletic, A.V.; Hother, C.; Ralfkiaer, E.; Rickert, R.C.; Gronbaek, K.; David, M. Onco-miR-155 targets SHIP1 to promote TNFalpha-dependent growth of B cell lymphomas. *EMBO Mol. Med.* **2009**, *1*, 288–295. [CrossRef] [PubMed]

150. Jung, I.; Aguiar, R.C.T. MicroRNA-155 expression and outcome in diffuse large B-cell lymphoma. *Br. J. Haematol.* **2009**, *144*, 138–140. [CrossRef] [PubMed]

151. Zhong, H.; Xu, L.; Zhong, J.-H.; Xiao, F.; Liu, Q.; Huang, H.-H.; Chen, F.-Y. Clinical and prognostic significance of miR-155 and miR-146a expression levels in formalin-fixed/paraffin-embedded tissue of patients with diffuse large B-cell lymphoma. *Exp. Ther. Med.* **2012**, *3*, 763–770. [CrossRef] [PubMed]

152. Lim, E.L.; Trinh, D.L.; Scott, D.W.; Chu, A.; Krzywinski, M.; Zhao, Y.; Robertson, A.G.; Mungall, A.J.; Schein, J.; Boyle, M.; et al. Comprehensive miRNA sequence analysis reveals survival differences in diffuse large B-cell lymphoma patients. *Genome Biol.* **2015**, *16*, 18. [CrossRef] [PubMed]

153. Lenz, G.; Wright, G.W.; Emre, N.C.T.; Kohlhammer, H.; Dave, S.S.; Davis, R.E.; Carty, S.; Lam, L.T.; Shaffer, A.L.; Xiao, W.; et al. Molecular subtypes of diffuse large B-cell lymphoma arise by distinct genetic pathways. *Proc. Natl. Acad. Sci. USA* **2008**, *105*, 13520–13525. [CrossRef] [PubMed]

154. Tan, L.P.; Wang, M.; Robertus, J.-L.; Schakel, R.N.; Gibcus, J.H.; Diepstra, A.; Harms, G.; Peh, S.-C.; Reijmers, R.M.; Pals, S.T.; et al. miRNA profiling of B-cell subsets: Specific miRNA profile for germinal center B cells with variation between centroblasts and centrocytes. *Lab. Investig.* **2009**, *89*, 708–716. [CrossRef] [PubMed]

155. Kim, S.-W.; Ramasamy, K.; Bouamar, H.; Lin, A.-P.; Jiang, D.; Aguiar, R.C.T. MicroRNAs miR-125a and miR-125b constitutively activate the NF-κB pathway by targeting the tumor necrosis factor alpha-induced protein 3 (TNFAIP3, A20). *Proc. Natl. Acad. Sci. USA* **2012**, *109*, 7865–7870. [CrossRef] [PubMed]

156. Craig, V.J.; Cogliatti, S.B.; Imig, J.; Renner, C.; Neuenschwander, S.; Rehrauer, H.; Schlapbach, R.; Dirnhofer, S.; Tzankov, A.; Müller, A. Myc-mediated repression of microRNA-34a promotes high-grade transformation of B-cell lymphoma by dysregulation of FOXP1. *Blood* **2011**, *117*, 6227–6236. [CrossRef] [PubMed]

157. Craig, V.J.; Tzankov, A.; Flori, M.; Schmid, C.A.; Bader, A.G.; Müller, A. Systemic microRNA-34a delivery induces apoptosis and abrogates growth of diffuse large B-cell lymphoma in vivo. *Leukemia* **2012**, *26*, 2421–2424. [CrossRef] [PubMed]

158. Lichtenstein, A.K.; Levine, A.; Taylor, C.R.; Boswell, W.; Rossman, S.; Feinstein, D.I.; Lukes, R.J. Primary mediastinal lymphoma in adults. *Am. J. Med.* **1980**, *68*, 509–514. [CrossRef]

159. Martelli, M.; Ferreri, A.; Di Rocco, A.; Ansuinelli, M.; Johnson, P.W.M. Primary mediastinal large B-cell lymphoma. *Crit. Rev. Oncol. Hematol.* **2017**, *113*, 318–327. [CrossRef] [PubMed]

160. Cazals-Hatem, D.; Lepage, E.; Brice, P.; Ferrant, A.; d'Agay, M.F.; Baumelou, E.; Brière, J.; Blanc, M.; Gaulard, P.; Biron, P.; et al. Primary mediastinal large B-cell lymphoma. A clinicopathologic study of 141 cases compared with 916 nonmediastinal large B-cell lymphomas, a GELA ("Groupe d'Etude des Lymphomes de l'Adulte") study. *Am. J. Surg. Pathol.* **1996**, *20*, 877–888. [CrossRef] [PubMed]

161. Falini, B.; Venturi, S.; Martelli, M.; Santucci, A.; Pileri, S.; Pescarmona, E.; Giovannini, M.; Mazza, P.; Martelli, M.F.; Pasqualucci, L.; et al. Mediastinal large B-cell lymphoma: Clinical and immunohistological findings in 18 patients treated with different third-generation regimens. *Br. J. Haematol.* **2008**, *89*, 780–789. [CrossRef]

162. Liu, P.-P.; Wang, K.-F.; Xia, Y.; Bi, X.-W.; Sun, P.; Wang, Y.; Li, Z.-M.; Jiang, W.-Q. Racial patterns of patients with primary mediastinal large B-cell lymphoma. *Medicine* **2016**, *95*, e4054. [CrossRef] [PubMed]

163. Lones, M.A.; Perkins, S.L.; Sposto, R.; Kadin, M.E.; Kjeldsberg, C.R.; Wilson, J.F.; Cairo, M.S. Large-cell lymphoma arising in the mediastinum in children and adolescents is associated with an excellent outcome: A Children's Cancer Group report. *J. Clin. Oncol.* **2000**, *18*, 3845–3853. [CrossRef] [PubMed]

164. Seidemann, K.; Tiemann, M.; Lauterbach, I.; Mann, G.; Simonitsch, I.; Stankewitz, K.; Schrappe, M.; Zimmermann, M.; Niemeyer, C.; Parwaresch, R.; et al. NHL Berlin-Frankfurt-Münster Group Primary mediastinal large B-cell lymphoma with sclerosis in pediatric and adolescent patients: Treatment and results from three therapeutic studies of the Berlin-Frankfurt-Münster Group. *J. Clin. Oncol.* **2003**, *21*, 1782–1789. [CrossRef] [PubMed]

165. Gerrard, M.; Waxman, I.M.; Sposto, R.; Auperin, A.; Perkins, S.L.; Goldman, S.; Harrison, L.; Pinkerton, R.; McCarthy, K.; Raphael, M.; et al. French-American-British/Lymphome Malins de Burkitt 96 (FAB/LMB 96) International Study Committee Outcome and pathologic classification of children and adolescents with mediastinal large B-cell lymphoma treated with FAB/LMB96 mature B-NHL therapy. *Blood* **2013**, *121*, 278–285. [CrossRef] [PubMed]

166. Burkhardt, B.; Oschlies, I.; Klapper, W.; Zimmermann, M.; Woessmann, W.; Meinhardt, A.; Landmann, E.; Attarbaschi, A.; Niggli, F.; Schrappe, M.; et al. Non-Hodgkin's lymphoma in adolescents: Experiences in 378 adolescent NHL patients treated according to pediatric NHL-BFM protocols. *Leukemia* **2011**, *25*, 153–160. [CrossRef] [PubMed]

167. Cairo, M.S.; Sposto, R.; Gerrard, M.; Auperin, A.; Goldman, S.C.; Harrison, L.; Pinkerton, R.; Raphael, M.; McCarthy, K.; Perkins, S.L.; et al. Advanced stage, increased lactate dehydrogenase, and primary site, but not adolescent age ($\geq$15 years), are associated with an increased risk of treatment failure in children and adolescents with mature B-cell non-Hodgkin's lymphoma: Results of the FAB LMB 96 study. *J. Clin. Oncol.* **2012**, *30*, 387–393. [CrossRef] [PubMed]

168. Kluiver, J.; Poppema, S.; de Jong, D.; Blokzijl, T.; Harms, G.; Jacobs, S.; Kroesen, B.-J.; van den Berg, A. BIC and miR-155 are highly expressed in Hodgkin, primary mediastinal and diffuse large B cell lymphomas. *J. Pathol.* **2005**, *207*, 243–249. [CrossRef] [PubMed]

169. Romero, M.; Gapihan, G.; Castro-Vega, L.J.; Acevedo, A.; Wang, L.; Li, Z.W.; Bouchtaoui, M.E.; Di Benedetto, M.; Ratajczak, P.; Feugeas, J.-P.; et al. Primary mediastinal large B-cell lymphoma: Transcriptional regulation by miR-92a through FOXP1 targeting. *Oncotarget* **2017**, *8*, 16243–16258. [CrossRef] [PubMed]

170. Malpeli, G.; Barbi, S.; Tosadori, G.; Greco, C.; Zupo, S.; Pedron, S.; Brunelli, M.; Bertolaso, A.; Scupoli, M.T.; Krampera, M.; et al. MYC-related microRNAs signatures in non-Hodgkin B-cell lymphomas and their relationships with core cellular pathways. *Oncotarget* **2018**, *9*, 29753–29771. [CrossRef] [PubMed]

171. Burkhardt, B.; Zimmermann, M.; Oschlies, I.; Niggli, F.; Mann, G.; Parwaresch, R.; Riehm, H.; Schrappe, M.; Reiter, A.; BFM Group. The impact of age and gender on biology, clinical features and treatment outcome of non-Hodgkin lymphoma in childhood and adolescence. *Br. J. Haematol.* **2005**, *131*, 39–49. [CrossRef] [PubMed]

172. Morris, S.W.; Kirstein, M.N.; Valentine, M.B.; Dittmer, K.G.; Shapiro, D.N.; Saltman, D.L.; Look, A.T. Fusion of a kinase gene, ALK, to a nucleolar protein gene, NPM, in non-Hodgkin's lymphoma. *Science* **1994**, *263*, 1281–1284. [CrossRef] [PubMed]

173. Liu, C.; Iqbal, J.; Teruya-Feldstein, J.; Shen, Y.; Dabrowska, M.J.; Dybkaer, K.; Lim, M.S.; Piva, R.; Barreca, A.; Pellegrino, E.; et al. MicroRNA expression profiling identifies molecular signatures associated with anaplastic large cell lymphoma. *Blood* **2013**, *122*, 2083–2092. [CrossRef] [PubMed]

174. Steinhilber, J.; Bonin, M.; Walter, M.; Fend, F.; Bonzheim, I.; Quintanilla-Martinez, L. Next-generation sequencing identifies deregulation of microRNAs involved in both innate and adaptive immune response in ALK$^+$ ALCL. *PLoS ONE* **2015**, *10*, e0117780. [CrossRef] [PubMed]

175. Dejean, E.; Renalier, M.H.; Foisseau, M.; Agirre, X.; Joseph, N.; de Paiva, G.R.; Al Saati, T.; Soulier, J.; Desjobert, C.; Lamant, L.; et al. Hypoxia-microRNA-16 downregulation induces VEGF expression in anaplastic lymphoma kinase (ALK)-positive anaplastic large-cell lymphomas. *Leukemia* **2011**, *25*, 1882–1890. [CrossRef] [PubMed]

176. Merkel, O.; Hamacher, F.; Laimer, D.; Sifft, E.; Trajanoski, Z.; Scheideler, M.; Egger, G.; Hassler, M.R.; Thallinger, C.; Schmatz, A.; et al. Identification of differential and functionally active miRNAs in both anaplastic lymphoma kinase (ALK)$^+$ and ALK$^-$ anaplastic large-cell lymphoma. *Proc. Natl. Acad. Sci. USA* **2010**, *107*, 16228–16233. [CrossRef] [PubMed]

177. Spaccarotella, E.; Pellegrino, E.; Ferracin, M.; Ferreri, C.; Cuccuru, G.; Liu, C.; Iqbal, J.; Cantarella, D.; Taulli, R.; Provero, P.; et al. STAT3-mediated activation of microRNA cluster 17~92 promotes proliferation and survival of ALK-positive anaplastic large cell lymphoma. *Haematologica* **2014**, *99*, 116–124. [CrossRef] [PubMed]

178. Merkel, O.; Hamacher, F.; Griessl, R.; Grabner, L.; Schiefer, A.-I.; Prutsch, N.; Baer, C.; Egger, G.; Schlederer, M.; Krenn, P.W.; et al. Oncogenic role of miR-155 in anaplastic large cell lymphoma lacking the t(2;5) translocation. *J. Pathol.* **2015**, *236*, 445–456. [CrossRef] [PubMed]

179. Cile Desjobert, C.; Ne Renalier, M.-H.; Bergalet, J.; Dejean, E.; Joseph, N.; Kruczynski, A.; Soulier, J.; Espinos, E.; Meggetto, F.; Rome Cavaillé, J.; et al. MiR-29a down-regulation in ALK-positive anaplastic large

cell lymphomas contributes to apoptosis blockade through MCL-1 overexpression. *Blood* **2011**. [CrossRef] [PubMed]

180. Vishwamitra, D.; Li, Y.; Wilson, D.; Manshouri, R.; Curry, C.V.; Shi, B.; Tang, X.M.; Sheehan, A.M.; Wistuba, I.I.; Shi, P.; et al. MicroRNA 96 is a post-transcriptional suppressor of anaplastic lymphoma kinase expression. *Am. J. Pathol.* **2012**, *180*, 1772–1780. [CrossRef] [PubMed]

181. Hoareau-Aveilla, C.; Valentin, T.; Daugrois, C.; Quelen, C.; Mitou, G.; Quentin, S.; Jia, J.; Spicuglia, S.; Ferrier, P.; Ceccon, M.; et al. Reversal of microRNA-150 silencing disadvantages crizotinib-resistant NPM-ALK(+) cell growth. *J. Clin. Investig.* **2015**, *125*, 3505–3518. [CrossRef] [PubMed]

182. Congras, A.; Caillet, N.; Torossian, N.; Quelen, C.; Daugrois, C.; Brousset, P.; Lamant, L.; Meggetto, F.; Hoareau-Aveilla, C. Doxorubicin-induced loss of DNA topoisomerase II and DNMT1- dependent suppression of MiR-125b induces chemoresistance in ALK-positive cells. *Oncotarget* **2018**, *9*, 14539–14551. [CrossRef] [PubMed]

183. Cortelazzo, S.; Ferreri, A.; Hoelzer, D.; Ponzoni, M. Lymphoblastic lymphoma. *Crit. Rev. Oncol. Hematol.* **2017**, *113*, 304–317. [CrossRef] [PubMed]

184. Shankland, K.R.; Armitage, J.O.; Hancock, B.W. Non-Hodgkin lymphoma. *Lancet* **2012**, *380*, 848–857. [CrossRef]

185. Burkhardt, B. Paediatric lymphoblastic T-cell leukaemia and lymphoma: One or two diseases? *Br. J. Haematol.* **2009**, *149*, 653–668. [CrossRef] [PubMed]

186. González-Gugel, E.; Villa-Morales, M.; Santos, J.; Bueno, M.J.; Malumbres, M.; Rodríguez-Pinilla, S.M.; Piris, M.Á.; Fernández-Piqueras, J. Down-regulation of specific miRNAs enhances the expression of the gene Smoothened and contributes to T-cell lymphoblastic lymphoma development. *Carcinogenesis* **2013**, *34*, 902–908. [CrossRef] [PubMed]

187. Mussolin, L.; Holmes, A.B.; Romualdi, C.; Sales, G.; D'Amore, E.S.G.; Ghisi, M.; Pillon, M.; Rosolen, A.; Basso, K. An aberrant microRNA signature in childhood T-cell lymphoblastic lymphoma affecting CDKN1B expression, NOTCH1 and growth factor signaling pathways. *Leukemia* **2014**, *28*, 1909–1912. [CrossRef] [PubMed]

188. Xi, Y.; Li, J.; Zhang, P.; Bai, W.; Gao, N.; Bai, W.; Zhang, Y.; Wu, Y.; Ning, Y. Upregulation of miRNA-17 and miRNA-19 is associated with unfavorable prognosis in patients with T-cell lymphoblastic lymphoma. *Exp. Mol. Pathol.* **2015**, *99*, 297–302. [CrossRef] [PubMed]

189. Fan, F.-Y.; Deng, R.; Yi, H.; Sun, H.-P.; Zeng, Y.; He, G.-C.; Su, Y. The inhibitory effect of MEG3/miR-214/AIFM2 axis on the growth of T-cell lymphoblastic lymphoma. *Int. J. Oncol.* **2017**, *51*, 316–326. [CrossRef] [PubMed]

190. Qian, D.; Chen, K.; Deng, H.; Rao, H.; Huang, H.; Liao, Y.; Sun, X.; Lu, S.; Yuan, Z.; Xie, D.; et al. MicroRNA-374b Suppresses Proliferation and Promotes Apoptosis in T-cell Lymphoblastic Lymphoma by Repressing AKT1 and Wnt-16. *Clin. Cancer Res.* **2015**, *21*, 4881–4891. [CrossRef] [PubMed]

191. López-Nieva, P.; Fernández-Navarro, P.; Vaquero-Lorenzo, C.; Villa-Morales, M.; Graña-Castro, O.; Cobos-Fernández, M.Á.; López-Lorenzo, J.L.; Llamas, P.; González-Sanchez, L.; Sastre, I.; et al. RNA-Seq reveals the existence of a CDKN1C-E2F1-TP53 axis that is altered in human T-cell lymphoblastic lymphomas. *BMC Cancer* **2018**, *18*, 430. [CrossRef] [PubMed]

192. Kelly, K.M. Hodgkin lymphoma in children and adolescents: Improving the therapeutic index. *Blood* **2015**, *126*, 2452–2458. [CrossRef] [PubMed]

193. Meti, N.; Esfahani, K.; Johnson, N. The Role of Immune Checkpoint Inhibitors in Classical Hodgkin Lymphoma. *Cancers* **2018**, *10*, 204. [CrossRef] [PubMed]

194. Bair, S.M.; Mato, A.; Svoboda, J. Immunotherapy for the Treatment of Hodgkin Lymphoma: An Evolving Paradigm. *Clin. Lymphoma Myeloma Leuk.* **2018**, *18*, 380–391. [CrossRef] [PubMed]

195. Sherief, L.M.; Elsafy, U.R.; Abdelkhalek, E.R.; Kamal, N.M.; Elbehedy, R.; Hassan, T.H.; Sherbiny, H.S.; Beshir, M.R.; Saleh, S.H. Hodgkin lymphoma in childhood: Clinicopathological features and therapy outcome at 2 centers from a developing country. *Medicine* **2015**, *94*, e670. [CrossRef] [PubMed]

196. Navarro, A.; Diaz, T.; Martinez, A.; Gaya, A.; Pons, A.; Gel, B.; Codony, C.; Ferrer, G.; Martinez, C.; Montserrat, E.; et al. Regulation of JAK2 by miR-135a: Prognostic impact in classic Hodgkin lymphoma. *Blood* **2009**, *114*, 2945–2951. [CrossRef] [PubMed]

197. Navarro, A.; Gaya, A.; Martinez, A.; Urbano-Ispizua, A.; Pons, A.; Balague, O.; Gel, B.; Abrisqueta, P.; Lopez-Guillermo, A.; Artells, R.; et al. MicroRNA expression profiling in classic Hodgkin lymphoma. *Blood* **2008**, *111*, 2825–2832. [CrossRef] [PubMed]

198. Huang, X.; Zhou, X.; Wang, Z.; Li, F.; Liu, F.; Zhong, L.; Li, X.; Han, X.; Wu, Z.; Chen, S.; et al. CD99 triggers upregulation of miR-9-modulated PRDM1/BLIMP1 in Hodgkin/Reed-Sternberg cells and induces redifferentiation. *Int. J. Cancer* **2012**, *131*, E382–E394. [CrossRef] [PubMed]

199. Khare, D.; Goldschmidt, N.; Bardugo, A.; Gur-Wahnon, D.; Ben-Dov, I.Z.; Avni, B. Plasma microRNA profiling: Exploring better biomarkers for lymphoma surveillance. *PLoS ONE* **2017**, *12*, e0187722. [CrossRef] [PubMed]

200. Paydas, S.; Acikalin, A.; Ergin, M.; Celik, H.; Yavuz, B.; Tanriverdi, K. Micro-RNA (miRNA) profile in Hodgkin lymphoma: Association between clinical and pathological variables. *Med. Oncol.* **2016**, *33*, 34. [CrossRef] [PubMed]

201. Gibcus, J.H.; Kroesen, B.-J.; Koster, R.; Halsema, N.; de Jong, D.; de Jong, S.; Poppema, S.; Kluiver, J.; Diepstra, A.; van den Berg, A. MiR-17/106b seed family regulates p21 in Hodgkin's lymphoma. *J. Pathol.* **2011**, *225*, 609–617. [CrossRef] [PubMed]

202. Zhu, M.; Xu, Z.; Wang, K.; Wang, N.; Zhu, M.; Wang, S. MicroRNA and gene networks in human Hodgkin's lymphoma. *Mol. Med. Rep.* **2013**, *8*, 1747–1754. [CrossRef] [PubMed]

203. Yuan, Y.; Kluiver, J.; Koerts, J.; de Jong, D.; Rutgers, B.; Abdul Razak, F.R.; Terpstra, M.; Plaat, B.E.; Nolte, I.M.; Diepstra, A.; et al. miR-24-3p Is Overexpressed in Hodgkin Lymphoma and Protects Hodgkin and Reed-Sternberg Cells from Apoptosis. *Am. J. Pathol.* **2017**, *187*, 1343–1355. [CrossRef] [PubMed]

204. Leucci, E.; Zriwil, A.; Gregersen, L.H.; Jensen, K.T.; Obad, S.; Bellan, C.; Leoncini, L.; Kauppinen, S.; Lund, A.H. Inhibition of miR-9 de-represses HuR and DICER1 and impairs Hodgkin lymphoma tumour outgrowth in vivo. *Oncogene* **2012**, *31*, 5081–5089. [CrossRef] [PubMed]

205. Van Vlierberghe, P.; De Weer, A.; Mestdagh, P.; Feys, T.; De Preter, K.; De Paepe, P.; Lambein, K.; Vandesompele, J.; Van Roy, N.; Verhasselt, B.; et al. Comparison of miRNA profiles of microdissected Hodgkin/Reed-Sternberg cells and Hodgkin cell lines versus CD77⁺ B-cells reveals a distinct subset of differentially expressed miRNAs. *Br. J. Haematol.* **2009**, *147*, 686–690. [CrossRef] [PubMed]

206. Gibcus, J.H.; Tan, L.P.; Harms, G.; Schakel, R.N.; de Jong, D.; Blokzijl, T.; Möller, P.; Poppema, S.; Kroesen, B.-J.; van den Berg, A. Hodgkin lymphoma cell lines are characterized by a specific miRNA expression profile. *Neoplasia* **2009**, *11*, 167–176. [CrossRef] [PubMed]

207. Lawrie, C.H.; Gal, S.; Dunlop, H.M.; Pushkaran, B.; Liggins, A.P.; Pulford, K.; Banham, A.H.; Pezzella, F.; Boultwood, J.; Wainscoat, J.S.; et al. Detection of elevated levels of tumour-associated microRNAs in serum of patients with diffuse large B-cell lymphoma. *Br. J. Haematol.* **2008**, *141*, 672–675. [CrossRef] [PubMed]

208. Pezuk, J.A. The Importance of Circulating miRNAs and Its Limitation on the Clinic. *Hum. J. Rev. Artic. Novemb. Ijsrm. Hum.* **2017**, *8*, 278–283.

209. Stamatopoulos, B.; Van Damme, M.; Crompot, E.; Dessars, B.; El Housni, H.; Mineur, P.; Meuleman, N.; Bron, D.; Lagneaux, L. Opposite Prognostic Significance of Cellular and Serum Circulating MicroRNA-150 in Patients with Chronic Lymphocytic Leukemia. *Mol. Med.* **2015**, *21*, 123–133. [CrossRef] [PubMed]

210. Jones, K.; Nourse, J.P.; Keane, C.; Bhatnagar, A.; Gandhi, M.K. Plasma MicroRNA Are Disease Response Biomarkers in Classical Hodgkin Lymphoma. *Clin. Cancer Res.* **2014**, *20*, 253–264. [CrossRef] [PubMed]

211. Swellam, M.; El-Khazragy, N. Clinical impact of circulating microRNAs as blood-based marker in childhood acute lymphoblastic leukemia. *Tumor Biol.* **2016**, *37*, 10571–10576. [CrossRef] [PubMed]

212. Swellam, M.; Hashim, M.; Mahmoud, M.S.; Ramadan, A.; Hassan, N.M. Aberrant Expression of Some Circulating miRNAs in Childhood Acute Lymphoblastic Leukemia. *Biochem. Genet.* **2018**, *56*, 283–294. [CrossRef] [PubMed]

213. Fayyad-Kazan, H.; Bitar, N.; Najar, M.; Lewalle, P.; Fayyad-Kazan, M.; Badran, R.; Hamade, E.; Daher, A.; Hussein, N.; ElDirani, R.; et al. Circulating miR-150 and miR-342 in plasma are novel potential biomarkers for acute myeloid leukemia. *J. Transl. Med.* **2013**, *11*, 31. [CrossRef] [PubMed]

214. Lin, X.; Wang, Z.; Zhang, R.; Feng, W. High serum microRNA-335 level predicts aggressive tumor progression and unfavorable prognosis in pediatric acute myeloid leukemia. *Clin. Transl. Oncol.* **2015**, *17*, 358–364. [CrossRef] [PubMed]

215. Guo, H.-Q.; Huang, G.-L.; Guo, C.-C.; Pu, X.-X.; Lin, T.-Y. Diagnostic and prognostic value of circulating miR-221 for extranodal natural killer/T-cell lymphoma. *Dis. Mark.* **2010**, *29*, 251–258. [CrossRef]

216. Zhao, Q.; Li, J.; Chen, S.; Shen, K.; Ai, G.; Dai, X.; Xie, B.; Shi, Y.; Jiang, S.; Feng, J.; et al. Decreased miR-144 expression as a non-invasive biomarker for acute myeloid leukemia patients. *Pharmazie* **2017**, *72*, 232–235. [CrossRef] [PubMed]

217. Huang, Y.; Zou, Y.; Lin, L.; Ma, X.; Chen, H. Identification of serum miR-34a as a potential biomarker in acute myeloid leukemia. *Cancer Biomark.* **2018**, *22*, 799–855. [CrossRef] [PubMed]

218. Lin, X.; Wang, Z.; Wang, Y.; Feng, W. Serum MicroRNA-370 as a potential diagnostic and prognostic biomarker for pediatric acute myeloid leukemia. *Int. J. Clin. Exp. Pathol.* **2015**, *8*, 14658–14666. [PubMed]

219. Hong, Z.; Zhang, R.; Qi, H. Diagnostic and prognostic relevance of serum miR-195 in pediatric acute myeloid leukemia. *Cancer Biomark.* **2018**, *21*, 269–275. [CrossRef] [PubMed]

220. Chen, W.; Wang, H.; Chen, H.; Liu, S.; Lu, H.; Kong, D.; Huang, X.; Kong, Q.; Lu, Z. Clinical significance and detection of microRNA-21 in serum of patients with diffuse large B-cell lymphoma in Chinese population. *Eur. J. Haematol.* **2014**, *92*, 407–412. [CrossRef] [PubMed]

221. Lopez-Santillan, M.; Larrabeiti-Etxebarria, A.; Arzuaga-Mendez, J.; Lopez-Lopez, E.; Garcia-Orad, A. Circulating miRNAs as biomarkers in diffuse large B-cell lymphoma: A systematic review. *Oncotarget* **2018**, *9*, 22850–22861. [CrossRef] [PubMed]

222. Ahmadvand, M.; Eskandari, M.; Pashaiefar, H.; Yaghmaie, M.; Manoochehrabadi, S.; Khakpour, G.; Sheikhsaran, F.; Montazer Zohour, M. Over expression of circulating miR-155 predicts prognosis in diffuse large B-cell lymphoma. *Leuk. Res.* **2018**, *70*, 45–48. [CrossRef] [PubMed]

223. Marchesi, F.; Regazzo, G.; Palombi, F.; Terrenato, I.; Sacconi, A.; Spagnuolo, M.; Donzelli, S.; Marino, M.; Ercolani, C.; Di Benedetto, A.; et al. Serum miR-22 as potential non-invasive predictor of poor clinical outcome in newly diagnosed, uniformly treated patients with diffuse large B-cell lymphoma: An explorative pilot study. *J. Exp. Clin. Cancer Res.* **2018**, *37*, 95. [CrossRef] [PubMed]

224. Yuan, W.X.; Gui, Y.X.; Na, W.N.; Chao, J.; Yang, X. Circulating microRNA-125b and microRNA-130a expression profiles predict chemoresistance to R-CHOP in diffuse large B-cell lymphoma patients. *Oncol. Lett.* **2016**, *11*, 423–432. [CrossRef] [PubMed]

225. Beta, M.; Krishnakumar, S.; Elchuri, S.V.; Salim, B.; Narayanan, J. A comparative fluorescent beacon-based method for serum microRNA quantification. *Anal. Sci.* **2015**, *31*, 231–235. [CrossRef] [PubMed]

226. Li, J.; Zhai, X.-W.; Wang, H.-S.; Qian, X.-W.; Miao, H.; Zhu, X.-H. Circulating MicroRNA-21, MicroRNA-23a, and MicroRNA-125b as Biomarkers for Diagnosis and Prognosis of Burkitt Lymphoma in Children. *Med. Sci. Monit.* **2016**, *22*, 4992–5002. [CrossRef] [PubMed]

227. Babar, I.A.; Cheng, C.J.; Booth, C.J.; Liang, X.; Weidhaas, J.B.; Saltzman, W.M.; Slack, F.J. Nanoparticle-based therapy in an in vivo microRNA-155 (miR-155)-dependent mouse model of lymphoma. *Proc. Natl. Acad. Sci. USA* **2012**, *109*, E1695–E1704. [CrossRef] [PubMed]

228. Tivnan, A.; Orr, W.S.; Gubala, V.; Nooney, R.; Williams, D.E.; McDonagh, C.; Prenter, S.; Harvey, H.; Domingo-Fernández, R.; Bray, I.M.; et al. Inhibition of neuroblastoma tumor growth by targeted delivery of microRNA-34a using anti-disialoganglioside GD2 coated nanoparticles. *PLoS ONE* **2012**, *7*, e38129. [CrossRef] [PubMed]

229. Hsu, S.-H.; Yu, B.; Wang, X.; Lu, Y.; Schmidt, C.R.; Lee, R.J.; Lee, L.J.; Jacob, S.T.; Ghoshal, K. Cationic lipid nanoparticles for therapeutic delivery of siRNA and miRNA to murine liver tumor. *Nanomedicine* **2013**, *9*, 1169–1180. [CrossRef] [PubMed]

230. Gill, S.-L.; O'Neill, H.; McCoy, R.J.; Logeswaran, S.; O'Brien, F.; Stanton, A.; Kelly, H.; Duffy, G.P. Enhanced delivery of microRNA mimics to cardiomyocytes using ultrasound responsive microbubbles reverses hypertrophy in an in-vitro model. *Technol. Health Care* **2014**, *22*, 37–51. [CrossRef] [PubMed]

231. Liu, J.; Dang, L.; Li, D.; Liang, C.; He, X.; Wu, H.; Qian, A.; Yang, Z.; Au, D.W.T.; Chiang, M.W.L.; et al. A delivery system specifically approaching bone resorption surfaces to facilitate therapeutic modulation of microRNAs in osteoclasts. *Biomaterials* **2015**, *52*, 148–160. [CrossRef] [PubMed]

232. Wang, S.; Cao, M.; Deng, X.; Xiao, X.; Yin, Z.; Hu, Q.; Zhou, Z.; Zhang, F.; Zhang, R.; Wu, Y.; et al. Degradable hyaluronic acid/protamine sulfate interpolyelectrolyte complexes as miRNA-delivery nanocapsules for triple-negative breast cancer therapy. *Adv. Healthc. Mater.* **2015**, *4*, 281–290. [CrossRef] [PubMed]

233. Janssen, H.L.A.; Reesink, H.W.; Lawitz, E.J.; Zeuzem, S.; Rodriguez-Torres, M.; Patel, K.; van der Meer, A.J.; Patick, A.K.; Chen, A.; Zhou, Y.; et al. Treatment of HCV infection by targeting microRNA. *N. Engl. J. Med.* **2013**, *368*, 1685–1694. [CrossRef] [PubMed]

234. Simonson, B.; Das, S. MicroRNA Therapeutics: The Next Magic Bullet? *Mini Rev. Med. Chem.* **2015**, *15*, 467–474. [CrossRef] [PubMed]

Int. J. Mol. Sci. **2018**, *19*, 2688

235. Cai, C.-K.; Zhao, G.-Y.; Tian, L.-Y.; Liu, L.; Yan, K.; Ma, Y.-L.; Ji, Z.-W.; Li, X.-X.; Han, K.; Gao, J.; et al. miR-15a and miR-16-1 downregulate CCND1 and induce apoptosis and cell cycle arrest in osteosarcoma. *Oncol. Rep.* **2012**, *28*, 1764–1770. [CrossRef] [PubMed]

236. Beg, M.S.; Brenner, A.J.; Sachdev, J.; Borad, M.; Kang, Y.-K.; Stoudemire, J.; Smith, S.; Bader, A.G.; Kim, S.; Hong, D.S. Phase I study of MRX34, a liposomal miR-34a mimic, administered twice weekly in patients with advanced solid tumors. *Investig. New Drugs* **2017**, *35*, 180–188. [CrossRef] [PubMed]

International Journal of
Molecular Sciences

MDPI

Article

Identification of Endogenous Control miRNAs for RT-qPCR in T-Cell Acute Lymphoblastic Leukemia

Monika Drobna [1], **Bronisława Szarzyńska-Zawadzka** [1], **Patrycja Daca-Roszak** [1], **Maria Kosmalska** [1], **Roman Jaksik** [2], **Michał Witt** [1] and **Małgorzata Dawidowska** [1,*]

[1] Institute of Human Genetics, Polish Academy of Sciences, 60-479 Poznań, Poland; monika.drobna@igcz.poznan.pl (M.D.); bronislawa.szarzynska-zawadzka@igcz.poznan.pl (B.S.-Z.); patrycja.daca-roszak@igcz.poznan.pl (P.D.-R.); maria.kosmalska@igcz.poznan.pl (M.K.); michal.witt@igcz.poznan.pl (M.W.)

[2] Department, Silesian University of Technology, 44-100 Gliwice, Poland; roman.jaksik@polsl.pl

[*] Correspondence: malgorzata.dawidowska@igcz.poznan.pl; Tel.: +48-61-65-79158

Received: 6 August 2018; Accepted: 18 September 2018; Published: 20 September 2018

Abstract: Optimal endogenous controls enable reliable normalization of microRNA (miRNA) expression in reverse-transcription quantitative PCR (RT-qPCR). This is particularly important when miRNAs are considered as candidate diagnostic or prognostic biomarkers. Universal endogenous controls are lacking, thus candidate normalizers must be evaluated individually for each experiment. Here we present a strategy that we applied to the identification of optimal control miRNAs for RT-qPCR profiling of miRNA expression in T-cell acute lymphoblastic leukemia (T-ALL) and in normal cells of T-lineage. First, using NormFinder for an iterative analysis of miRNA stability in our miRNA-seq data, we established the number of control miRNAs to be used in RT-qPCR. Then, we identified optimal control miRNAs by a comprehensive analysis of miRNA stability in miRNA-seq data and in RT-qPCR by analysis of RT-qPCR amplification efficiency and expression across a variety of T-lineage samples and T-ALL cell line culture conditions. We then showed the utility of the combination of three miRNAs as endogenous normalizers (hsa-miR-16-5p, hsa-miR-25-3p, and hsa-let-7a-5p). These miRNAs might serve as first-line candidate endogenous controls for RT-qPCR analysis of miRNAs in different types of T-lineage samples: T-ALL patient samples, T-ALL cell lines, normal immature thymocytes, and mature T-lymphocytes. The strategy we present is universal and can be transferred to other RT-qPCR experiments.

Keywords: miRNA (microRNA); T-cell acute lymphoblastic leukemia (T-ALL); normalization of miRNA expression in RT-qPCR; endogenous controls; reference genes; tissue analysis; cell lines

1. Introduction

1.1. Background

MicroRNAs (miRNAs) belong to the class of small noncoding RNAs, serving as negative regulators of gene expression at the posttranscriptional level [1,2]. They bind to 3′ untranslated regions (3′ UTRs) of their target mRNAs, leading to translational repression and thus gene silencing [3]. Aberrant expression of miRNAs contributes to diseases, including malignancies. In cancer, miRNAs may serve as proto-oncogenes, which target and silence tumor-suppressor genes (these miRNAs attain oncogenic properties when overexpressed), and as tumor-suppressor miRNAs, which negatively regulate the expression of oncogenes (the suppressive function of these miRNAs is lost or diminished due to their downregulation) [4]. Research on cancer-related miRNAs provides insight into the molecular pathogenesis of cancer and forms the basis for using miRNAs as diagnostic and prognostic markers and even as therapeutic targets. Thus, intracellular and extracellular (circulating exosomal) miRNAs are extensively studied in various cancer types [5–7], including hematologic malignancies [8–10].

The successful implementation of miRNAs as biomarkers in standardized diagnostic and treatment-stratification strategies depends on many factors, including the choice of a reliable method for miRNA profiling and an optimal strategy for normalization of miRNA expression [11,12].

1.2. Challenges of Normalization in RT-qPCR-Based miRNA Profiling

Despite a rapid increase in the use of high-throughput deep sequencing of the miRNA-transcriptome (miRNA-seq), RT-qPCR still remains the gold standard of miRNA profiling, used either as a primary method for expression analysis or for the validation of miRNA-seq results [13–15]. In contrast to miRNA-seq, which, due to a large number of features, allows for the use of distribution-based normalization techniques, RT-qPCR often requires the choice of appropriate endogenous controls to normalize miRNA expression levels. An optimal endogenous normalizer (EN) should be a gene or a combination of genes exhibiting stable and relatively abundant expression across all samples examined by RT-qPCR, regardless of their tissue of origin, the preanalytical procedures, or the time points analyzed [16].

The use of miRNAs (and not other types of RNAs) as ENs is currently the most commonly advocated strategy in RT-qPCR-based miRNA profiling [9,17–20]. The length of miRNA molecules excludes the possibility of using housekeeping gene transcripts, which are standard ENs for mRNA expression. Differences in length between miRNAs and mRNAs affect isolation yield, reverse transcription, and amplification efficiencies, while these should be similar for ENs and the studied transcripts [11]. Other small RNAs, such as small nuclear RNA (snRNA) and small nucleolar RNA (snoRNA), have frequently been used as endogenous controls in miRNA studies [8,9,21–23]. Yet, the length of these small RNAs is also different from miRNAs (60–200 nt for snoRNAs [24] and 150 nt for snRNAs [25], as compared to 20–24 for miRNAs [1]). miRNAs also differ structurally. Unlike other small RNA types, miRNAs contain 5′-phosphate and 3′-hydroxyl groups at their ends [1]. For these reasons, some commercially available technologies for RT-qPCR experiments, based on hydrolysis probes, implement solutions that hamper the use of ENs other than miRNAs. For example, the TaqMan Advanced miRNA technology provided by Thermo Fisher Scientific makes it possible to reverse-transcribe only the miRNA fraction during cDNA synthesis, excluding other small RNA types.

However, the question remains as to which miRNAs will serve as optimal ENs in a particular experiment. Universal endogenous control miRNAs are lacking, due to a large variation of miRNA expression profiles across various cell and tissue types [18]. This problem is particularly valid in research on miRNAs as potential cancer biomarkers. Therefore, instead of aiming to identify universal EN miRNAs, the focus should be on applying a universal methodology for the identification of optimal ENs for each particular study. Such a methodology should allow for a comprehensive evaluation of stability (using *in silico* and wet-lab approaches) of candidate ENs, selected and tested in a pilot phase, preceding the RT-qPCR expression analysis [12].

Here we present a strategy we applied for the identification of optimal EN miRNAs for miRNA profiling in T-cell acute lymphoblastic leukemia (T-ALL), an aggressive and highly heterogeneous type of hematologic cancer [26,27]. The study was driven by the shortage of data on comprehensive assessment of miRNAs as optimal ENs, specifically for cells of T-lineage.

1.3. miRNA Expression Profiling in T-Cell Acute Lymphoblastic Leukemia

In recent years, miRNAs have become an object of growing interest in the field of T-ALL research, due to their involvement in oncogenesis and their potential as candidate biomarkers [28,29]. Most of the data on miRNA expression in T-ALL patients and T-ALL cell lines comes from RT-qPCR-based studies [13,30,31], with only a few studies exploiting miRNA-seq technology [32,33].

One of the essential challenges in T-ALL research is the choice of a proper control material. The use of thymocytes separated from the thymus gland is widely approved [34–36] but technically demanding. Thus, other types of cells are often used as controls, including CD34+-enriched cells [37,38] or mature T-lymphocytes from peripheral blood or bone marrow [39,40]. Additionally, T-ALL cell lines serving as an *in vitro* model of this disease are often used for miRNA expression profiling. Importantly, miRNA expression in cell lines might be affected by culturing conditions, such as time of culture or medium composition [41,42]. Considering the diversity of cell and tissue types used in miRNA research in T-ALL, identifying stably expressed miRNAs to be used as optimal ENs for RT-qPCR is highly necessary.

The endogenous control miRNAs we analyzed can serve as the first-line choice for those dealing with miRNA expression in cells of T-lineage: T-ALL patient samples, T-ALL cell lines, normal thymocytes, and normal mature T-lymphocytes. The workflow we applied for the identification of optimal EN miRNAs was successfully used to validate our miRNA-seq results. Additionally, the strategy is universal and can be adapted to other RT-qPCR experiments for miRNA profiling in cancer samples and cell lines, and is potentially transferable to the identification of cancer biomarkers, cancer diagnostics, and therapy.

2. Results

Here we present a strategy for the identification of miRNAs as optimal endogenous normalizers for RT-qPCR miRNA expression profiling used in T-lymphoid cells. The results of an miRNA-seq experiment were the starting point of our study, including 34 T-ALL samples and 5 samples of normal mature T-lymphocytes from bone marrow used as controls [43]. Based on miRNA-seq results, we identified a set of miRNAs that were differentially expressed between T-ALL samples and controls, including known and new potential oncogenic and tumor-suppressive miRNAs. We selected several of these miRNAs for validation by RT-qPCR, using TaqMan Advanced miRNA Assays (Thermo Fisher Scientific, Waltham, MA USA). To identify optimal EN miRNAs, we applied a stepwise strategy for the selection and evaluation of candidate ENs. The scheme of the workflow with respect to the type of analysis and material used in each step is shown in Figure 1.

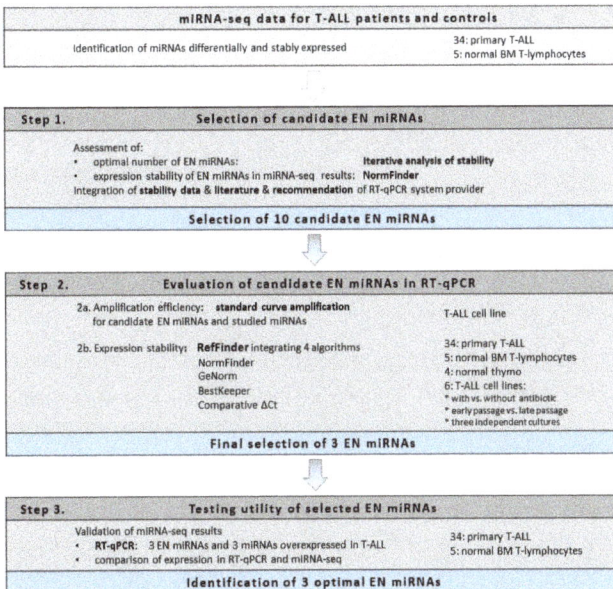

Figure 1. Workflow diagram illustrating strategy for identification of endogenous normalizer microRNAs (miRNAs) for RT-qPCR. BM, bone marrow; EN, endogenous normalizer; miRNA-seq, miRNA sequencing; T-ALL, T-cell acute lymphoblastic leukemia; thymo, thymocytes.

2.1. Selection of Candidate Endogenous Normalizer miRNAs (Step 1)

In order to test how many reference miRNAs should be used, we created an iterative algorithm based on NormFinder (further referred to as iterative analysis of stability), which we used to select the best set of up to 100 miRNAs from our miRNA-seq data (Figure 1, Step 1). First, we conducted a standard NormFinder analysis and selected the most stable miRNAs out of a total of 1503 expressed in the analyzed cells. Then, we averaged the signals of the selected miRNAs with each of the remaining N - 1 miRNAs individually, each time repeating the NormFinder analysis. Based on that, we selected the best miRNA pair in terms of the stability index. This process was repeated by adding an additional miRNA from the remaining pool to the previously selected pair until we obtained a set of 100 unique miRNAs, which is already beyond any practical application in small-scale experiments such as RT-qPCR. The file including the code for this iterative algorithm, written in R programming language, is available in the Supplementary Materials (IterativeStability_v1.0.R).

In Figure 2 we present the plots of the average NormFinder scores obtained for all sets in each iteration, and the minimum score that describes the best miRNA set of specific size. By adding additional miRNAs, we were able to reduce the score in general. However, the minimal score shows that the gain of incorporating one additional miRNA very quickly becomes negligible, being the highest for the first three iterations. Thus, in our dataset we show that the use of three miRNAs as ENs is reasonable, both for a fair representation of stably expressed miRNAs and for the cost of an RT-qPCR experiment.

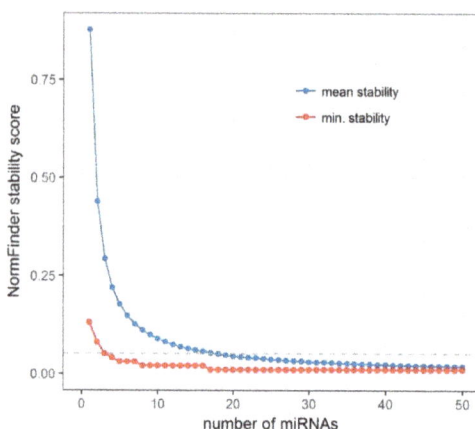

Figure 2. NormFinder stability scores relative to the number of miRNAs tested. For visualization purposes, we limited the plot to the results of the first 50 iterations. Beyond that point, no clear changes could be observed in the linear scale.

In our experimental setting, the most optimal combination of three miRNAs identified by this iterative analysis of ENs was hsa-miR-1301-3p, hsa-miR-185-5p, and hsa-miR-30d-5p. Their stability scores according to NormFinder and mean read counts in miRNA-seq are shown in Supplementary Table S1. Due to relatively low read counts for two of these (hsa-miR-1301-3p and hsa-miR-185-5p), we decided to aim for three EN miRNAs in our RT-qPCR experiments, but searched for additional suitable candidates.

To accomplish this, we compared the list of the most stably expressed miRNAs in our miRNA-seq data with miRNAs recommended as suitable ENs for TaqMan Advanced miRNA Assays (www.thermofisher.com/advancedmirna). In addition, we searched through literature data regarding different tissues, including those of malignant origin [44–48]. Importantly, no such data regarding T-ALL and cells of T-lineage specifically are available so far. Thus, by integrating our miRNA-seq results with the recommendations for TaqMan assays and literature review, we selected 10 candidate EN miRNAs based on their stability scores and mean read counts in miRNA-seq, as illustrated in Figure 1, Step 1 and presented in Table 1. These 10 candidates were further comprehensively evaluated by RT-qPCR (Figure 1, Step 2).

Table 1. Candidate endogenous normalizer miRNAs.

Candidate EN miRNAs	Criteria for Selection			
miRNA Name	Stability in miRNA-Seq		Thermo Fisher Scientific Recommendation	Literature Data
	Stability Score *	Mean Read Count		
hsa-let-7a-5p	0.25	330,280	–	[44]
hsa-miR-30d-5p	0.25	98,732	–	–
hsa-miR-92a-3p	0.31	630,854	Suitable endogenous control for tissue samples	–
hsa-miR-93-5p	0.36	9838	Suitable endogenous control	[45]
hsa-let-7f-5p	0.37	326,028	–	–
hsa-miR-25-3p	0.45	108,466	Suitable endogenous control	[45]
hsa-miR-26a-5p	0.5	176,895	Suitable endogenous control for breast and heart tissue	–
hsa-miR-21-5p	0.5	116,490	Suitable endogenous control for tissue samples	–
hsa-miR-16-5p	0.52	5746	Suitable endogenous control	[46–48]
hsa-let-7g-5p	0.56	194,895	Suitable endogenous control	–

* Stability score according to NormFinder software. The stability score value is inversely proportional to the stability of gene expression.

2.2. Evaluation of Candidate Endogenous Normalizer miRNAs in RT-qPCR (Step 2)

Ten candidate EN miRNAs and three overexpressed miRNAs were tested by RT-qPCR for amplification efficiency of the assays and expression stability (Figure 1, Steps 2a and 2b, respectively). The slope and amplification efficiency data are shown in Table 2. Out of 13 assays, 11 exhibited amplification efficiency in the desired range of 90–110%, except for hsa-let-7f-5p and hsa-miR-181a-5p, with amplification efficiencies of 89% and 81%, respectively.

Table 2. Amplification efficiency of miRNA assays.

miRNA Name	TaqMan Advanced miRNA Assay Name	Standard Curve		Amplification Efficiency (%)
		Slope	R2 (Correlation Coefficient)	
Candidate EN miRNAs				
hsa-let-7a-5p	478575_mir	−3.292	0.991	101
hsa-miR-30d-5p	478606_mir	−3.371	0.996	98
hsa-miR-92a-3p	477827_mir	−3.549	0.994	91
hsa-miR-93-5p	478210_mir	−3.371	0.996	98
hsa-let-7f-5p	478578_mir	−3.626	0.985	89
hsa-miR-25-3p	477994_mir	−3.533	0.994	92
hsa-miR-26a-5p	477995_mir	−3.558	0.997	91
hsa-miR-21-5p	477975_mir	−3.516	0.994	92
hsa-miR-16-5p	477860_mir	−3.582	0.997	90
hsa-let-7g-5p	478580_mir	−3.338	0.959	99
Selected miRNAs Overexpressed in T-ALL vs. Control				
hsa-miR-181a-5p	477857_mir	−3.872	0.966	81
hsa-miR-128-3p	477892_mir	−3.406	0.99	97
hsa-miR-20b-5p	477804_mir	−3.570	0.997	91

Mean raw Cq values and mean standard deviation (SD) across biological groups representing different types of material (T-ALL patient samples, T-ALL cell lines, normal mature T-lymphocytes of bone marrow, and normal thymocytes) are shown in Table 3. Mean raw Cq values and mean SD values obtained for T-ALL cell lines analyzed with the application of different culture types and conditions are shown in Supplementary Table S2. The variability of Cq values across all samples with respect to the type of material analyzed is presented for all candidate EN miRNAs in Figure 3.

Table 3. Mean raw Cq and standard deviation (SD) values for candidate endogenous normalizer miRNA across analyzed samples with respect to biological groups.

miRNA	All Samples		T-ALL Samples		Normal BM T-Lymphocytes		Thymocytes		T-ALL Cell Lines		*p-adj*
	Cq	SD	Cq	SD	Cq	SD	Cq	SD	Cq	SD	
hsa-miR-92a-3p	21.54	1.68	21.38	1.61	20.37	0.92	23.22	1.78	21.65	1.53	0.173
hsa-miR-16-5p	22.58	1.66	22.18	1.42	23.61	1.79	22.89	2.42	23.40	1.82	0.192
hsa-miR-25-3p	23.00	1.64	22.70	1.51	24.11	1.76	22.60	1.90	23.67	1.83	0.240
hsa-let-7a-5p	23.08	1.91	22.66	1.83	23.98	1.99	24.22	2.15	23.73	1.89	0.240
hsa-miR-26a-5p	23.17	2.07	22.48	1.42	23.57	2.21	24.51	2.70	24.99	2.50	0.028
hsa-let-7f-5p	24.15	1.96	23.69	1.77	24.26	2.07	24.88	2.20	25.43	2.15	0.192
hsa-miR-93-5p	24.89	1.55	24.76	1.54	26.10	1.64	25.31	1.77	24.63	1.47	0.308
hsa-let-7g-5p	25.46	1.95	24.98	1.72	25.18	2.12	26.55	2.46	26.82	2.02	0.163
hsa-miR-21-5p	26.16	2.20	25.55	1.61	26.62	2.36	28.13	2.19	27.58	2.80	0.064
hsa-miR-30d-5p	26.18	1.49	25.91	1.38	26.98	1.62	26.83	1.90	26.59	1.56	0.308

Cq and SD values represent mean values across biological replicates. Significance of Cq differences between biological groups was tested with one-way ANOVA and Benjamini and Hochberg correction to adjust for multiple testing (*p-adj*). BM, bone marrow.

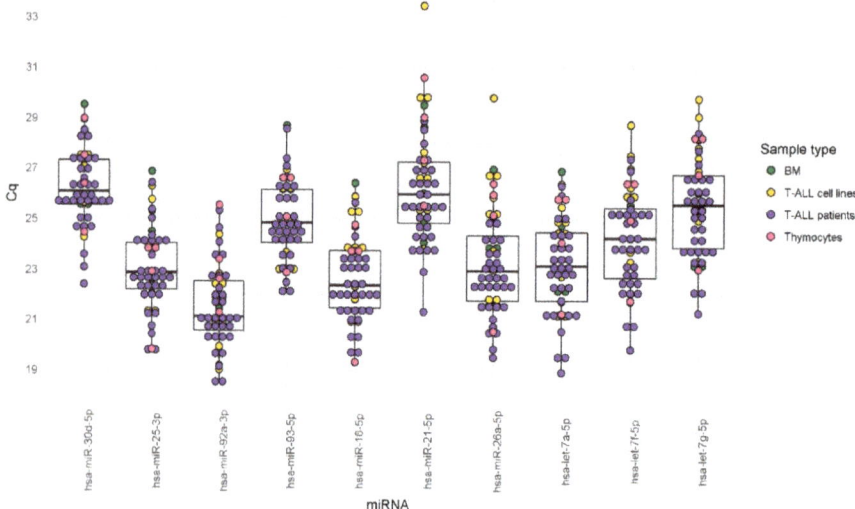

Figure 3. Overview of Cq values obtained by RT-qPCR for all samples with respect to the type of sample. BM, mature T-lymphocytes from normal bone marrow; thymocytes, normal precursors of T-cells. Dots represent mean raw Cq values for technical replicates of individual samples. Boxes correspond to the interquartile range (IQR) for each miRNA. Lines inside boxes indicate median Cq values. Candidate endogenous control miRNAs are ranked from left to right, according to increasing IQR value.

Next, for these 10 candidate EN miRNAs, we analyzed expression stability by RT-qPCR across all samples using the RefFinder tool (http://leonxie.esy.es/RefFinder) (Figure 1, Step 2b) [49]. This online open source tool integrates four algorithms for expression stability assessment: NormFinder [50], geNorm [16], BestKeeper [51], and a comparative Delta C_T method [52]. Thus, RefFinder generates a comprehensive ranking of candidate ENs. The comprehensive stability ranking is shown in Table 4. The individual ranks generated by each algorithm separately are shown in Supplementary Tables

S3–S6. The overlap between four stability-testing algorithms is presented in a Venn diagram in Figure 4. Out of the 10 candidate ENs, we finally selected three miRNAs: hsa-miR-16-5p, hsa-miR-25-3p, and hsa-let-7a-5p. The selection was based on the highest stability according to the RefFinder tool (Table 4), high overlapping between the four algorithms (Figure 4), and relatively high expression (low mean Cq values; equivalent to C_T values in Delta C_T method) across all samples (Table 3 and Figure 3).

Table 4. RefFinder comprehensive ranking score of miRNA expression stability.

miRNA Name	Comprehensive Ranking Stability Score
hsa-miR-16-5p	2.11
hsa-miR-30d-5p	2.71
hsa-miR-25-3p	2.99
hsa-let-7g-5p	3.98
hsa-let-7a-5p	4.36
hsa-miR-93-5p	4.53
hsa-let-7f-5p	4.58
hsa-miR-92a-3p	5.66
hsa-miR-21-5p	8.74
hsa-miR-26a-5p	10

The stability score value is inversely proportional to the stability of gene expression.

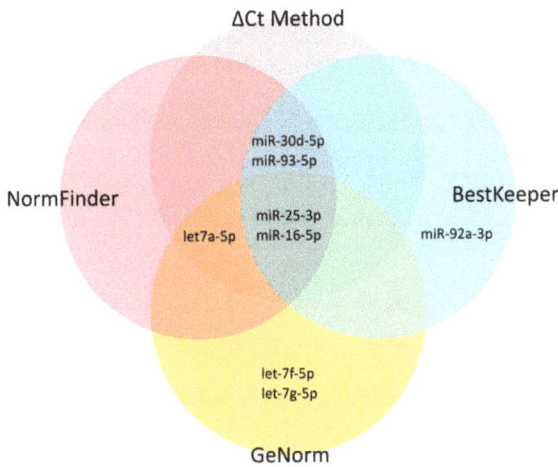

Figure 4. Venn diagram presenting the overlap between the five most stable candidate miRNAs indicated by each of four stability-testing algorithms.

2.3. Testing the Utility of Selected Candidate Endogenous Normalizer miRNAs (Step 3)

To test for the applicability of the three selected EN miRNAs, we used them in the RT-qPCR validation of our miRNA-seq results (Figure 1, Step 3). For the purposes of this study, we report the validation results for three miRNAs with an already-reported role in T-ALL biology: hsa-miR-128-3p, hsa-miR-181a-5p, and hsa-miR-20b-5p. We identified these miRNAs to be overexpressed in T-ALL (see Table 5), which is in line with their reported role as oncogenic miRNAs in this disease. In the RT-qPCR, we used the same samples that were used in miRNA-seq. Normalized relative expression levels of hsa-miR-20b-5p, hsa-miR-128-3p, and hsa-miR-181a-5p in patient samples and in normal controls are presented in Figure 5. For all three miRNAs tested by RT-qPCR, we observed statistically significant overexpression in T-ALL samples. The comparison of logarithmic fold change and *p*-values obtained by miRNA-seq and RT-qPCR is shown in Table 5.

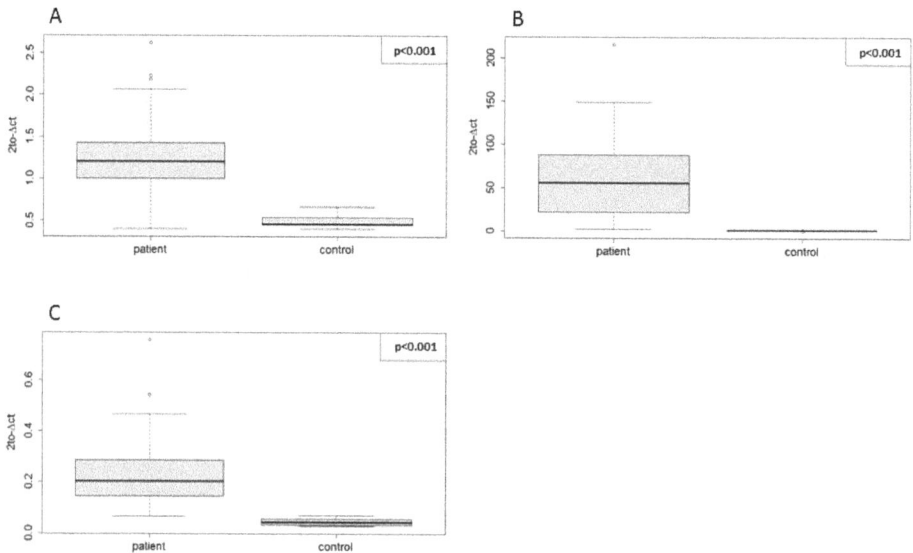

Figure 5. Normalized relative expression levels of miRNAs overexpressed in T-ALL vs. controls. (**A**) hsa-miR-20b-5p, (**B**) hsa-miR-181a-5p, and (**C**) hsa-miR-128-3p in patients (34 T-ALL samples) and in controls (5 samples of mature T-lymphocytes obtained from the bone marrow of healthy donors). Dots represent relative gene expression in individual samples. Upper and lower edges of boxes correspond to first (Q1) and third (Q3) quartiles, respectively. Lines inside boxes indicate median expression values. Whiskers extend to the smallest and largest observations within the 1.5-times interquartile range (IQR) from the box.

Table 5. Validation of expression levels for selected miRNAs with oncogenic role in T-ALL.

miRNA Name	miRNA-Seq		RT-qPCR	
	Log2 Fold Change	*p*-Value	Log2 Fold Change	*p*-Value
hsa-miR-128-3p	2.814	<0.001	2.373	<0.001
hsa-miR-181a-5p	2.362	<0.001	5.951	<0.001
hsa-miR-20b-5p	3.522	<0.001	1.329	<0.001

To explain the possible reason for discrepancies between log2 fold changes of both methods, most striking in the case of hsa-miR-181a-5p, we tested for similarities between miRNA sequences. These may cause the nonspecific binding of primers and probes used in RT-qPCR. We performed a computational analysis of the global similarity between mature miRNA sequences, their seed sequences, and the Pearson's correlation of expression based on our miRNA-seq results. The outcome of this analysis for hsa-miR-181a-5p is presented in Supplementary Table S7. The similarity of hsa-miR-181a-5p and its isoforms (isomiRs) was considerable, which may cause nonspecific annealing of primers and probes. It is noteworthy that the mean read counts of hsa-miR-181b-5p and hsa-miR-181d-5p in our miRNA-seq results were high enough (see Supplementary Table S7) to produce potential bias in the quantification of hsa-miR-181a-5p by RT-qPCR and explain the higher fold change value than that in miRNA-seq.

Regardless of the fold change discrepancies, the direction of expression changes and the statistical significance were consistent between both methods for all three miRNAs (see Table 5). Thus, the overexpression of hsa-miR-20b-5p, hsa-miR-128-3p, and hsa-miR-181a-5p in T-ALL patients relative to normal controls that we observed in the miRNA-seq experiment was successfully validated by RT-qPCR with the use of the selected EN miRNAs.

Taken together, the combination of criteria we adopted (i.e., comprehensively tested stability of expression and relative abundance of expression) allowed us to identify three miRNAs (hsa-miR-16-5p, hsa-miR-25-3p, and hsa-let-7a-5p) as optimal endogenous normalizers for RT-qPCR experiments in T-ALL samples.

3. Discussion

The choice of appropriate endogenous controls to normalize expression levels is one of the key factors greatly influencing the results of RT-qPCR expression profiling [53]. Optimal endogenous normalizers are crucial for the comparison of miRNA expression levels from different RT-qPCR experiments. Such interexperimental reproducibility of expression data is particularly important when miRNAs are considered as candidate biomarkers to be implemented in standardized diagnostic or treatment stratification procedures [54].

3.1. Strategy for the Identification of Optimal Endogenous Normalizer miRNAs for RT-qPCR

Here we present the strategy used for the identification of optimal endogenous normalizer (EN) miRNAs for RT-qPCR miRNA profiling in cells of T-lineage. The study design and experimental workflow are presented in Figure 1. The starting point of our study was the results of miRNA transcriptome profiling with use of next-generation sequencing (miRNA-seq) performed in pediatric T-ALL samples and normal controls [43]. First, using an iterative analysis of the stability of candidate ENs, we established three as an applicable number of miRNAs to be combined as ENs. Based on the expression stability in our miRNA-seq data, we selected 10 candidate EN miRNAs. Out of these, seven were recommended by Thermo Fisher Scientific as endogenous controls for TaqMan Advanced miRNA Assays (www.thermofisher.com/advancedmirna). Four miRNAs were also reported in the literature as suitable endogenous controls for different cancer samples [44–48]. To get insight into the potential utility of these candidate EN miRNAs in other experimental settings relative to cells of T-lineage, we included in our study four types of material commonly used in this research: primary T-ALL samples, normal mature T-lymphocytes from bone marrow, immature thymocyte samples, and six T-ALL cell lines. We also included varying cell line culture conditions for a more in-depth analysis of miRNA stability. Next, we analyzed the expression of these candidate EN miRNAs by RT-qPCR, and with the use of four algorithms we assessed the expression stability across all samples with respect to different material types. We selected three EN miRNAs (hsa-miR-16-5p, hsa-miR-25-3p, and hsa-let-7a-5p) exhibiting stable and abundant expression across all samples under study and successfully used them in the RT-qPCR validation of our miRNA-seq results.

Note that two of the three miRNAs we identified as the most optimal ENs for T-lineage cells, hsa-miR-25-3p and hsa-miR-16-5p, were among those proposed as universal ENs for TaqMan Advanced miRNA Assays (www.thermofisher.com/advancedmirna). These two miRNAs were also reported in the literature as optimal ENs used for miRNA expression profiling in cancer cells. Specifically, hsa-miR-25-3p was shown to be stably expressed in several human cancer cell lines, including cervical, breast, and colorectal cancer, acute lymphoblastic leukemia, and testicular embryonal carcinoma [45]. hsa-miR-16-5p was reported to be a suitable EN for miRNA expression profiling in malignant, benign, and normal breast tissues [44], and was also used as an EN for the profiling of exosomal miRNA expression in blood serum of patients with breast and gastric cancer [46–48]. The third miRNA we identified as an optimal EN, hsa-let-7a-5p, was reported to be suitable for miRNA expression analysis in human breast cancer [44]. Here we show that these three miRNAs are also optimal ENs for miRNA profiling in T-ALL.

Below, we discuss several important aspects of the selection of optimal normalizers for miRNA expression in light of the proposed strategy.

3.2. Number of miRNAs to Be Used as Endogenous Normalizers in RT-qPCR

There are several approaches to the normalization of expression data in RT-qPCR, such as the use of a single endogenous control gene, the geometric mean of expression of several endogenous controls, and the global mean of expression of all transcripts under study. Below, we discuss the issue of the number of endogenous normalizers to be used in RT-qPCR.

One of the approaches to normalization in RT-qPCR is the use of a single endogenous control gene. However, this approach may generate a large normalization error [16]. Therefore, normalization against a geometric mean of expression of several validated endogenous control genes is a more appropriate approach and greatly reduces the error rate [53,55].

Another possible strategy for RT-qPCR normalization is the use of a global mean expression value of all miRNAs under study (or genes, in the case of mRNA profiling) [56–58]. This approach was demonstrated by Mestdagh et al. to be accurate and reliable for miRNA profiling in RT-qPCR experiments [56]. In this study, the stability of a global mean was higher than the stability of commonly used endogenous control small RNAs. However, this approach is not suitable for every dataset, especially those in which a significant proportion of miRNAs show a positive correlation of signal intensities, which will limit the effectiveness of the approach. Additionally, a limitation of this method is the need to analyze a large and unbiased group of examined miRNAs or genes [56]. Another option for normalization of miRNA expression is a method based on the use of weight mean of miRNA expression to generate an artificial endogenous control used to calculate ΔCq values. The standard deviation of miRNA expression across all samples is used as a measure of stability, and the expression of each miRNA is weighted by its stability [59]. Yet, the utility of these approaches is limited to high- and medium-throughput expression profiling experiments like RT-qPCR-based arrays, and they are not suitable for small numbers of studied miRNAs or genes. Thus, the valid question is, what number of miRNAs should be used as ENs in a particular experiment?

To address this question, we created an algorithm for an iterative analysis of miRNA expression stability (Supplementary Material, IterativeStability_v1.0.R) and applied it to our miRNA-seq data. Several aspects should be considered when translating findings from miRNA-seq to an RT-qPCR experimental setting. The stability of miRNAs as measured by miRNA-seq depends not only on the intersample variability in miRNA concentrations, but also on the accuracy of the measurement method. The accuracy depends on the RNA isolation, the library preparation, the process of sequencing, and the data processing algorithms, including the data standardization strategy and its assumptions [17]. Furthermore, the expression of low-level miRNAs might be significantly affected by the sampling process, caused by the limited capacity of the flow-cell used in the sequencing process, as in RNA-seq [60]. For these reasons, even the most stable miRNAs will show significant variance between samples, caused by noise of technical origin. In order to reduce this effect, it is often beneficial to combine several endogenous reference RNAs for the normalization of expression data in RT-qPCR. The goal is to establish the number and optimal set of miRNAs to be combined. Using our miRNA-seq data, we selected the best k-number of miRNAs. However, in other experiments, a $k + 1$ number of combined miRNAs might provide a more stable signal. In general, in this iterative approach the stability is expected to increase as k increases, until we reach the point at which we start to incorporate differentially expressed miRNAs. However, in most cases, this will happen long after we exceed the practical limits of the number of miRNAs that can be used as reference in RT-qPCR experiments.

In our iterative analysis of miRNA expression stability, we demonstrated that the use of more than three EN miRNAs could provide only a limited increase of stability. The advantage of combining more miRNAs is likely disproportionate in view of additional cost. The use of three miRNAs seems to give a reasonable balance between good stability for endogenous normalization and cost-effectiveness of the study.

3.3. Expression Stability Relative to Sample Type and Culture Conditions

In our study, we analyzed the expression stability of 10 candidate EN miRNAs in several types of material commonly used in research on T-lineage cells, both normal and malignant. We included T-ALL samples, normal thymocytes, normal mature T-lymphocytes, and six T-ALL cell lines, cultured in varying conditions (as presented in Figure 1 and described in detail in the Materials section). We did not observe significant differences in the expression levels of candidate EN miRNAs between different types of biological material tested, as presented in Table 3. This observation also applies to T-ALL cell line samples cultured in different conditions, as presented in Supplementary Table S2. The variability of expression of these miRNAs is rather sample-dependent, as illustrated in Supplementary Figure S1.

We also demonstrated that our three selected EN miRNAs were suitable for miRNA profiling across different T-ALL cell lines and varying culture conditions by evaluating the relative expression of hsa-miR-128-3p in these samples in reference to normal bone marrow T-lymphocytes. As shown in Supplementary Figure S2, the hsa-miR-128-3p expression levels were very close in the MOLT-4 cells cultured with and without antibiotic. Similarly, the time of culture (early vs. late passage analyzed in the case of the CCRF-CEM cell line) did not affect the expression level of this miRNA. We observed the difference in fold change in one of the three independent cultures of Jurkat cells and, naturally, differences of fold change values across different cell lines. Nevertheless, using our three EN miRNAs, we showed that hsa-miR-128-3p was overexpressed in all six T-ALL cell lines, which is in line with literature data [35,61].

3.4. RT-qPCR Validation of miRNA-Seq Results

One of the discrepancies between miRNA-seq and RT-qPCR results when used for validation purposes is the difference of fold change values observed for differentially expressed miRNAs [62]. This can be explained by the fact that the operating principles for the two methods of quantification are different. miRNA-seq is based on the read count number of a particular transcript, whereas RT-qPCR is based on hybridization with primers and hydrolysis probe and amplification. The design of amplification primers and probes is particularly challenging in the case of miRNAs due to the fact that mature miRNA sequences are short. In combination with the high similarity between sequences of isomiRs, this limits the options for the optimal design of a miRNA assay to ensure the highest sensitivity and specificity [17]. These factors can potentially lead to nonspecific hybridization and eventually to biased RT-qPCR results as compared to miRNA-seq [62]. In our experimental setting, the difference in the fold change values was most striking in hsa-miR-181a-5p, as shown in Table 5. Based on the analysis of similarity between mature miRNA sequences, their seed sequences, and the Pearson's correlation of expression in our miRNA-seq data, we concluded that this discrepancy might be attributed to high similarity of hsa-miR-181a-5p and its isomiRs, as presented in Supplementary Table S7.

Nevertheless, the direction of expression change and statistical significance were concordant between the two methods for all three miRNAs, which proves the positive validation of miRNA-seq findings with RT-qPCR using our EN miRNAs.

4. Materials and Methods

4.1. Materials

Bone marrow samples of 34 pediatric T-ALL patients and 5 age-related healthy donors were collected in Polish pediatric hemato-oncology centers. The study was approved by the Ethics Committee of the Medical University of Silesia (KNW/0022/KB1/153/I/16/17, 3 Octorber 2017). Informed consent was obtained in accordance with the Declaration of Helsinki. T-ALL cells and normal T-lymphocytes were selected from the samples by immunomagnetic separation. Four RNA samples obtained from CD34+ and CD4+CD8+ normal thymocyte subsets were a kind gift from Prof. Pieter van Vlierberghe (Center for Medical Genetics Ghent, Ghent University, Belgium).

The samples were obtained and processed as described previously [32]. Six T-ALL cell lines—Jurkat, DND-41, CCRF-CEM, BE-13, P12-Ichikawa, and MOLT-4—were purchased from the Leibniz Institute DSMZ–German Collection of Microorganisms and Cell Cultures. To gain insight into miRNA stability under different culture conditions, we applied culture/harvesting settings as follows. All cell lines were harvested for RNA extraction at the fifth (early) passage. Additionally, CCRF-CEM cells were also harvested in the twentieth (late) passage. MOLT-4 cells were harvested after culture in antibiotic-free medium as well as after standard antibiotic-including culture, applied to all cell lines. The Jurkat cell line was harvested after 3 independent cultures in standard conditions. Details of T-ALL and bone marrow sample preparation, T-ALL cell line culture conditions, and RNA isolation are included in the Supplementary Materials and Methods.

4.2. miRNA-Seq

miRNA-seq was conducted by Exiqon (NGS Services Exiqon, Denmark) using NextSeq500 Illumina, with 10 million reads per sample, read length: 51 bp single-end; reference annotation: GRCh37. Raw sequencing reads were adapter-trimmed using Cutadapt (version 1.11) and aligned with Bowtie (version 1.2.2) to a modified version of miRBase (version 21) created according to the miRge specifications [63]. Differentially expressed miRNAs were selected using edgeR [64].

4.3. Reverse-Transcription and RT-qPCR Amplification Conditions

Total RNA, including miRNA fraction, was reverse-transcribed using an adapter-based TaqMan Advanced miRNA cDNA Synthesis Kit (Thermo Fisher Scientific, Waltham, MA, USA) with preamplification step according to the manufacturer's protocol. Predesigned hydrolysis probe-based TaqMan Advanced miRNA Assays and TaqMan Fast Advanced Master Mix (Thermo Fisher Scientific) were used. The TaqMan assay IDs for the miRNAs are shown in Table 2. The reactions were performed using a 7900HT Fast Real-Time PCR System with 96-well block module (Applied Biosystems, currently part of Thermo Fisher Scientific, Waltham, MA, USA). All RT-qPCR reactions were performed in 3 technical replicates for each sample.

4.4. Amplification Efficiency

Standard curve analysis was applied to test the amplification efficiency of all assays (candidate EN miRNAs and miRNAs overexpressed in T-ALL). Standard curves were generated by 10-fold serial dilutions (10^0–10^{-5}) of cDNA, reverse-transcribed from RNA obtained from the DND-41 cell line.

4.5. Analysis of Gene Expression

For the relative quantification of gene expression, a comparative C_T method ($\Delta\Delta C_T$) was used [65]. Statistical analysis was performed with Data Assist Software (v. 3.01; Thermo Fisher Scientific, Waltham, MA USA). The significance of differences in expression level between patients and controls was tested using a 2-tailed Student's *t*-test, with $p < 0.05$ indicating statistical significance. Results are presented as boxplots. Whenever relevant, C_T values are further referred to as Cq values, in accordance with the Minimum Information for Publication of Quantitative Real-Time PCR Experiments guidelines [11]. Significance of Cq differences between biological groups representing different types of biological material used in the study were tested with one-way ANOVA and Benjamini and Hochberg correction to adjust for multiple testing (*p-adj*).

4.6. Analysis of Expression Stability

NormFinder software (https://moma.dk/normfinder-software) [50] was used to assess the stability of expression across all samples in miRNA-seq results. For stability testing by RT-qPCR, the Cq data collected for each candidate EN miRNA for each sample were used as input data for the RefFinder tool (http://leonxie.esy.es/RefFinder), integrating 4 algorithms for the assessment of

Int. J. Mol. Sci. **2018**, *19*, 2858

expression stability: NormFinder [50], geNorm [16], BestKeeper [51], and a comparative Delta C_T method [52]. Stability ranks were generated comprehensively and individually from the 4 independent stability-assessing algorithms.

4.7. Analysis of the Similarity of Mature miRNA Sequences

Mature miRNA sequences were downloaded from miRBase v.21 and compared pairwise using the Needleman–Wunsch global alignment algorithm, as implemented in the pairwise Alignment function of the Biostrings R package (version 2.48). Seed sequences, defined as the nucleotides at positions 2–7 of the mature miRNA sequences, were compared separately, retaining only information about ideal matches of the compared sequences. Expression levels between individual miRNAs were compared pairwise using Pearson's correlation coefficient.

5. Conclusions

Herein we present a strategy for the identification of miRNAs as optimal endogenous normalizers for the RT-qPCR profiling of miRNA expression in T-ALL cells and normal cells of T-lineage. Our strategy includes a computational approach to assess the applicable number of miRNAs to be used for the normalization of RT-qPCR results. The identification of optimal endogenous normalizers was based on a comprehensive analysis of miRNA expression stability in both miRNA-seq data and RT-qPCR, complemented by an analysis of RT-qPCR amplification efficiency and an analysis of expression across a variety of sample types and cell culture conditions. We showed the utility of the combination of three endogenous normalizers (hsa-miR-16-5p, hsa-miR-25-3p, and hsa-let-7a-5p) for successful validation of our miRNA-seq results.

The panel of 10 miRNAs, particularly the three miRNAs we identified as optimal endogenous normalizers, might serve as the first-line choice for those dealing with RT-qPCR expression analysis of miRNAs in different types of T-lineage samples: patient samples, T-ALL cell lines, normal thymocytes, and mature T-lymphocytes. So far, no data have been available on the comprehensive evaluation of candidate endogenous normalizer miRNAs regarding T-ALL and normal T-cells. Thus, our results fill this research gap, while the strategy we propose, including the algorithm we created for the assessment of expression stability, might be used more universally and transferred to other experimental settings concerning miRNA profiling with the use of RT-qPCR.

Supplementary Materials: Supplementary materials can be found at http://www.mdpi.com/1422-0067/19/10/2858/s1.

Author Contributions: M.D. (Małgorzata Dawidowska) conceived the study and designed the experiments. M.D. (Monika Drobna), B.S.-Z., and M.K. carried out the experiments. M.D. (Monika Drobna) and P.D.-R. performed statistical analysis and relevant graphics. R.J. carried out the bioinformatics analysis. M.D. (Monika Drobna) and M.D. (Małgorzata Dawidowska) interpreted the results of the experiments and drafted the manuscript. M.W. was in charge of overall direction and funding. M.D. (Małgorzata Dawidowska) supervised the study and was in charge of the final version of the manuscript. All authors critically revised the manuscript and approved the submitted version.

Funding: This research was funded by grants from the National Science Centre, Poland (2014/15/B/NZ2/03394 and 2017/25/N/NZ2/01132), and the National Centre of Research and Development (STRATEGMED3/304586/5/NCBR/2017). Bronisława Szarzyńska-Zawadzka received financial resources as part of financing of a doctoral scholarship from the National Science Center (ETIUDA scholarship number: 2017/24/T/NZ5/00359).

Acknowledgments: The authors wish to acknowledge Prof. Pieter van Vlierberghe (Center for Medical Genetics Ghent, Ghent University, Belgium), who kindly provided RNA samples obtained from CD34+ and CD4+CD8+ normal thymocyte subsets. The authors wish to thank Paweł Śledziński for technical assistance in data visualization.

Conflicts of Interest: The authors declare no conflict of interest.

Abbreviations

BM	bone marrow
Cq	quantification cycle (nomenclature in accordance with Minimum Information for Publication of Quantitative Real-Time PCR Experiments, MIQE guidelines); Cq is equivalent to C_T (threshold cycle) which is only used, when referred to $\Delta\Delta\,C_T$ method implemented in Data Assist Software version 3.01 (Thermo Fisher Scientific) or Comparative Delta C_T method; these methods has retained their original names
EN	endogenous normalizer
IQR	interquartile range
isomiR	miRNA isoform
miRNA-seq	next-generation sequencing of miRNA-transcriptome
Q1	first quartile
Q3	third quartile
SD	standard deviation
snoRNA	small nucleolar RNA
snRNA	small nuclear RNA
T-ALL	T-cell acute lymphoblastic leukemia
thymo	thymocytes

References

1. MacFarlane, L.A.; Murphy, P.R. MicroRNA: Biogenesis, Function and Role in Cancer. *Curr. Genom.* **2010**, *11*, 537–561. [CrossRef] [PubMed]
2. Bartel, D.P. MicroRNAs: Target Recognition and Regulatory Functions. *Cell* **2009**, *136*, 215–233. [CrossRef] [PubMed]
3. Krol, J.; Loedige, I.; Filipowicz, W. The widespread regulation of microRNA biogenesis, function and decay. *Nat. Rev. Genet.* **2010**, *11*, 597–610. [CrossRef] [PubMed]
4. Esquela-Kerscher, A.; Slack, F.J. Oncomirs-microRNAs with a role in cancer. *Nat. Rev. Cancer* **2006**, *6*, 259–269. [CrossRef] [PubMed]
5. Ogata-Kawata, H.; Izumiya, M.; Kurioka, D.; Honma, Y.; Yamada, Y.; Furuta, K.; Gunji, T.; Ohta, H.; Okamoto, H.; Sonoda, H.; et al. Circulating Exosomal microRNAs as Biomarkers of Colon Cancer. *PLoS ONE* **2014**, *9*, e92921. [CrossRef] [PubMed]
6. Wang, J.; Yan, F.; Zhao, Q.; Zhan, F.; Wang, R.; Wang, L.; Zhang, Y.; Huang, X. Circulating exosomal miR-125a-3p as a novel biomarker for early-stage colon cancer. *Sci. Rep.* **2017**, *7*, 4150. [CrossRef] [PubMed]
7. Bhome, R.; Del Vecchio, F.; Lee, G.-H.; Bullock, M.D.; Primrose, J.N.; Sayan, A.E.; Mirnezami, A.H. Exosomal microRNAs (exomiRs): Small molecules with a big role in cancer. *Cancer Lett.* **2018**, *420*, 228–235. [CrossRef] [PubMed]
8. Hornick, N.I.; Huan, J.; Doron, B.; Goloviznina, N.A.; Lapidus, J.; Chang, B.H.; Kurre, P. Serum Exosome MicroRNA as a Minimally-Invasive Early Biomarker of AML. *Sci. Rep.* **2015**, *5*, 11295. [CrossRef] [PubMed]
9. Elhamamsy, A.R.; El Sharkawy, M.S.; Zanaty, A.F.; Mahrous, M.A.; Mohamed, A.E.; Abushaaban, E.A. Circulating miR-92a, miR-143 and miR-342 in Plasma are Novel Potential Biomarkers for Acute Myeloid Leukemia. *Int. J. Mol. Cell. Med.* **2017**, *6*, 77–86. [CrossRef] [PubMed]
10. Manier, S.; Liu, C.-J.; Avet-Loiseau, H.; Park, J.; Shi, J.; Campigotto, F.; Salem, K.Z.; Huynh, D.; Glavey, S.V.; Rivotto, B.; et al. Prognostic role of circulating exosomal miRNAs in multiple myeloma. *Blood* **2017**, *129*, 2429–2436. [CrossRef] [PubMed]
11. Bustin, S.A.; Benes, V.; Garson, J.A.; Hellemans, J.; Huggett, J.; Kubista, M.; Mueller, R.; Nolan, T.; Pfaffl, M.W.; Shipley, G.L.; et al. The MIQE guidelines: Minimum information for publication of quantitative real-time PCR experiments. *Clin. Chem.* **2009**, *55*, 611–622. [CrossRef] [PubMed]
12. Marabita, F.; de Candia, P.; Torri, A.; Tegnér, J.; Abrignani, S.; Rossi, R.L. Normalization of circulating microRNA expression data obtained by quantitative real-time RT-PCR. *Brief. Bioinform.* **2016**, *17*, 204–212. [CrossRef] [PubMed]

13. Mavrakis, K.J.; Wolfe, A.L.; Oricchio, E.; Palomero, T.; de Keersmaecker, K.; McJunkin, K.; Zuber, J.; James, T.; Chang, K.; Khan, A.A.; et al. Genome-wide RNA-mediated interference screen identifies miR-19 targets in Notch-induced T-cell acute lymphoblastic leukaemia. *Nat. Cell Biol.* **2010**, *12*, 372. [CrossRef] [PubMed]

14. Kroh, E.M.; Parkin, R.K.; Mitchell, P.S.; Tewari, M. Analysis of circulating microRNA biomarkers in plasma and serum using quantitative reverse transcription-PCR (qRT-PCR). *Methods* **2010**, *50*, 298–301. [CrossRef] [PubMed]

15. Farr, R.J.; Januszewski, A.S.; Joglekar, M.V.; Liang, H.; McAulley, A.K.; Hewitt, A.W.; Thomas, H.E.; Loudovaris, T.; Kay, T.W.H.; Jenkins, A.; et al. A comparative analysis of high-throughput platforms for validation of a circulating microRNA signature in diabetic retinopathy. *Sci. Rep.* **2015**, *5*, 10375. [CrossRef] [PubMed]

16. Vandesompele, J.; De Preter, K.; Pattyn, F.; Poppe, B.; Van Roy, N.; De Paepe, A.; Speleman, F. Accurate normalization of real-time quantitative RT-PCR data by geometric averaging of multiple internal control genes. *Genome Biol.* **2002**, *3*, 12. [CrossRef]

17. Chugh, P.; Dittmer, D.P. Potential pitfalls in microRNA profiling. *Wiley Interdiscip. Rev. RNA* **2012**, *3*, 601–616. [CrossRef] [PubMed]

18. Schwarzenbach, H.; da Silva, A.M.; Calin, G.; Pantel, K. Data Normalization Strategies for MicroRNA Quantification. *Clin. Chem.* **2015**, *61*, 1333–1342. [CrossRef] [PubMed]

19. Gee, H.E.; Buffa, F.M.; Camps, C.; Ramachandran, A.; Leek, R.; Taylor, M.; Patil, M.; Sheldon, H.; Betts, G.; Homer, J.; et al. The small-nucleolar RNAs commonly used for microRNA normalisation correlate with tumour pathology and prognosis. *Br. J. Cancer* **2011**, *104*, 1168–1177. [CrossRef] [PubMed]

20. Lamba, V.; Ghodke-Puranik, Y.; Guan, W.; Lamba, J.K. Identification of suitable reference genes for hepatic microRNA quantitation. *BMC Res. Notes* **2014**, *7*, 129. [CrossRef] [PubMed]

21. Serafin, A.; Foco, L.; Blankenburg, H.; Picard, A.; Zanigni, S.; Zanon, A.; Pramstaller, P.P.; Hicks, A.A.; Schwienbacher, C. Identification of a set of endogenous reference genes for miRNA expression studies in Parkinson's disease blood samples. *BMC Res. Notes* **2014**, *7*, 715. [CrossRef] [PubMed]

22. Correia, N.C.; Melao, A.; Povoa, V.; Sarmento, L.; de Cedron, M.G.; Malumbres, M.; Enguita, F.J.; Barata, J.T. microRNAs regulate TAL1 expression in T-cell acute lymphoblastic leukemia. *Oncotarget* **2016**, *7*, 8268–8281. [CrossRef] [PubMed]

23. Ortega, M.; Bhatnagar, H.; Lin, A.P.; Wang, L.; Aster, J.C.; Sill, H.; Aguiar, R.C.T. A microRNA-mediated regulatory loop modulates NOTCH and MYC oncogenic signals in B- and T-cell malignancies. *Leukemia* **2015**, *29*, 968–976. [CrossRef] [PubMed]

24. Scott, M.S.; Ono, M. From snoRNA to miRNA: Dual function regulatory non-coding RNAs. *Biochimie* **2011**, *93*, 1987–1992. [CrossRef] [PubMed]

25. Delpu, Y.; Larrieu, D.; Gayral, M.; Arvanitis, D.; Dufresne, M.; Cordelier, P.; Torrisani, J. Noncoding RNAs: Clinical and Therapeutic Applications. In *Drug Discovery in Cancer Epigenetics*; Academic Press: Boston, MA, USA, 2016; Chapter 12; pp. 305–326. ISBN 978-0-12-802208-5.

26. Belver, L.; Ferrando, A. The genetics and mechanisms of T cell acute lymphoblastic leukaemia. *Nat. Rev. Cancer* **2016**, *16*, 494–507. [CrossRef] [PubMed]

27. Van Vlierberghe, P.; Pieters, R.; Beverloo, H.B.; Meijerink, J.P.P. Molecular-genetic insights in paediatric T-cell acute lymphoblastic leukaemia. *Br. J. Haematol.* **2008**, *143*, 153–168. [CrossRef] [PubMed]

28. Wallaert, A.; Durinck, K.; Taghon, T.; Van Vlierberghe, P.; Speleman, F. T-ALL and thymocytes: A message of noncoding RNAs. *J. Hematol. Oncol.* **2017**, *10*, 17. [CrossRef] [PubMed]

29. Drobna, M.; Szarzyńska-Zawadzka, B.; Dawidowska, M. T-cell acute lymphoblastic leukemia from miRNA perspective: Basic concepts, experimental approaches, and potential biomarkers. *Blood Rev.* **2018**. [CrossRef] [PubMed]

30. Mets, E.; Van der Meulen, J.; Van Peer, G.; Boice, M.; Mestdagh, P.; Van de Walle, I.; Lammens, T.; Goossens, S.; De Moerloose, B.; Benoit, Y.; et al. MicroRNA-193b-3p acts as a tumor suppressor by targeting the MYB oncogene in T-cell acute lymphoblastic leukemia. *Leukemia* **2015**, *29*, 798–806. [CrossRef] [PubMed]

31. Correia, N.C.; Durinck, K.; Leite, A.P.; Ongenaert, M.; Rondou, P.; Speleman, F.; Enguita, F.J.; Barata, J.T. Novel TAL1 targets beyond protein-coding genes: Identification of TAL1-regulated microRNAs in T-cell acute lymphoblastic leukemia. *Leukemia* **2013**, *27*, 1603–1606. [CrossRef] [PubMed]

32. Wallaert, A.; Van Loocke, W.; Hernandez, L.; Taghon, T.; Speleman, F.; Van Vlierberghe, P. Comprehensive miRNA expression profiling in human T-cell acute lymphoblastic leukemia by small RNA-sequencing. *Sci. Rep.* **2017**, *7*, 7901. [CrossRef] [PubMed]

33. Schotte, D.; Moqadam, F.A.; Lange-Turenhout, E.A.M.; Chen, C.; van Ijcken, W.F.J.; Pieters, R.; den Boer, M.L. Discovery of new microRNAs by small RNAome deep sequencing in childhood acute lymphoblastic leukemia. *Leukemia* **2011**, *25*, 1389–1399. [CrossRef] [PubMed]

34. Mavrakis, K.J.; Van der Meulen, J.; Wolfe, A.L.; Liu, X.P.; Mets, E.; Taghon, T.; Khan, A.A.; Setti, M.; Rondou, P.; Vandenberghe, P.; et al. A cooperative microRNA-tumor suppressor gene network in acute T-cell lymphoblastic leukemia (T-ALL). *Nat. Genet.* **2011**, *43*, 673. [CrossRef] [PubMed]

35. Mets, E.; Van Peer, G.; Van der Meulen, J.; Boice, M.; Taghon, T.; Goossens, S.; Mestdagh, P.; Benoit, Y.; De Moerloose, B.; Van Roy, N.; et al. MicroRNA-128-3p is a novel oncomiR targeting PHF6 in T-cell acute lymphoblastic leukemia. *Haematologica* **2014**, *99*, 1326–1333. [CrossRef] [PubMed]

36. Sanghvi, V.R.; Mavrakis, K.J.; Van der Meulen, J.; Boice, M.; Wolfe, A.L.; Carty, M.; Mohan, P.; Rondou, P.; Socci, N.D.; Benoit, Y.; et al. Characterization of a set of tumor suppressor microRNAs in T cell acute lymphoblastic leukemia. *Sci. Signal.* **2014**, *7*, ra111. [CrossRef] [PubMed]

37. Oliveira, L.H.; Schiavinato, J.L.; Fraguas, M.S.; Lucena-Araujo, A.R.; Haddad, R.; Araujo, A.G.; Dalmazzo, L.F.; Rego, E.M.; Covas, D.T.; Zago, M.A.; et al. Potential roles of microRNA-29a in the molecular pathophysiology of T-cell acute lymphoblastic leukemia. *Cancer Sci.* **2015**, *106*, 1264–1277. [CrossRef] [PubMed]

38. Coskun, E.; von der Heide, E.K.; Schlee, C.; Kuhnl, A.; Gokbuget, N.; Hoelzer, D.; Hofmann, W.K.; Thiel, E.; Baldus, C.D. The role of microRNA-196a and microRNA-196b as ERG regulators in acute myeloid leukemia and acute T-lymphoblastic leukemia. *Leuk. Res.* **2011**, *35*, 208–213. [CrossRef] [PubMed]

39. Lv, M.; Zhang, X.; Jia, H.; Li, D.; Zhang, B.; Zhang, H.; Hong, M.; Jiang, T.; Jiang, Q.; Lu, J.; et al. An oncogenic role of miR-142-3p in human T-cell acute lymphoblastic leukemia (T-ALL) by targeting glucocorticoid receptor-alpha and cAMP/PKA pathways. *Leukemia* **2012**, *26*, 769–777. [CrossRef] [PubMed]

40. Nemes, K.; Csoka, M.; Nagy, N.; Mark, A.; Varadi, Z.; Danko, T.; Kovacs, G.; Kopper, L.; Sebestyen, A. Expression of Certain Leukemia/Lymphoma Related microRNAs and its Correlation with Prognosis in Childhood Acute Lymphoblastic Leukemia. *Pathol. Oncol. Res.* **2015**, *21*, 597–604. [CrossRef] [PubMed]

41. Ikari, J.; Smith, L.M.; Nelson, A.J.; Iwasawa, S.; Gunji, Y.; Farid, M.; Wang, X.Q.; Basma, H.; Feghali-Bostwick, C.; Liu, X.D.; et al. Effect of culture conditions on microRNA expression in primary adult control and COPD lung fibroblasts in vitro. *In Vitro Cell. Dev. Biol.-Anim.* **2015**, *51*, 390–399. [CrossRef] [PubMed]

42. Wagner, W.; Horn, P.; Castoldi, M.; Diehlmann, A.; Bork, S.; Saffrich, R.; Benes, V.; Blake, J.; Pfister, S.; Eckstein, V.; et al. Replicative Senescence of Mesenchymal Stem Cells: A Continuous and Organized Process. *PLoS ONE* **2008**, *3*, e2213. [CrossRef] [PubMed]

43. Dawidowska, M.; Szarzyńska-Zawadzka, B.; Jaksik, R.; Drobna, M.; Kosmalska, M.; Łukasz, S.; Szczepańskiand, T.; Witt, M. miRNA Profiling in Pediatric T-ALL with Use of Next-Generation Sequencing: Focus on T-ALL Pathobiology and Heterogeneity. *Blood* **2017**, *130*, 1443.

44. Davoren, P.A.; McNeill, R.E.; Lowery, A.J.; Kerin, M.J.; Miller, N. Identification of suitable endogenous control genes for microRNA gene expression analysis in human breast cancer. *BMC Mol. Biol.* **2008**, *9*, 76. [CrossRef] [PubMed]

45. Das, M.K.; Andreassen, R.; Haugen, T.B.; Furu, K. Identification of Endogenous Controls for Use in miRNA Quantification in Human Cancer Cell Lines. *Cancer Genom. Proteom.* **2016**, *13*, 63–68.

46. Muller, V.; Gade, S.; Steinbach, B.; Loibl, S.; von Minckwitz, G.; Untch, M.; Schwedler, K.; Lubbe, K.; Schem, C.; Fasching, P.A.; et al. Changes in serum levels of miR-21, miR-210, and miR-373 in HER2-positive breast cancer patients undergoing neoadjuvant therapy: A translational research project within the Geparquinto trial. *Breast Cancer Res. Treat.* **2014**, *147*, 61–68. [CrossRef] [PubMed]

47. Song, J.N.; Bai, Z.G.; Han, W.; Zhang, J.; Meng, H.; Bi, J.T.; Ma, X.M.; Han, S.W.; Zhang, Z.T. Identification of Suitable Reference Genes for qPCR Analysis of Serum microRNA in Gastric Cancer Patients. *Dig. Dis. Sci.* **2012**, *57*, 897–904. [CrossRef] [PubMed]

48. McDermott, A.M.; Kerin, M.J.; Miller, N. Identification and Validation of miRNAs as Endogenous Controls for RQ-PCR in Blood Specimens for Breast Cancer Studies. *PLoS ONE* **2013**, *8*, e83718. [CrossRef] [PubMed]

49. Xie, F.L.; Xiao, P.; Chen, D.L.; Xu, L.; Zhang, B.H. miRDeepFinder: A miRNA analysis tool for deep sequencing of plant small RNAs. *Plant Mol. Biol.* **2012**, *80*, 75–84. [CrossRef] [PubMed]

50. Andersen, C.L.; Jensen, J.L.; Orntoft, T.F. Normalization of real-time quantitative reverse transcription-PCR data: A model-based variance estimation approach to identify genes suited for normalization, applied to bladder and colon cancer data sets. *Cancer Res.* **2004**, *64*, 5245–5250. [CrossRef] [PubMed]

51. Pfaffl, M.W.; Tichopad, A.; Prgomet, C.; Neuvians, T.P. Determination of stable housekeeping genes, differentially regulated target genes and sample integrity: BestKeeper—Excel-based tool using pair-wise correlations. *Biotechnol. Lett.* **2004**, *26*, 509–515. [CrossRef] [PubMed]

52. Silver, N.; Best, S.; Jiang, J.; Thein, S.L. Selection of housekeeping genes for gene expression studies in human reticulocytes using real-time PCR. *BMC Mol. Biol.* **2006**, *7*, 33. [CrossRef] [PubMed]

53. Derveaux, S.; Vandesompele, J.; Hellemans, J. How to do successful gene expression analysis using real-time PCR. *Methods* **2010**, *50*, 227–230. [CrossRef] [PubMed]

54. Henry, N.L.; Hayes, D.F. Cancer biomarkers. *Mol. Oncol.* **2012**, *6*, 140–146. [CrossRef] [PubMed]

55. Thellin, O.; Zorzi, W.; Lakaye, B.; De Borman, B.; Coumans, B.; Hennen, G.; Grisar, T.; Igout, A.; Heinen, E. Housekeeping genes as internal standards: Use and limits. *J. Biotechnol.* **1999**, *75*, 291–295. [CrossRef]

56. Mestdagh, P.; Van Vlierberghe, P.; De Weer, A.; Muth, D.; Westermann, F.; Speleman, F.; Vandesompele, J. A novel and universal method for microRNA RT-qPCR data normalization. *Genome Biol.* **2009**, *10*, R64. [CrossRef] [PubMed]

57. Kang, K.; Peng, X.; Luo, J.; Gou, D. Identification of circulating miRNA biomarkers based on global quantitative real-time PCR profiling. *J. Anim. Sci. Biotechnol.* **2012**, *3*, 4. [CrossRef] [PubMed]

58. Chang, K.H.; Mestdagh, P.; Vandesompele, J.; Kerin, M.J.; Miller, N. MicroRNA expression profiling to identify and validate reference genes for relative quantification in colorectal cancer. *BMC Cancer* **2010**, *10*, 173. [CrossRef] [PubMed]

59. Qureshi, R.; Sacan, A. A novel method for the normalization of microRNA RT-PCR data. *BMC Med. Genom.* **2013**, *6* (Suppl. 1), S14. [CrossRef]

60. McIntyre, L.M.; Lopiano, K.K.; Morse, A.M.; Amin, V.; Oberg, A.L.; Young, L.J.; Nuzhdin, S.V. RNA-seq: Technical variability and sampling. *BMC Genom.* **2011**, *12*, 293. [CrossRef] [PubMed]

61. Yamada, N.; Noguchi, S.; Kumazaki, M.; Shinohara, H.; Miki, K.; Naoe, T.; Akao, Y. Epigenetic regulation of microRNA-128a expression contributes to the apoptosis-resistance of human T-cell leukaemia Jurkat cells by modulating expression of Fas-associated protein with death domain (FADD). *Biochim. Biophys. Acta Mol. Cell Res.* **2014**, *1843*, 590–602. [CrossRef] [PubMed]

62. Git, A.; Dvinge, H.; Salmon-Divon, M.; Osborne, M.; Kutter, C.; Hadfield, J.; Bertone, P.; Caldas, C. Systematic comparison of microarray profiling, real-time PCR, and next-generation sequencing technologies for measuring differential microRNA expression. *RNA* **2010**, *16*, 991–1006. [CrossRef] [PubMed]

63. Baras, A.S.; Mitchell, C.J.; Myers, J.R.; Gupta, S.; Weng, L.-C.; Ashton, J.M.; Cornish, T.C.; Pandey, A.; Halushka, M.K. miRge-A Multiplexed Method of Processing Small RNA-Seq Data to Determine MicroRNA Entropy. *PLoS ONE* **2015**, *10*, e0143066. [CrossRef] [PubMed]

64. Robinson, M.D.; McCarthy, D.J.; Smyth, G.K. edgeR: A Bioconductor package for differential expression analysis of digital gene expression data. *Bioinformatics* **2010**, *26*, 139–140. [CrossRef] [PubMed]

65. Schmittgen, T.D.; Livak, K.J. Analyzing real-time PCR data by the comparative C(T) method. *Nat. Protoc.* **2008**, *3*, 1101–1108. [CrossRef] [PubMed]

International Journal of
Molecular Sciences

MDPI

Review

Relevance of MicroRNAs as Potential Diagnostic and Prognostic Markers in Colorectal Cancer

Grzegorz Hibner *, Małgorzata Kimsa-Furdzik and Tomasz Francuz

Department of Biochemistry, School of Medicine in Katowice, Medical University of Silesia in Katowice, St. Medyków 18, 40-752 Katowice, Poland; malgorzata.kimsa@sum.edu.pl (M.K.-F.); tfrancuz@sum.edu.pl (T.F.)
* Correspondence: ghibner@sum.edu.pl; Tel.: +48-32-252-5088

Received: 3 July 2018; Accepted: 25 September 2018; Published: 27 September 2018

Abstract: Colorectal cancer (CRC) is currently the third and the second most common cancer in men and in women, respectively. Every year, more than one million new CRC cases and more than half a million deaths are reported worldwide. The majority of new cases occur in developed countries. Current screening methods have significant limitations. Therefore, a lot of scientific effort is put into the development of new diagnostic biomarkers of CRC. Currently used prognostic markers are also limited in assessing the effectiveness of CRC therapy. MicroRNAs (miRNAs) are a promising subject of research especially since single miRNA can recognize a variety of different mRNA transcripts. MiRNAs have important roles in epigenetic regulation of basic cellular processes, such as proliferation, apoptosis, differentiation, and migration, and may serve as potential oncogenes or tumor suppressors during cancer development. Indeed, in a large variety of human tumors, including CRC, significant distortions in miRNA expression profiles have been observed. Thus, the use of miRNAs as diagnostic and prognostic biomarkers in cancer, particularly in CRC, appears to be an inevitable consequence of the advancement in oncology and gastroenterology. Here, we review the literature to discuss the potential usefulness of selected miRNAs as diagnostic and prognostic biomarkers in CRC.

Keywords: colorectal; cancer; microRNA; biomarkers

---

1. Introduction

Colorectal cancer (CRC) accounts for about 10% of all cancer cases worldwide. The latest Global Cancer Observatory data from 2012 estimates that nearly 1.4 million new cases of CRC and over 694,000 deaths were reported. CRC is more common in developed countries, with over 65% of cases. In Europe, about 400,000 new patients are diagnosed with CRC each year, and more than 200,000 die annually. The majority of CRCs occur in patients over the age of 50, with more than 75% being those over 60 years old. The risk of CRC increases with age and is 1.5–2 times higher in men than in women. Over the past three decades, in the male population, there has been a steady increase in mortality. In the female population, the increase in mortality stabilized in the mid-1990s and since then, mortality has remained relatively constant [1]. Currently used screening methods, including the fecal occult blood test (FOBT), stool DNA test, double-contrast barium enema (DCBE), and colonoscopy, have significant limitations. Colonoscopy is an invasive procedure that carries the risk of bowel perforation. Additionally, due to the nature of this test, some patients deny its implementation. Furthermore, the other tests mentioned are limited by either insufficient sensitivity or high percentage of false positive results. Therefore, currently, the development of new biomarkers for CRC screening tests is the subject of intensive research. However, most of the results of recent studies require confirmation on larger groups of patients before being implemented in clinical practice [2].

Currently, tumor-node-metastasis (TNM) classification is the primary prognostic marker in CRC therapy. The TNM system describes the size of the primary tumor, the degree of invasion of the intestinal wall, nearby lymph nodes and distant organs [3]. Although the TNM classification is the basis for CRC prognosis, this system has important caveats. Insufficient analysis of lymph node status may lead to underestimation of tumor progression, which in turn may result in ineffective treatment [3]. In addition, histologically indistinguishable CRCs may have various genetic and epigenetic backgrounds that contribute to different progressions and responses to treatment. For example, patients with TNM stage II CRC without lymph node metastasis have relatively good survival rates, but still around 25% of these patients have a high risk of relapse after surgical removal of the tumor. Unfortunately, currently used CRC prognostic markers have limited use in identification of patients with increased risk of recurrence and are still not highly accurate in assessing effectiveness of treatment.

MicroRNAs (miRNAs), which are short, single-stranded non-coding RNA sequences of approximately 21–23 nucleotides, are interesting and promising targets in CRC therapy and diagnostics. MiRNAs are products of intracellular processing of hairpin precursor molecules. Firstly, a miRNA gene is transcribed into a primary miRNA (pri-miRNA), which is then processed in the nucleus by Drosha RNase to form a precursor miRNA (pre-miRNA). The pre-miRNA is then transported to the cytoplasm via the activity of the exportin 5, which interacts with the Ran protein. In the cytoplasm, Dicer RNase is responsible for further processing of pre-miRNA. Subsequently, mature miRNA is complexed with Argonaute family proteins to form the RISC complex. Physiologically, miRNAs are responsible for epigenetic regulation of translation by attaching to the 3'-untranslated region (3'-UTR) of the target messenger RNA (mRNA) (Figure 1). In this way, miRNA mediates mRNA degradation and the repression of translation [4].

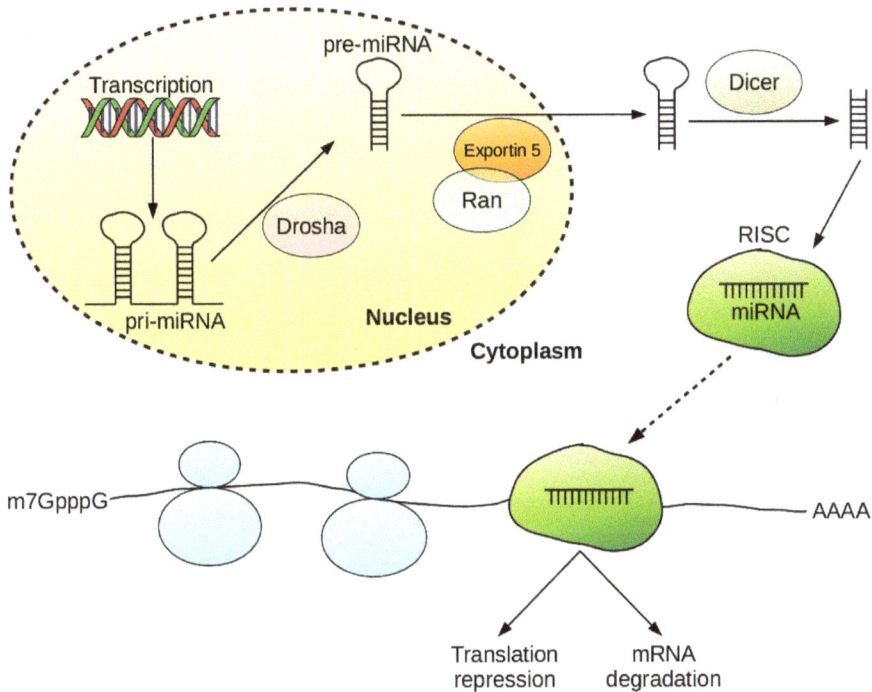

Figure 1. Synthesis and mechanism of miRNA activity.

One type of miRNA molecules can regulate the expression of multiple target genes and activation of different signaling pathways, and their crosstalk. The miRNA expression is unique for different tissues (including tumor tissue), and it participates in maintaining a differentiated cellular state. Therefore, disorders of miRNA expression may shift the cell to the undifferentiated proliferative phenotype. MiRNA expression disorders may be caused by a variety of molecular processes, such as deletions, amplifications, mutations within the miRNA *locus*, epigenetic silencing, or abnormal regulation of transcription factor activity. It is estimated that miRNAs are responsible for the regulation of translation of about 30–60% of human genes [5,6]. Because miRNAs recognize many different mRNA transcripts, these molecules are important in epigenetic regulation of basic cell processes, such as proliferation, apoptosis, differentiation, and migration. Therefore, miRNAs may act as potential oncogenes or suppressors in tumor development processes. Indeed, for many different human cancers, including CRC, significant abnormalities in miRNA expression have been observed. In CRCs, the altered expression of a large number of miRNAs associated with the development, progression and response to the treatment of this cancer was observed [7–9]. Changes in miRNA expression are associated with more frequent metastases, promotion of tumor mass growth, and increased malignancy of tumor cells. MiRNA expression is also associated with the risk of recurrence, response to the therapeutic regimen, and survival time in different cancers. Therefore, miRNA profiling may be a new and valuable tool in the diagnosis and prognosis of many types of cancer, including CRC. However, knowledge in this area still remains fragmented, and previous studies were conducted only on small groups of patients. In addition, the vast majority of studies conducted so far have evaluated the expression of miRNAs only in tumor tissues. However, miRNA expression profiling in tumor tissue has significant disadvantages. The heterogeneity of tumor cells produces variable results. Obtaining reliable data in this case is problematic and requires the isolation of individual cancer cells, e.g., using the laser microdissection method. This method, however, is costly, requires specialized equipment and is therefore not commonly used in clinical practice. Furthermore, studying tumor tissues does not allow for the assessment of changes in miRNA expression during anti-cancer therapy. Alternatively, choosing serum or blood plasma as the test material is a more practical and cheaper approach to miRNA profiling. Stool is also suitable for miRNA expression analysis. In both cases, the material is readily available, which makes it possible to identify potential biomarkers at any stage during and after the therapy, e.g., for early detection of cancer recurrence. MiRNA in plasma and serum is encapsulated in exosomes, which makes it highly stable. Therefore, the material does not require any special storage or protection operations prior to analysis. Potential applications of miRNA expression analysis for diagnostic, therapeutic and prognostic purposes are summarized in Table 1. The use of miRNAs as diagnostic and prognostic markers in neoplastic diseases, particularly in CRC, appears to be an inevitable consequence of advances in clinical oncology and gastroenterology.

Table 1. Possibilities and advantages of using miRNA for diagnostic, therapeutic and prognostic purposes in cancer diseases.

Possible Applications of miRNA:
Predictors of response to therapy and prognostic biomarkers
Use of miRNA as a drug—modulation of gene expression
Prediction and detection of metastases or non-invasive tumor phenotypes
Cancer diagnostics—detection of tumor specific miRNA signatures
Advantages of Using miRNA:
Easy detection in various biological materials (serum/plasma, cerebrospinal fluid, faeces)
High stability of miRNA molecules
Ability to determine specific types of cancer, and predict response to therapy and prognosis based on miRNA expression profile
Potential for use as antagonists in cancer therapy

2. MiRNAs as Diagnostic Biomarkers

Circulating miRNAs are stable due to encapsulation in exosomes [10], but the mechanism of their formation and secretion has not yet been fully understood. There is increasing evidence that miRNAs contained in exosomes may be involved in the exchange of information between distant tissues [11]. Serum miRNA expression is known to be altered in CRC patients and current studies attempt to investigate the correlation of the expression of certain types of miRNAs, both in serum and tumor tissue. Due to minimal invasiveness of miRNA harvesting procedures, circulating miRNAs may be potentially used as diagnostic biomarkers of various cancers, including CRC. A significant limitation of miRNA expression analysis is its poor utility in diagnosis, due to non-specificity and high variability of expression of a single miRNA type. Therefore, recent studies have attempted to evaluate the expression of miRNA sets. Such studies were conducted using microarray technology as well as quantitative real-time reverse transcriptase polymerase chain reaction (qRT-PCR) [12]. Insufficient sensitivity resulting from low concentrations of miRNA in serum or plasma of the patients is the principal limitation of a microarray experiment. QRT-PCR is characterized by better sensitivity, but evaluation of expression of numerous miRNAs using this method is a difficult and time-consuming task. Next-generation sequencing (NGS) is a novel and promising method that may be applied to evaluate the expression profiles of many miRNAs simultaneously [13]. Moreover, novel isolation techniques allow obtaining more miRNA for analysis, which, in combination with specialized NGS protocols, can increase the utility of this method in diagnostics of cancer, including CRC in the near future.

Some non-invasive screening methods used in CRC diagnosis are based on stool testing. Endogenous miRNAs encapsulated in exosomes are protected against RNases, in contrast to mRNAs or proteins that are rapidly degraded. For this reason, the detection of miRNA in stool is relatively easier. However, in order to ensure sensitivity and replicability, appropriate protocols, including material preparation, extraction, and quantitative analysis of miRNA, are required in this case [14]. Stool tests allow for earlier detection of tumor cells and most tumor markers, as compared to peripheral blood tests. Therefore, stool miRNA assays may be useful in detecting precancerous lesions [15]. Stool miRNA purification kits are commercially available, making it possible to obtain high-quality and high-purity nucleic acids for further analysis. However, methods that use stool miRNA molecules as biomarkers of CRC are still in their infancy. Although many recent studies indicate that stool miRNA tests have higher specificity, higher sensitivity and higher reproducibility than peripheral blood assays, no particular stool test has yet passed the preclinical phase. Therefore, it is necessary to carry out further studies and validation of methods based on miRNA derived from this material.

3. MiRNAs as Prognostic Biomarkers and Therapeutic Agents

MiRNA molecules also appear to be promising prognostic biomarkers, as has been shown so far in many preclinical and clinical studies [16–22]. The profiling of miRNA expression for prognostic purposes has been demonstrated in many human tumors, including: colorectal, pancreatic, ovarian, breast cancer and glioblastoma [15,23–25]. Since one type of miRNA molecule can influence the regulation of expression of many different genes, the use of these molecules in anti-cancer therapy also seems promising. Currently, there are two potential strategies: (1) the inhibition of oncogenic miRNAs and (2) the activation of suppressive miRNAs. Both strategies can be effective, as shown in preclinical studies. Direct inhibition involves antisense oligonucleotides used to sequester a given miRNAs. Modified antisense oligonucleotides used to inhibit miRNA in vivo are often referred to in the literature as antagomirs [26]. For example, a study conducted by Lanford et al. [27] published in Science in 2010 showed that the use of anti-miR-122 in chimpanzees chronically infected with hepatitis C virus (HCV) contributes to an improvement in liver disease. Currently, anti-miR-122 is in phase II clinical trials of HCV therapy in humans, and this miRNA-based therapy may possibly be included in clinical treatments in the coming years. The use of miRNA antagonists seems to be a promising form of therapy, as evidenced by the successful treatment of patients with chronic HCV infection [28]. Additionally, miRNAs can also be inhibited indirectly using a variety of chemical compounds [29]. Moreover, there are also studies on genetic knockout of miRNAs in cancer cells. These studies can provide valuable information about the role of miRNAs in cancer and contribute to the development of novel anti-cancer strategies. For example, Shi et al. [30] showed in mouse model that knockout of oncogenic *miR-21* causes an attenuated proliferation of colitis-associated CRC. In turn, Jiang and Hermeking [31] and Rokavec et al. [32] performed studies on suppressive miR-34 in mouse model. In these studies, the authors showed that genetic knockout of *miR-34a* and *miR-34b/c* can contribute to CRC progression.

4. MiR-21

One of the most intensively studied miRNA molecules is miR-21, which is often overexpressed in CRC [23,33]. MiR-21 lowers the expression of several different suppressor genes that influence various biological functions, such as proliferation, adhesion, angiogenesis, migration, metabolism, and apoptosis [34]. Therefore, aberrant miR-21 has potentially oncogenic properties. It is worth noting that some colorectal polyps transform into malignant tumors as a result of successive, consecutive genetic events. Many studies have shown that miR-21 is associated with the progression of polyps into malignant tumor, and that expression of this miRNA may be increased in CRC [35,36]. One study evaluated the expression of miR-21 in different stages of CRC in 39 surgically removed tumors and 34 polyps after endoscopic resection. Using in situ hybridization (ISH) of nucleic acids, expression of miR-21 was shown to be increased in non-malignant polyps in comparison with controls and was highest in advanced CRC tumors and also in adjacent stromal fibroblasts [36]. In another study, Bastaminejad et al. [37], using the qRT-PCR method, investigated the expression level of miR-21 in serum and stool samples from 40 patients with CRC and 40 healthy controls. The expression level of this miRNA was significantly up-regulated in serum (12.1-fold) and stool (10.0-fold) in CRC patients, compared to the control group. The sensitivity and specificity of serum miR-21 expression level were found to be 86.05% and 72.97%, respectively, and the sensitivity and specificity of stool miR-21 expression were 86.05% and 81.08%, respectively. The expression level of miR-21 was able to significantly distinguish CRC stages III–IV from stages I–II (according to the American Joint Committee on Cancer (AJCC) TNM staging system) in stool samples with a sensitivity and a specificity of 88.1% and 81.6%, respectively, and in serum samples with a sensitivity and a specificity of 88.10% and 73.68%, respectively. Significantly increased miR-21 expression was also demonstrated in stool samples of 88 CRC patients compared to control group of 101 healthy volunteers [38]. Similar results were obtained in 29 patients with CRC and eight healthy patients [39]. Therefore, the expression of this miRNA in tumor tissues as well as in serum and stool may be a potential and minimally invasive diagnostic biomarker of CRC.

MiR-21 overexpression is closely related to proliferation and lymph node metastases in CRC, which are important prognostic factors in this type of cancer. Analysis of the expression of miR-21 derived from CRC tissues may also be helpful in prognosis. Fukushima et al. [40] assessed the prognostic value of miR-21 in a group of 306 CRC patients. The authors found that high miR-21 expression was correlated with low overall survival (OS), as well as low disease-free survival (DFS) in CRC patients in Dukes stages B, C, and D [40]. In another study, the prognostic value of miR-21 was also considered in patients classified in the TNM staging system. After tumor tissues from 301 patients with varying degrees of CRC were investigated, a statistically significant correlation between miR-21 expression and prognosis was observed [23]. Moreover, high expression of this miRNA was associated with low OS. Oue et al. [23] demonstrated that miR-21 expression in tumor tissues is significantly increased in patients with tumors infiltrating adjacent organs (T4), compared to patients with tumors limited to the colon (T1–T3). The study also presented a similar relationship in patients with regional lymph node metastases present (N1) compared to patients without cancer in the lymph nodes (N0). Furthermore, high miR-21 expression in CRC patients was correlated with insensitivity to 5-fluorouracil (5-FU) treatment. Oue et al. [23] analyzed the expression of miR-21 in German (stage II, *n* = 145) and

Japanese (stage I-IV, n = 156) cohorts of patients using the qRT-PCR method. MiR-21 overexpression was associated with poor prognosis in both Japanese (stage II/III) and German patients (stage II). These correlations were not dependent on other clinical data in a multivariate model. In addition, the use of adjuvant chemotherapy did not benefit patients with high miR-21 expression in both cohorts. Similar correlations were also obtained in other studies [16,19,41–43]. Moreover, Schetter et al. [16], using the ISH method, observed a high expression level of miR-21 in colonic epithelial cells in tumor tissues compared to adjacent non-tumor tissues. In turn, Nielsen et al. [42] detected miR-21 expression mainly in stromal fibroblasts adjacent to tumors. On the other hand, Xia et al. [44] showed in a meta-analysis of miR-21 expression profiles of 1174 CRC tissue samples that overexpression of this miRNA is associated with low OS, but there was no correlation with the carcinoembryonic antigen (CEA) level. Additionally, Chen et al. [45] and Peng et al. [46] in their meta-analysis studies showed that miR-21 expression in tumor tissues is associated with poor DFS and OS in CRC patients. However, Chen et al. [45] revealed the significant correlation between miR-21 expression and poor OS in studies based on the qRT-PCR analysis but not the fluorescence in situ hybridization (FISH) method. Nonetheless, a few previous studies showed that a higher miR-21 expression level detected with the use of ISH method is associated with poor recurrence-free survival of CRC patients in stage II [41,47]. Moreover, in these studies, miR-21 expression was also detected mainly in stromal fibroblasts adjacent to tumors and only in a few samples in cancer cells. All studies discussed above indicated that the analysis of miR-21 expression in tumor tissues may be a potential, but certainly not an ideal biomarker of CRC prognosis.

The prognostic value of miR-21 expression in blood and stool of CRC patients is also the subject of intensive research. Kanaan et al. [48] observed significantly increased plasma levels of miR-21 in CRC patients. In turn, Toiyama et al. [33] evaluated the expression of miR-21 in a cell culture medium from two different CRC lines, in serum collected from 12 CRC patients and 12 healthy volunteers separately, and confirmed that this miRNA belongs to the secretory group of the miRNAs. The same research group subsequently measured miR-21 expression in 246 blood samples from CRC patients, 53 healthy volunteers, and 43 colorectal polyps. They also compared the expression of miR-21 in serum and tumor tissues in 166 paired samples. Statistically significant increase in serum miR-21 expression was observed in patients with benign polyps and in those with CRC. Moreover, a decrease in the expression of this miRNA in serum was observed in patients after surgical removal of the tumor [33]. Furthermore, many studies showed that expression of miR-21 both in tissue and serum samples of CRC patients is associated with lower OS and DFS [33,49]. However, Chen et al. [45] showed no significant association of serum miR-21 expression with poor OS of CRC patients in their meta-analysis studies (421 patients). In another study, it was shown that the increase in miR-21 expression in stool of CRC patients may be correlated with TNM classification [50]. Similarly, Bastaminejad et al. [37] revealed that increased expression of miR-21 was associated with AJCC TNM staging, more related with III and IV compared to I and II stages, in both serum and stool samples of 40 CRC patients. The above-mentioned studies show that miR-21 expression is associated with tumor size, metastases and low patient survival. Therefore, the expression of this miRNA in tumor tissues as well as in serum and stool may be a potential and minimally invasive prognostic biomarker of CRC.

5. MiR-29 Family

Another promising potential CRC biomarker is the miR-29 family, which includes three related miRNAs: miR-29a, miR-29b, and miR-29c. This family is associated with various molecular functions, such as the regulation of cell proliferation, cell senescence and tumor metastasis. Therefore, these molecules can participate in carcinogenesis and progression. It has been shown that the expression of miRNAs belonging to this family is altered in many different cancers [51]. Wang et al. [52] performed a study in 114 patients with CRC—56 subjects without metastasis and 58 with liver metastasis, which are commonly found in this type of cancer. The authors demonstrated that the expression of miR-29a in serum of patients with liver metastasis was significantly increased. In addition, a significantly increased expression of miR-29a was also observed in patients in stage T4, compared to T2. The authors concluded that miR-29a enables the early detection of liver metastasis in CRC [52]. In addition to early metastasis detection, miR-29a was also tested for potential use as a diagnostic biomarker for CRC. Huang et al. [53], using qRT-PCR, studied the expression of 12 miRNAs (miR-17-3p, miR-25, miR-29a, miR-92a, miR-134, miR-146a, miR-181d, miR-191, miR-221, miR-222, miR-223, and miR-320a) in the plasma of patients with advanced stages of colorectal neoplasia (CRC and advanced adenomas), compared to a group of healthy volunteers. It was shown that the expression of two miRNAs, miR-29a and miR-92a, can have a significant diagnostic value in CRC. For miR-29a, the area under the curve (AUC) was 0.844, while for miR-92a, the AUC was 0.838 in differentiating patients with CRC from healthy volunteers. The utility of both miRNAs was also demonstrated in differentiation of advanced adenomas and normal tissues. In this case, the AUC value for miR-29a was 0.769, while for miR-92a, it was 0.749. Overall, a receiver operating characteristic (ROC) analysis for both miRNAs showed an AUC of 0.883 (sensitivity = 83.0% and specificity = 84.7%) in differentiation of CRC and an AUC of 0.773 (sensitivity = 73.0% and specificity = 79.7%) in differentiation of advanced adenomas. Similarly, Al-Sheikh et al. [54] revealed up-regulation of miR-29 and miR-92, and down-regulation of miR-145 and miR-195 in 20 CRC patients (both in tissue and plasma) compared to controls. The above-mentioned results suggest that the determination of miR-29a and miR-92a expression in plasma may be a novel and potential biomarker in the diagnosis of CRC.

MiR-29b is also the member of miR-29 family. This miRNA inhibits proliferation and induces apoptosis in CRC cells. MiR-29b mediates the inhibition of the epithelial–mesenchymal transition. In many tumors that originate from epithelium, including CRC, the epithelial–mesenchymal transition is considered to be a key processes in initiation of metastasis. Li et al. [55] showed decreased expression of miR-29b in tissue and plasma samples obtained from CRC patients compared to controls. In addition, Basati et al. [56] showed down-regulation of miR-29b and miR-194 in serum samples obtained from 55 CRC patients compared to controls. Moreover, these authors showed a negative correlation between these miRNA expression and TNM stages.

MiR-29 is also a potential prognostic biomarker of CRC. Tang et al. [57] analyzed the expression of miR-29a and *KLF4* mRNA in 85 tumor tissues of CRC patients and CRC cell lines using the qRT-PCR method. It was shown that reduced expression of *KLF4* mRNA is associated with the presence of metastasis. Moreover, increased miR-29a expression indicated presence of metastasis and worsened prognosis of patients with CRC. It is known that *KLF4* is a target of miR-29a and that it acts to inhibit metastasis by reducing *MMP-2* and increasing E-cadherin expression [57]. The study mentioned above showed that high expression of miR-29a is associated with metastasis and poor prognosis. The predictive value of miR-29a was also shown in stage II CRC. Weissmann-Brenner et al. [58] performed studies on 110 CRC patients (51 with stage I cancer and 59 patients with stage II cancer according to the AJCC TNM staging system) using miRNA microarrays and verified the microarray results using the qRT-PCR technique afterwards. RNA was extracted from formalin-fixed paraffin-embedded tumor tissues. The authors defined a poor prognosis as a recurrence of the disease within 36 months of surgery. Patients with a good prognosis ($n = 87$; 79%) and a poor prognosis ($n = 23$; 21%) were compared. There were no statistically significant differentially expressed miRNAs between good- and poor-prognosis stage I CRC patients, among the set of 903 analyzed miRNAs. On the other hand, the expression of miR-29a was significantly higher in stage II CRC patients with good prognosis compared to the poor-prognosis group. High expression of this miRNA was associated with longer DFS in both univariate and multivariate analysis. In case of miR-29a expression, a positive predictive value (PPV) for non-recurrence of the disease was 94% (two cases out of 31). Differences in the miR-29a expression were confirmed using qRT-PCR, and this method showed the effect of overexpression of this miRNA on prolonging of the DFS. This study demonstrated a significant association of the miR-29a expression level with the risk of CRC recurrence in stage II patients.

For the patients in stage I, no such correlation was demonstrated. Moreover, significantly decreased expressions of miR-29a and miR-29c were reported in tumor tissues in 43 early-recurrence patients compared to the control group [59]. Although increased expressions of both miR-29a and miR-29c were associated with better outcome after 12 months of therapy, the authors suggested that only miR-29a may be used as a predictor marker for an early recurrence of the disease. The low PPV of miR-29c in this case may result from short follow-up time of the patients and the small study group [59]. In another study on 245 patients by Inoue et al. [60], the expression of miR-29b level in tumor tissues was used to divide the patients into two groups. The reference value was the median expression of this miRNA. The authors concluded that higher expression of miR-29b is associated with higher five-year DFS and OS values. Analysis of the severity of the disease showed that the miR-29b expression reflects the five-year DFS and has a significant prognostic value, but only in patients with stage III CRC. In addition, the low level of miR-29b expression was also a predictor of lymph node metastasis. These findings confirmed the prognostic value of this miRNA in stage III CRC patients. In another study, Yuan et al. [61] studied the expression of miR-29b in tumor tissue and adjacent normal mucosa samples of 41 patients using the qRT-PCR method. The authors found a significant decrease in the expression of this miRNA in CRC and concluded that the level of miR-29b may be associated with the size of a tumor, clinical status and lymph node metastasis. Basati et al. [56] studied 55 serum samples of CRC patients and revealed that lower expression of miR-29b is correlated with poor prognosis. In turn, Ulivi et al. [62] showed that higher miR-29b level in plasma samples is associated with longer progression-free survival (PFS) and OS of metastatic CRC patients treated with bevacizumab-based chemotherapy. As indicated by numerous studies, analysis of changes of miR-29 expression may be helpful in assessing an early recurrence and evaluating DFS in CRC patients.

6. MiR-34 Family

There are also studies on miR-34, a group of miRNAs that includes miR-34a, miR-34b and miR-34c. These molecules show suppressor properties and are regulated by p53 protein and DNA hypermethylation. The miR-34 group influences various processes that occur in tumor cells such as differentiation, drug resistance and metastasis [63]. For example, miR-34a overexpression inhibits NOTCH signaling and suppresses symmetric cell division, which prevents expansion of the CRC stem cell niche [64]. Wu et al. [65] studied the possibility of using miR-34 as a potential diagnostic biomarker for CRC. The authors showed abnormal methylation of miR-34a, miR-34b, and miR-34c genes in tissue and stool samples of 82 CRC patients. In turn, Aherne et al. [66] reported higher expression of miR-34a in plasma samples obtained from CRC patients compared to controls.

MiR-34 expression has also been found to be useful for CRC prognosis. The usefulness of miR-34 as a biomarker of a recurrence of the disease in two independent groups of 268 CRC patients was evaluated. It was shown that the miR-34a expression in CRC tissues is directly proportional to DFS, and therefore, this molecule may be a good prognostic factor in assessing the risk of the recurrence of CRC. In addition, the expression of miR-34a was significantly higher in patients with high expression of p53 compared to those with low expression of this protein. The authors suggested that miR-34 inhibits the growth and invasiveness of CRC in p53-dependent manner, which allows this miRNA to be used as a potential biomarker for a recurrence in patients with stage II and stage III CRC [67]. The PAR2 receptor also plays an important role in the progression of CRC. Previous reports have indicated that miR-34a expression is inhibited by PAR2, which results in increased synthesis of cyclin D1 and transforming growth factor β (*TGF-β*) in CRC cells [68]. Furthermore, silencing of miR-34a expression through promoter methylation in CRC tissues is associated with the occurrence of metastases [69]. In other studies, a lower expression of this miRNA was observed in some patients with CRC in tissue [70] and serum/plasma samples [71], which suggests that miR-34a may play a role in the progression of this cancer. Li et al. [72] showed that lower expression of miR-34a in CRC tissues is correlated with the lymph node metastasis and TNM stage. Zhang et al. [73] performed studies on 103 CRC tissue samples and showed that miR-34a expression was down-regulated compared to adjacent normal mucosa samples with the use of ISH technique. Moreover, these authors indicated that the lower expression of this molecule is correlated with more distant metastasis and shorter OS time. Studies on the effect of the miR-34 group on the prognosis of CRC have also been performed. The purpose of these studies was to determine the relationship between the miR-34b and miR-34c expression in CRC tissues and the development of this disease. Samples were obtained from 159 American and 113 Chinese patients with stages I–IV CRC. Using the qRT-PCR method, Hiyoshi et al. [74] showed an increased expression of miR-34b and miR-34c in advanced CRC, which was associated with poor prognosis in both study groups. Similarly to miR-34a, the expressions of miR-34b and miR-34c are also regulated by p53 protein at the transcriptional level. The results of all of these studies show that miR-34 may be an interesting prognostic tool and may be used to assess the risk of a recurrence in patients with CRC.

Furthermore, recent studies have determined the possibility of using miR-34 as a potential therapeutic agent. A low level of this miRNA was observed in DLD-1 CRC cell line that was resistant to 5-FU. Restoration of the miR-34a expression caused the sensitization of cells to 5-FU treatment and resulted in an inhibition of cell growth [75]. Moreover, Sun et al. [76] showed that miR-34a expression was down-regulated in blood samples of CRC patients after oxaliplatin-based chemotherapy. The negative correlation between miR-34a and *TGF-β/SMAD4* expression is also noted. The authors suggested that miR-34a and *TGF-β/SMAD4* expression changes can lead to activation of macroautophagy and oxaliplatin resistance in CRC cells.

7. MiR-124

MiR-124 is known to inhibit cell proliferation, metastasis and invasion in CRC. This miRNA not only down-regulates rho-associated protein kinase 1 (ROCK1) [77], which functions as an oncogene, but also inhibits the activity of cyclin-dependent kinase 4 (CDK4), which is responsible for cell cycle progression at the G1/S checkpoint [78]. Studies have shown that, in CRC cells, the expression of miR-124 is silenced through DNA methylation [79,80]. Since *miR-124* gene is more likely to be methylated in CRCs compared to other tumors, DNA methylation status of this miRNA may be used as a specific marker for CRC. Harada et al. [79] performed the detection of DNA methylation in bowel lavage fluid for CRC screening. These authors analyzed DNA methylation status in a total of 508 patients—56 with CRC, 53 with advanced adenoma, 209 with minor polyp, and 190 healthy individuals. Three genes showed the greatest sensitivity for CRC detection (*miR-124-3, LOC386758*, and *SFRP1*) after training set analysis ($n = 345$). A scoring system based on the methylation of these three genes achieved 82% sensitivity and 79% specificity, and the AUC was 0.834. These results were subsequently validated in an independent test set ($n = 153$; AUC = 0.808). In another study, Xi et al. [77] investigated the expression of ROCK1 and miR-124 in CRC patients using 68 paired tissue specimens (38 cases of non-metastatic CRC and 30 cases of metastatic CRC). The use of qRT-PCR revealed that expression of miR-124 was higher in normal compared with CRC tissues, and in non-metastatic compared to metastatic CRC tissues. In contrast, ROCK1 was significantly overexpressed in CRC compared with control tissues and between metastatic tissues and non-metastatic CRC tissues. The above-mentioned results suggest that miR-124, as a tumor suppressor, may play a role in tumor growth and metastasis. Recently, the miR-124a level has also been studied in 40 patients with ulcerative colitis (without colorectal cancer), four patients with CRC or inflammatory dysplasia, eight patients with CRC (without inflammatory background), and 12 healthy volunteers. It was found that miR-124a-1, miR-124a-2, and miR-124a-3 genes are methylated in tumor tissues. The authors suggested that the methylation of miR-124a-3 occurring during oncogenesis in patients with ulcerative colitis can be used to evaluate the individual's risk of developing cancer [81].

MiR-124 may also be a prognostic marker for CRC. Wang et al. [82] studied 96 tumor tissue samples and showed that down-regulated expression of miR-124 is correlated with poor OS and DFS in CRC patients. Moreover, Jinushi et al. [83] showed that higher expression of miR-124-5p (both in plasma and tissues) was associated with better prognosis of CRC patients. In turn, Slattery et al. [84] performed studies in 1893 patients with CRC. The authors found that miR-124-3p belongs to infrequently highly expressed miRNAs in tumor tissues and showed that up-regulation of miR-124-3p can worsen prognosis of these CRC patients.

8. MiR-130b

The direct target of miR-130b is the PPARγ receptor, of which inhibition results in the regulation of the expression of cadherin E, vascular endothelial growth factor (VEGF) and phosphatase and tensin homolog (PTEN). Since tissue miR-130b overexpression was observed in stage III and IV CRCs, it is suggested that miR-130b-PPARγ signaling may play a significant role in increasing tumor malignancy. Furthermore, evaluating the expression of proteins in this pathway, including miR-130b, may be a promising prognostic biomarker in CRC [85]. Finally, a study performed in 53 cancer and non-cancerous samples [86] suggested miR-130a as a good biomarker of CRC because of correlation with TNM staging and lymph node metastasis.

9. MiR-155

The sequence of this miRNA is located in a non-coding region of MIR155 host gene (*BIC*). Altered expression of miR-155 has been observed in many different tumors, and is associated with severity of disease, progression, and response to treatment. Interestingly, Sempere et al. [87] using the ISH method observed that miR-155 expression is detected mainly in tumor-infiltrating immune cells. Lv et al. [88] examined the possibility of using serum miR-155 expression as a diagnostic tool. Using qRT-PCR, they measured the expression levels of miR-155 in 146 CRC patients and 60 healthy controls. Serum miR-155 was up-regulated in CRC patients compared with the matched healthy controls. Moreover, ROC curve analysis indicated that miR-155 is a suitable marker for discriminating CRC patients from healthy controls, with an AUC of 0.776. Therefore, this molecule can be used as a potential tumor biomarker in the diagnosis of CRC.

MiR-155 can also play an important role in CRC prognosis. Shibuya et al. [43] showed that patients with an increased expression of miR-155 in tumor tissues were characterized by shorter OS and DFS, compared to those with lower expression of this miRNA. In another study, multivariate analysis also demonstrated a relationship between the level of miR-155 expression and poor prognosis in CRC, depending on the severity of the disease. The control group consisted of 60 healthy volunteers, while the experimental group consisted of 146 patients. The authors did not observe changes in the miR-155 serum level in patients with stage I CRC, but reported a statistically significant overexpression of this miRNA in patients with stages II–IV of the disease [88]. In another study, the serum CEA level and miR-155 expression were measured in tissues that were obtained before and after surgery of 84 CRC patients. It is well known that CEA is used for determining the prognosis, evaluating the effectiveness of therapy and monitoring the recurrence of CRC. In this study, the miR-155 expression was observed to be significantly increased in patients with CRC. This is associated with metastases and a recurrence of the disease [89]. Therefore, the evaluation of miR-155 expression in association with serum CEA may provide additional diagnostic information and enable a more accurate assessment of the risk of the metastasis of CRC. A statistically significant increase in miR-155 expression in tumor tissues, compared to normal samples, was also observed by Zhang et al. [90] in a study of 76 patients with CRC. In addition, the authors observed a correlation between miR-155 expression and lymph node and distant metastases and disease progression. Moreover, they observed that miR-155 overexpression inhibited E-cadherin expression and positively regulated zinc finger E-box binding homeobox 1 protein (ZEB-1), which resulted in an increased cell migration and metastases. Similarly, Qu et al. [91] revealed association of miR-155-5p expression in tumor tissues with location, grade of tumor, TNM stage and distant metastasis. Ulivi et al. [62] in multivariate analysis also showed that increased expression of circulating miR-155 is associated with shorter PFS and OS in metastatic CRC patients treated with bevacizumab-based chemotherapy. The above-mentioned observations indicate that miR-155 may play a role in the development and metastasis of CRC.

10. MiR-224

Recent reports have revealed that miR-224 may influence many processes that are associated with tumor cell growth and development, such as proliferation, growth, differentiation and apoptosis [92]. Some groups have investigated the expression of miR-224 in CRC patients. Zhu et al. [93] found significantly lower miR-224 levels in feces from CRC patients than these from normal volunteers in their retrospective analysis of miR-224 levels in fecal samples from 80 CRC patients and 51 normal controls. The authors suggested that the miR-224 expression level in feces can be used for screening and early diagnosis of CRC.

MiR-224 is also a potential prognostic biomarker in CRC. Zhang et al. [94] evaluated the clinical and pathological information of 40 patients with a recurrence and 68 patients without a recurrence within three years after a surgical intervention. Moreover, using the qRT-PCR and Western blot methods, the authors analyzed samples from all 108 patients with stages I and II CRC. They showed that miR-224 is involved in the regulation of *SMAD4* protein, which is involved in cell signaling. *SMAD4*, together with other proteins from this family, forms a DNA-binding complex that acts as a transcription factor. Furthermore, a significant increase in miR-224 expression in CRC tissues was observed in the study and this change was associated with a higher risk of a recurrence and a shorter DFS. In another study, Ling et al. [95] showed that miR-224 is an activator of metastasis and that the target of this miRNA is *SMAD4*. The authors concluded that an evaluation of miR-224 alone or an evaluation of miR-224 together with *SMAD4* may be an independent prognostic marker in patients with CRC. MiR-224 expression in tumor tissues and its association with the survival of patients were evaluated in 449 CRC patients. In this study, the patients were divided into two groups. These two groups were characterized by low and high levels of miR-224 expression, respectively. A shorter OS and survival time without metastases were observed in patients with miR-224 overexpression. Moreover, Wang et al. [96] showed an inverse correlation between *SMAD4* and miR-224 expression in tumor tissues of 40 CRC patients. These authors also revealed that miR-224 can regulate USP3 expression and its higher expression is associated with poor prognosis. In turn, Liao et al. [97] observed a statistically significant increase in the expression of this miRNA in tumor tissues of patients with an aggressive CRC phenotype and poor prognosis. In another study, miR-224 expression was evaluated in 79 patients with CRC and 18 healthy volunteers. The authors observed a significant inhibition of miR-224 expression in tumor tissues. Since the molecular target of miR-224 is *CDC42*, a lower expression of this miRNA reduces tumor cells migration. In general, the study indicated an important role of miR-224 in inhibiting migration of CRC cells. The authors concluded that miR-224 may be a promising biomarker in evaluating development of CRC [98]. Furthermore, Zhang et al. [99] showed in their meta-analysis that the increased level of miR-224-5p is correlated with poor OS of CRC patients.

11. MiR-378

MiR-378 is known to play a significant role in development of different types of cancer, including CRC. Current studies have shown that miR-378 is overexpressed in CRC cells and its targets include *FUS-1* and *SUFU* suppressor genes [100,101]. In addition, miR-378 is involved in tumor progression by promoting cell survival, migration and angiogenesis [102]. Significant differences in the level of blood and tumor miR-378 expression between oncological and healthy subjects were observed. Zanutto et al. [103] analyzed miRNA expression in serum samples from 65 CRC patients and 70 healthy volunteers, and found significantly increased miR-378 levels in CRC patients compared to the control group. At the same time, the authors observed a statistically significant decrease in the expression of this miRNA after the surgical removal of the tumor. Similar results were obtained in patients who had no recurrence of the disease within four to six months after surgery. The results suggest that serum miR-378 levels may be useful not only for differentiating CRC patients from healthy subjects, but that miR-378 is synthesized in the tumor tissue and its concentration is associated with tumor mass and possible recurrence. In turn, Wang et al. [104] qualified miR-378 as a tumor suppressor after analyzing miRNA expression in 47 CRC samples that were matched with normal tissue samples. In another study, Zhang et al. [105] observed an inhibition of miR-378 expression in 84 CRC samples compared to normal mucosal samples. Similarly, Zeng et al. [106] showed lower miR-378 expression in 27 CRC samples compared to paired adjacent normal samples. There have also been studies that showed an association of a reduced expression of miR-378 in cancer tissues with increased tumor size, metastasis and shorter OS in patients with CRC [105,107]. The above-mentioned results suggest that miR-378 may play an important role in carcinogenesis and may have clinical value as a potential biomarker of CRC.

12. Other miRNAs

Many recent studies attempted to identify other miRNAs in tumor tissues or plasma/serum samples as the potential diagnostic, prognostic or predictive biomarkers of CRC, e.g., miR-17, miR-19a, miR-20a [108], miR-22 [109], miR-24-3p [110], miR-26a, miR-26b [111], miR-9 [112–114], miR-106a [115], miR-122, miR-200 [116], miR-125a-3p [117], miR-126-3p [118], miR-139-3p [119], miR-139-5p [120], miR-148a, miR-625-3p [121], miR-181a, miR-181b [122], miR-181c [123], miR-181d [124], miR-193a-3p [125], miR-200c [126,127], miR-196b-5p [126], miR-223 [108,113], miR-375, miR-760 [112], miR-506, miR-4316 [128], miR-1290 [129]. In addition, Kiss et al. [130] using microarray and qRT-PCR techniques revealed that miR-92b-3p, miR-3156-5p, miR-10a-5p, miR-125a-5p may be used as predictive biomarkers of response to bevacizumab/FOLFOX therapy of CRC patients with metastasis. Similarly, Fiala et al. [118] showed that miR-126-3p expression is correlated with response to bevacizumab/FOLFOX treatment of CRC patients with metastasis. On the other hand, some studies showed only altered expression of various miRNAs in CRC tissues/serum samples compared to controls [131,132]. The significance of miRNAs as diagnostic and prognostic biomarkers in CRC is summarized in Table 2.

Table 2. The significance of miRNAs as diagnostic and prognostic biomarkers of colorectal cancer.

MiRNA	Biomarker Type	Regulation in CRC	Source of miRNA	Cohort Size	Correlation/ Differentiation	Detection Method	Authors
miR-17	diagnostic	up-regulation	serum	$n = 190$	control vs. CRC	qRT-PCR	Zekri et al. [108]
miR-17-5p	diagnostic	up-regulation	serum	$n = 39$	control vs. CRC	qRT-PCR	Fu et al. [114]
miR-19a	diagnostic	up-regulation	serum	$n = 190$	control vs. CRC	qRT-PCR	Zekri et al. [108]
miR-20a	diagnostic	up-regulation	serum	$n = 190$	control vs. CRC	qRT-PCR	Zekri et al. [108]
miR-21	diagnostic	up-regulation	plasma, tissues	$n = 116$	control vs. CRC	microfluidic array technology, qRT-PCR	Kanaan et al. [48]
miR-21	diagnostic	up-regulation	tissues	$n = 73$	control vs. CRC vs. polyps	ISH	Yamamichi et al. [36]
miR-21	diagnostic	up-regulation	serum, stool	$n = 80$	TNM stages	qRT-PCR	Bastaminejad et al. [37]
miR-21	diagnostic	up-regulation	stool, tissues	$n = 246$	control vs. CRC vs. polyps	qRT-PCR	Wu et al. [38]
miR-21	diagnostic	up-regulation	stool	$n = 55$	TNM stages	miRNA microarrays, qRT-PCR	Ahmed et al. [50]
miR-21	diagnostic	up-regulation	stool, tissues	$n = 37$	control vs. CRC vs. adenomas	qRT-PCR, miRNA microarrays	Link et al. [39]
miR-21	diagnostic, prognostic	up-regulation	serum, tissues	$n = 279$	TNM stages, tumor size, distant metastasis, poor survival	qRT-PCR	Toiyama et al. [33]
miR-21	prognostic	up-regulation	tissues	$n = 301$	poor survival	qRT-PCR	Oue et al. [23]
miR-21	prognostic	up-regulation	tissues	$n = 306$	liver metastasis, Dukes' stage, shorter OS, shorter DFS	qRT-PCR	Fukushima et al. [40]

Table 2. *Cont.*

MiRNA	Biomarker Type	Regulation in CRC	Source of miRNA	Cohort Size	Correlation/ Differentiation	Detection Method	Authors
miR-21	prognostic	up-regulation	tissues	$n = 197$	poor survival, poor therapeutic outcome	miRNA microarrays, qRT-PCR, ISH	Schetter et al. [16]
miR-21	prognostic	up-regulation	tissues	$n = 46$	shorter DFS	qRT-PCR	Kulda et al. [19]
miR-21	prognostic	up-regulation	tissues	$n = 520$	inferior recurrence-free cancer-specific survival	ISH	Kjaer-Frifeldt et al. [41]
miR-21	prognostic	up-regulation	tissues	$n = 234$	shorter DFS	ISH	Nielsen et al. [42]
miR-21	prognostic	up-regulation	tissues	$n = 156$	liver metastasis, shorter OS, shorter DFS	qRT-PCR	Shibuya et al. [43]
miR-21	prognostic	up-regulation	tissues	$n = 277$	shorter RFS	ISH	Kang et al. [47]
miR-22	prognostic	down-regulation	tissues	$n = 193$	shorter OS	qRT-PCR	Li et al. [109]
miR-24-3p	prognostic	up-regulation	tissues	$n = 268$	shorter OS, shorter DFS	qRT-PCR	Kerimis et al. [110]
miR-26a/b	prognostic	down-regulation	tissues	$n = 58$	shorter OS	qRT-PCR	Li et al. [111]
miR-29	diagnostic	up-regulation	tissues, plasma	$n = 40$	control vs. CRC	qRT-PCR	Al-Sheikh et al. [54]
miR-29a	diagnostic	up-regulation	plasma	$n = 209$	control vs. CRC	qRT-PCR	Huang et al. [53]
miR-29a	diagnostic	down-regulation	stool	$n = 131$	control vs. CRC	qRT-PCR	Zhu et al. [93]
miR-29a	diagnostic	up-regulation	serum	$n = 74$	liver metastasis	qRT-PCR	Wang and Gu [52]
miR-29a	prognostic	down-regulation	tissues	$n = 110$	poor prognosis	miRNA microarrays, qRT-PCR	Weissmann-Brenner et al. [58]
miR-29b	diagnostic	down-regulation	tissues, plasma	$n = 600$	control vs. CRC	qRT-PCR	Li et al. [55]
miR-29b	diagnostic, prognostic	down-regulation	serum	$n = 110$	TNM stages	qRT-PCR	Basati et al. [56]
miR-29b	prognostic	down-regulation	tissues	$n = 245$	shorter OS, shorter DFS	qRT-PCR	Inoue et al. [60]
miR-29b	prognostic	down-regulation	plasma	$n = 52$	shorter OS, shorter PFS	qRT-PCR	Ulivi et al. [62]
miR-34a	diagnostic	up-regulation	plasma	$n = 185$	polyps vs. advanced cancer	qRT-PCR	Aherne et al. [66]
miR-34a	diagnostic, prognostic	methylation	tissues, stool	$n = 142$	lymph metastasis, control vs. CRC	methylation-specific PCR	Wu et al. [65]
miR-34a	prognostic	down-regulation	tissues	$n = 268$	shorter DFS, recurrence	qRT-PCR	Gao et al. [67]
miR-34a	prognostic	methylation	tissues	$n = 94$	liver metastasis	methylation-specific PCR	Siemens et al. [69]
miR-34a	prognostic	down-regulation	tissues	$n = 176$	poor prognosis	qRT-PCR	Li et al. [72]
miR-34a	prognostic	down-regulation	tissues	$n = 103$	distant metastasis	qRT-PCR, ISH	Zhang et al. [73]
miR-34a/b/c	prognostic	up-regulation	tissues	$n = 272$	poor prognosis	qRT-PCR	Hiyoshi et al. [74]
miR-34b/c	diagnostic	methylation	tissues, stool	$n = 142$	control vs. CRC	methylation-specific PCR	Wu et al. [65]
miR-92a	diagnostic	up-regulation	tissues, stool, plasma	$n = 907$	control vs. CRC	qRT-PCR	Chang et al. [113]
miR-92a	diagnostic, prognostic	up-regulation	serum	$n = 91$	control vs. CRC, TNM stages, poor prognosis	qRT-PCR	Elshafei et al. [112]
miR-92a-3p	diagnostic	up-regulation	serum	$n = 39$	control vs. CRC	qRT-PCR	Fu et al. [114]
miR-106a	diagnostic	up-regulation	tissues, plasma	$n = 84$	control vs. CRC	qRT-PCR	He et al. [115]
miR-122	prognostic	up-regulation	plasma	$n = 543$	shorter RFS, shorter OS	qRT-PCR	Maierthaler et al. [116]

Table 2. *Cont.*

MiRNA	Biomarker Type	Regulation in CRC	Source of miRNA	Cohort Size	Correlation/ Differentiation	Detection Method	Authors
miR-124	prognostic	down-regulation	tissues	$n = 96$	shorter OS, shorter DFS	qRT-PCR	Wang et al. [82]
miR-124-3p	prognostic	up-regulation	tissues	$n = 1893$	increased likelihood of dying	miRNA microarrays	Slattery et al. [84]
miR-124-5p	prognostic	down-regulation	tissues	$n = 71$	shorter OS	qRT-PCR	Jinushi et al. [83]
miR-125a-3p	diagnostic	down-regulation	tissues	$n = 35$	control vs. CRC	qRT-PCR	Liang et al. [117]
miR-126-3p	prognostic	down-regulation	tissues	$n = 63$	shorter OS, shorter PFS	qRT-PCR	Fiala et al. [118]
miR-126-5p	prognostic	down-regulation	tissues	$n = 63$	shorter OS, shorter PFS	qRT-PCR	Fiala et al. [118]
miR-130b	prognostic	up-regulation	tissues	$n = 80$	poor prognosis	miRNA microarrays	Colangelo et al. [85]
miR-139-3p	diagnostic	down-regulation	tissues, serum	$n = 249$	control vs. CRC	qRT-PCR	Ng et al. [119]
miR-139-5p	prognostic	up-regulation	tissues, serum	$n = 433$	shorter RFS	miRNA microarrays, qRT-PCR	Miyoshi et al. [120]
miR-148a	prognostic	down-regulation	tissues	$n = 54$	shorter PFS	qRT-PCR	Baltruskeviciene et al. [121]
miR-150	diagnostic	down-regulation	plasma	$n = 185$	adenomas vs. advanced cancer	qRT-PCR	Aherne et al. [66]
miR-155	prognostic	up-regulation	serum	$n = 206$	shorter OS, shorter PFS	qRT-PCR	Lv et al. [88]
miR-155	prognostic	up-regulation	tissues	$n = 156$	shorter OS, shorter DFS	qRT-PCR	Shibuya et al. [43]
miR-155	prognostic	up-regulation	tissues	$n = 84$	poor prognosis	qRT-PCR	Hongliang et al. [89]
miR-155	prognostic	up-regulation	tissues	$n = 76$	lymph node and distant metastases	qRT-PCR	Zhang et al. [90]
miR-155-5p	prognostic	up-regulation	tissues	$n = 372$	metastasis	qRT-PCR	Qu et al. [91]
miR-155-5p	prognostic	down-regulation	plasma	$n = 52$	shorter OS, shorter PFS	qRT-PCR	Ulivi et al. [62]
miR-181c	prognostic	up-regulation	tissues	$n = 147$	shorter RFS	miRNA microarrays, qRT-PCR	Yamazaki et al. [123]
miR-181d	prognostic	up-regulation	tissues	$n = 40$	metastasis	qRT-PCR	Guo et al. [124]
miR-193a-3p	prognostic	down-regulation	tissues	$n = 96$	shorter OS	miRNA microarrays, qRT-PCR	Lin et al. [125]
miR-196-5p	prognostic	down-regulation	tissues	$n = 48$	shorter OS	miRNA microarrays, qRT-PCR	Li et al. [126]
miR-200	prognostic	down-regulation	plasma	$n = 543$	shorter RFS, shorter OS	qRT-PCR	Maierthaler et al. [116]
miR-200c	prognostic	down-regulation	tissues	$n = 48$	shorter OS	miRNA microarrays, qRT-PCR	Li et al. [126]
miR-200c	prognostic	up-regulation	serum	$n = 90$	shorter OS	qRT-PCR	Tayel et al. [127]
miR-223	diagnostic	down-regulation	stool	$n = 131$	control vs. CRC	qRT-PCR	Zhu et al. [93]
miR-223	diagnostic	up-regulation	serum	$n = 190$	control vs. CRC	qRT-PCR	Zekri et al. [108]
miR-223	diagnostic	up-regulation	tissues, stool, plasma	$n = 907$	control vs. CRC	qRT-PCR	Chang et al. [113]
miR-224	diagnostic	down-regulation	stool	$n = 131$	control vs. CRC	qRT-PCR	Zhu et al. [93]
miR-224	prognostic	up-regulation	tissues	$n = 108$	shorter DFS	qRT-PCR	Zhang et al. [94]
miR-224	prognostic	up-regulation	tissues	$n = 621$	shorter OS	miRNA microarrays, qRT-PCR	Ling et al. [95]

Table 2. *Cont.*

MiRNA	Biomarker Type	Regulation in CRC	Source of miRNA	Cohort Size	Correlation/ Differentiation	Detection Method	Authors
miR-224	prognostic	up-regulation	tissues	n = 110	shorter OS	qRT-PCR	Liao et al. [97]
miR-375	diagnostic, prognostic	down-regulation	serum	n = 91	control vs. CRC, TNM stages, poor prognosis	qRT-PCR	Elshafei et al. [112]
miR-378	diagnostic	up-regulation	plasma	n = 65	control vs. CRC	qRT-PCR	Zanutto et al. [103]
miR-378	prognostic	down-regulation	tissues	n = 84	shorter OS	qRT-PCR	Zhang et al. [105]
miR-378a-3p	prognostic	down-regulation	tissues	n = 96	shorter OS	qRT-PCR	Li et al. [107]
miR-378a-5p	prognostic	down-regulation	tissues	n = 96	shorter OS	qRT-PCR	Li et al. [107]
miR-506	diagnostic	up-regulation	plasma	n = 126	control vs. CRC	qRT-PCR	Krawczyk et al. [128]
miR-625-3p	diagnostic	down-regulation	tissues	n = 54	control vs. CRC	qRT-PCR	Baltruskeviciene et al. [121]
miR-664-3p	prognostic	down-regulation	tissues	n = 63	shorter OS, shorter PFS	qRT-PCR	Fiala et al. [118]
miR-760	diagnostic, prognostic	down-regulation	serum	n = 91	control vs. CRC, TNM stages, poor prognosis	qRT-PCR	Elshafei et al. [112]
miR-1290	prognostic	up-regulation	tissues	n = 291	shorter OS, shorter DFS	miRNA microarrays, qRT-PCR	Ye et al. [129]
miR-4316	diagnostic	up-regulation	plasma	n = 126	control vs. CRC	qRT-PCR	Krawczyk et al. [128]

RFS—relapse-free survival.

13. MiRNA Panels

More researchers are focusing on finding miRNA sets that may be used as potential diagnostic or prognostic markers due to the fact that the expression of a single miRNA may not have sufficient specificity and sensitivity to distinguish CRC stages or CRC patients from healthy controls. Recently, interesting study was performed by Pan et al. [133], in which the expression level of 30 miRNAs in plasma samples was analyzed with the use of qRT-PCR. These authors showed that analysis of plasma expression level of five miRNAs, such as miR-15b, miR-17, miR-21, miR-26b, and miR-145, together with CEA, can improve diagnostic accuracy of CRC (AUC = 0.85 in the training cohort, AUC = 0.818 in the validation cohort). In turn, Guo et al. [134] selected a 5-miRNA set (miR-1246, miR-202-3p, miR-21-3p, miR-1229-3p, and miR-532-3p) from 528 miRNAs in serum and revealed that these miRNA panels have good sensitivity and specificity to distinguish CRC patients from colorectal adenoma patients (AUC = 0.951, sensitivity = 94.4%, specificity = 84.7%) and healthy controls (AUC = 0.960, sensitivity = 91.6%, specificity = 91.7%). Zhu et al. [135] showed that a 3-serum miRNA set (miR-19a-3p, miR-21-5p, and miR-425-5p) can be useful in diagnosis of CRC (AUC = 0.886 in the training phase, AUC = 0.768 in the validation phase, AUC = 0.783 in the combined training and validation phases, and AUC = 0.830 in the external validation phase). Wang et al. [136] also evaluated a 3-plasma miRNA set (miR-409-3p, miR-7, and miR-93) with diagnostic potential for CRC patients in Dukes stages A–D (AUC = 0.866 in the training phase, sensitivity = 91%, specificity = 88%; AUC = 0.897 in the validation phase, sensitivity = 82%, specificity = 89%) and also for non-metastatic CRC patients in Dukes stages A and B (AUC = 0.809 in the training phase, sensitivity = 85%, specificity = 88%; AUC = 0.892 in the validation phase, sensitivity = 82%, specificity = 89%), and for metastatic CRC patients in Dukes stages C and D (AUC = 0.917 in the training phase, sensitivity = 96%, specificity = 88%; AUC = 0.865 in the validation phase, sensitivity = 91%, specificity = 88%). Similar results were obtained in CRC patients when compared to age-matched healthy controls (for CRC patients in Dukes stages A–D: AUC = 0.894, sensitivity = 90%, specificity = 96%; in Dukes stages A and B: AUC = 0.850, sensitivity = 85%, specificity = 96%; and in Dukes stages C and D: AUC = 0.937, sensitivity = 95%, specificity = 96%). In turn, Kanaan et al. [137] observed that a 3-plasma miRNA set (miR-431, miR-15b, and miR-139-3p) can distinguish stage IV CRC from controls with high specificity and sensitivity (AUC = 0.896, sensitivity = 93%, specificity = 74%). They also found that a 5-miRNA set (miR-331, miR-15b, miR-21, miR-142-3p, and miR-339-3p) may be used to distinguish colorectal adenoma patients from CRC patients (AUC = 0.856, sensitivity = 91%, specificity = 69%), and CRC patients may be distinguished from healthy controls with the use of a 2-miRNA set (miR-431, and miR-139-3p) (AUC = 0.829, sensitivity = 91%, specificity = 57%). Wikberg et al. [138] used a 4-plasma miRNA set (miR-18a, miR-21, miR-22, and miR-25) to diagnose CRC (in all CRC stages, AUC = 0.93, sensitivity = 81% and 67% at 80% and 90% specificity, respectively; in CRC stages I–II, AUC = 0.92, sensitivity = 88% and 73% at 80% and 90% specificity, respectively; and in CRC stages, III–IV AUC = 0.85, sensitivity = 68% and 57% at 80% and 90% specificity, respectively). It is worth noticing that miR-21 is present in majority of miRNA sets discussed above. Additionally, Wikberg et al. [138] showed that miR-21 expression was increasing during three years before CRC diagnosis. Interestingly, Yang et al. [139] revealed that a 6-miRNA set (miR-7, miR-93, miR-195, miR-141, miR-494, and let-7b) in tumor tissues together with six clinicopathological factors (the Union for International Cancer Control (UICC) stage, location, type of tumor, vascular invasion, perineural invasion, and lymph node metastasis) can be used as potential prognostic markers of CRC recurrence within 12 months after surgery (AUC = 0.948, sensitivity = 89.4%, specificity = 88.9%). Moreover, based on literature data, Liu et al. [140] found 63 miRNAs, which can have diagnostic value for CRC. Then, using qRT-PCR technique, the authors analyzed the expression of five miRNAs: miR-21, miR-29a, miR-92a, miR-125b and miR-223 in serum samples obtained from 85 CRC patients and revealed that expression levels of these miRNAs were up-regulated in CRC samples compared to healthy controls. However, there were no differences in expression levels between TNM stages. The authors suggested that analysis of these five miRNAs together has higher diagnostic value than expression analysis of a single miRNA (AUC = 0.952, sensitivity = 84.7%, specificity = 98.7%). Interestingly, the authors also compared diagnostic value of CEA and CA19-9 with gas chromatography-mass spectrometry metabolomic data. The results discussed above show that miRNA panels in general have better sensitivity and similar specificity when compared to CRC screening tests currently used in clinical practice, such as FOBT (sensitivity = 64.3% (range: 35.6–86%), specificity = 90.1% (range: 89.3–90.8%)) and fecal immunochemical tests (FITs) (sensitivity = 81.8% (range: 47.8–96.8%), specificity = 96.9% (range: 96.4–97.4%)) [141,142]. The significance of miRNA panels as diagnostic and prognostic biomarkers in CRC is summarized in Table 3.

Table 3. The significance of miRNA panels as diagnostic and prognostic biomarkers of colorectal cancer.

MiRNA Panel	Biomarker Type	Regulation in CRC	Source of miRNA	Cohort Size	Correlation/ Differentiation	Detection Method	Authors
miR-15b miR-17 miR-21 miR-26b miR-145	diagnostic	up-regulation	plasma	$n = 280$	control vs. CRC	qRT-PCR	Pan et al. [133]
miR-1246 miR-202-3p miR-21-3p miR-1229-3p miR-532-3p	diagnostic	up-regulation (miR-1246, miR-1229-3p, miR-532-3p) down-regulation (miR-202-3p, miR-21-3p)	serum	$n = 575$	control vs. CRC vs. colorectal adenomas	qRT-PCR	Guo et al. [134]
miR-19a-3p miR-21-5p miR-425-5p	diagnostic	up-regulation	serum	$n = 334$	control vs. CRC	qRT-PCR	Zhu et al. [135]

Table 3. *Cont.*

MiRNA Panel	Biomarker Type	Regulation in CRC	Source of miRNA	Cohort Size	Correlation/ Differentiation	Detection Method	Authors
miR-409-3p miR-7 miR-93	diagnostic	up-regulation (miR-409-3p) down-regulation (miR-7, miR-93)	plasma	$n = 241$	control vs. CRC	miRNA microarrays, qRT-PCR	Wang et al. [136]
miR-431 miR-15b miR-139-3p	diagnostic	up-regulation	plasma	$n = 87$	control vs. stage IV CRC	microfluidic array technology, qRT-PCR	Kanaan et al. [137]
miR-431 miR-139-3p	diagnostic	up-regulation	plasma	$n = 87$	control vs. CRC	microfluidic array technology, qRT-PCR	Kanaan et al. [137]
miR-331 miR-15b miR-21 miR-142-3p miR-339-3p	diagnostic	up-regulation	plasma	$n = 87$	colorectal adenomas vs. CRC	microfluidic array technology, qRT-PCR	Kanaan et al. [137]
miR-18a miR-21 miR-22 miR-25	diagnostic	up-regulation (miR-18a, miR-21, miR-25) down-regulation (miR-22)	plasma	$n = 201$	control vs. CRC	semi-quantitative RT-PCR	Wikberg et al. [138]
miR-7 miR-93 miR-195 miR-141 miR-494 let-7b	prognostic	up-regulation (miR-7, miR-141, miR-494) down-regulation (miR-93, miR-195, let-7b))	tissues	$n = 104$	non-early relapsed CRC vs. early relapsed CRC	qRT-PCR	Yang et al. [139]
miR-21 miR-15b miR-29a miR-92a miR-125b miR-223	diagnostic	up-regulation	serum	$n = 163$	control vs. CRC	qRT-PCR	Liu et al. [140]

14. Conclusions

Our paper presents the latest reports on the diagnostic and prognostic values of selected miRNAs in CRC. Although the number of published papers that describe miRNAs as potential biomarkers for CRC has increased significantly over the past decade, clinical knowledge still remains fragmented. Only two miRNAs (miR-21 and miR-29) have been described in more detail in many previous studies. However, it is necessary to conduct further prospective validation studies before translating the knowledge into clinical use. Most of the current findings are from preliminary studies, which are often not free of methodological limitations such as small sample size, lack of detailed patient information, untested replicability, and statistical errors. It is also worth noticing that using the expression of a single miRNA as a diagnostic or prognostic biomarker of CRC is often limited due to insufficient specificity and sensitivity. Currently, many groups of researchers are investigating miRNA panels as CRC biomarkers, which appears to be a more promising strategy than the use of single miRNA tests. The development of panels containing many miRNA biomarkers seems to be essential and may enable more accurate diagnoses and prognoses of CRC in the future. However, the cost–benefit issue is also important in this case. In addition, for every potential miRNA biomarker, it is necessary to understand its molecular and biological functions as well as the mechanisms that are associated with its regulation. Understanding these processes is key to clinical application and identification of new therapeutic targets.

Author Contributions: Conceptualization, G.H. and M.K.-F.; Investigation, G.H. and M.K.-F.; Writing-Original Draft Preparation, G.H. and M.K.-F.; Writing-Review & Editing, T.F.; Visualization, G.H. and M.K.-F.; Supervision, T.F.; Funding Acquisition, T.F. All authors approved the final version of the article before submission.

Int. J. Mol. Sci. **2018**, *19*, 2944

Funding: This research was funded by Medical University of Silesia (grant number: KNW-1-002/K/7/0).

Conflicts of Interest: The authors declare no conflict of interest.

Abbreviations

3′-UTR	3′-untranslated region
5-FU	5-fluorouracil
AJCC	American Joint Committee on Cancer
AUC	area under the curve
BIC	MIR155 host gene
CDK6	cyclin-dependent kinase 6
CEA	carcinoembryonic antigen
CRC	colorectal cancer
DCBE	double-contrast barium enema
DFS	disease-free survival
FISH	fluorescence in situ hybrydization
FOBT	fecal occult blood test
HCV	hepatitis C virus
ISH	in situ hybridization
miRNAs	microRNAs
mRNA	messenger RNA
NGS	next-generation sequencing
OS	overall survival
PFS	progression-free survival
PPV	positive predictive value
PTEN	phosphatase and tensin homolog
RFS	relapse-free survival
qRT-PCR	quantitative real-time reverse transcriptase polymerase chain reaction
ROC	receiver operating characteristic
ROCK1	rho-associated protein kinase 1
TGF-β	transforming growth factor β
TNM	tumor-node-metastasis
UICC	Union for International Cancer Control
VEGF	vascular endothelial growth factor
ZEB-1	zinc finger E-box binding homeobox 1 protein

References

1. Global Cancer Observatory. Available online: http://gco.iarc.fr/today/home (accessed on 16 June 2018).
2. Maida, M.; Macaluso, F.S.; Ianiro, G.; Mangiola, F.; Sinagra, E.; Hold, G.; Maida, C.; Cammarota, G.; Gasbarrini, A.; Scarpulla, G. Screening of colorectal cancer: Present and future. *Expert Rev. Anticancer Ther.* **2017**, *17*, 1131–1146. [CrossRef] [PubMed]
3. Wolpin, B.M.; Mayer, R.J. Systemic treatment of colorectal cancer. *Gastroenterology* **2008**, *134*, 1296–1310. [CrossRef] [PubMed]
4. Tétreault, N.; De Guire, V. miRNAs: Their discovery, biogenesis and mechanism of action. *Clin. Biochem.* **2013**, *46*, 842–845. [CrossRef] [PubMed]
5. Lewis, B.P.; Burge, C.B.; Bartel, D.P. Conserved seed pairing, often flanked by adenosines, indicates that thousands of human genes are microRNA targets. *Cell* **2005**, *120*, 15–20. [CrossRef] [PubMed]
6. Friedman, R.C.; Farh, K.K.-H.; Burge, C.B.; Bartel, D.P. Most mammalian mRNAs are conserved targets of microRNAs. *Genome Res.* **2009**, *19*, 92–105. [CrossRef] [PubMed]
7. Volinia, S.; Calin, G.A.; Liu, C.-G.; Ambs, S.; Cimmino, A.; Petrocca, F.; Visone, R.; Iorio, M.; Roldo, C.; Ferracin, M.; et al. A microRNA expression signature of human solid tumors defines cancer gene targets. *Proc. Natl. Acad. Sci. USA* **2006**, *103*, 2257–2261. [CrossRef] [PubMed]

8. Lu, J.; Getz, G.; Miska, E.A.; Alvarez-Saavedra, E.; Lamb, J.; Peck, D.; Sweet-Cordero, A.; Ebert, B.L.; Mak, R.H.; Ferrando, A.A.; et al. MicroRNA expression profiles classify human cancers. *Nature* **2005**, *435*, 834–838. [CrossRef] [PubMed]
9. Tan, W.; Liu, B.; Qu, S.; Liang, G.; Luo, W.; Gong, C. MicroRNAs and cancer: Key paradigms in molecular therapy. *Oncol. Lett.* **2018**, *15*, 2735–2742. [CrossRef] [PubMed]
10. Chaffer, C.L.; Weinberg, R.A. A perspective on cancer cell metastasis. *Science* **2011**, *331*, 1559–1564. [CrossRef] [PubMed]
11. Ahlquist, D.A. Molecular detection of colorectal neoplasia. *Gastroenterology* **2010**, *138*, 2127–2139. [CrossRef] [PubMed]
12. Muniyappa, M.K.; Dowling, P.; Henry, M.; Meleady, P.; Doolan, P.; Gammell, P.; Clynes, M.; Barron, N. MiRNA-29a regulates the expression of numerous proteins and reduces the invasiveness and proliferation of human carcinoma cell lines. *Eur. J. Cancer* **2009**, *45*, 3104–3118. [CrossRef] [PubMed]
13. Eminaga, S.; Christodoulou, D.C.; Vigneault, F.; Church, G.M.; Seidman, J.G. Quantification of microRNA expression with next-generation sequencing. *Curr. Protoc. Mol. Biol.* **2013**. [CrossRef] [PubMed]
14. Slaby, O.; Svoboda, M.; Fabian, P.; Smerdova, T.; Knoflickova, D.; Bednarikova, M.; Nenutil, R.; Vyzula, R. Altered expression of miR-21, miR-31, miR-143 and miR-145 is related to clinicopathologic features of colorectal cancer. *Oncology* **2007**, *72*, 397–402. [CrossRef] [PubMed]
15. Pfeffer, U.; Romeo, F.; Noonan, D.M.; Albini, A. Prediction of breast cancer metastasis by genomic profiling: Where do we stand? *Clin. Exp. Metastasis* **2009**, *26*, 547–558. [CrossRef] [PubMed]
16. Schetter, A.J.; Leung, S.Y.; Sohn, J.J.; Zanetti, K.A.; Bowman, E.D.; Yanaihara, N.; Yuen, S.T.; Chan, T.L.; Kwong, D.L.W.; Au, G.K.H.; et al. MicroRNA expression profiles associated with prognosis and therapeutic outcome in colon adenocarcinoma. *JAMA* **2008**, *299*, 425–436. [CrossRef] [PubMed]
17. Wang, C.-J.; Zhou, Z.-G.; Wang, L.; Yang, L.; Zhou, B.; Gu, J.; Chen, H.-Y.; Sun, X.-F. Clinicopathological significance of microRNA-31, -143 and -145 expression in colorectal cancer. *Dis. Markers* **2009**, *26*, 27–34. [CrossRef] [PubMed]
18. Koga, Y.; Yasunaga, M.; Takahashi, A.; Kuroda, J.; Moriya, Y.; Akasu, T.; Fujita, S.; Yamamoto, S.; Baba, H.; Matsumura, Y. MicroRNA expression profiling of exfoliated colonocytes isolated from feces for colorectal cancer screening. *Cancer Prev. Res.* **2010**, *3*, 1435–1442. [CrossRef] [PubMed]
19. Kulda, V.; Pesta, M.; Topolcan, O.; Liska, V.; Treska, V.; Sutnar, A.; Rupert, K.; Ludvikova, M.; Babuska, V.; Holubec, L.; et al. Relevance of miR-21 and miR-143 expression in tissue samples of colorectal carcinoma and its liver metastases. *Cancer Genet. Cytogenet.* **2010**, *200*, 154–160. [CrossRef] [PubMed]
20. Sundaram, P.; Hultine, S.; Smith, L.M.; Dews, M.; Fox, J.L.; Biyashev, D.; Schelter, J.M.; Huang, Q.; Cleary, M.A.; Volpert, O.V.; et al. p53-responsive miR-194 inhibits thrombospondin-1 and promotes angiogenesis in colon cancers. *Cancer Res.* **2011**, *71*, 7490–7501. [CrossRef] [PubMed]
21. Wang, X.; Wang, J.; Ma, H.; Zhang, J.; Zhou, X. Downregulation of miR-195 correlates with lymph node metastasis and poor prognosis in colorectal cancer. *Med. Oncol.* **2012**, *29*, 919–927. [CrossRef] [PubMed]
22. Wu, X.; Somlo, G.; Yu, Y.; Palomares, M.R.; Li, A.X.; Zhou, W.; Chow, A.; Yen, Y.; Rossi, J.J.; Gao, H.; et al. De novo sequencing of circulating miRNAs identifies novel markers predicting clinical outcome of locally advanced breast cancer. *J. Transl. Med.* **2012**, *10*, 42. [CrossRef] [PubMed]
23. Oue, N.; Anami, K.; Schetter, A.J.; Moehler, M.; Okayama, H.; Khan, M.A.; Bowman, E.D.; Mueller, A.; Schad, A.; Shimomura, M.; et al. High miR-21 expression from FFPE tissues is associated with poor survival and response to adjuvant chemotherapy in colon cancer. *Int. J. Cancer* **2014**, *134*, 1926–1934. [CrossRef] [PubMed]
24. Zhao, G.; Zhang, J.; Shi, Y.; Qin, Q.; Liu, Y.; Wang, B.; Tian, K.; Deng, S.; Li, X.; Zhu, S.; et al. MiR-130b is a prognostic marker and inhibits cell proliferation and invasion in pancreatic cancer through targeting STAT3. *PLoS ONE* **2013**, *8*, e73803. [CrossRef]
25. Niyazi, M.; Zehentmayr, F.; Niemöller, O.M.; Eigenbrod, S.; Kretzschmar, H.; Schulze-Osthoff, K.; Tonn, J.-C.; Atkinson, M.; Mörtl, S.; Belka, C. MiRNA expression patterns predict survival in glioblastoma. *Radiat. Oncol.* **2011**, *6*, 153. [CrossRef] [PubMed]
26. Krützfeldt, J.; Rajewsky, N.; Braich, R.; Rajeev, K.G.; Tuschl, T.; Manoharan, M.; Stoffel, M. Silencing of microRNAs in vivo with "antagomirs". *Nature* **2005**, *438*, 685–689. [CrossRef]

27. Lanford, R.E.; Hildebrandt-Eriksen, E.S.; Petri, A.; Persson, R.; Lindow, M.; Munk, M.E.; Kauppinen, S.; Ørum, H. Therapeutic silencing of microRNA-122 in primates with chronic hepatitis C virus infection. *Science* **2010**, *327*, 198–201. [CrossRef] [PubMed]

28. Janssen, H.L.A.; Reesink, H.W.; Lawitz, E.J.; Zeuzem, S.; Rodriguez-Torres, M.; Patel, K.; van der Meer, A.J.; Patick, A.K.; Chen, A.; Zhou, Y.; et al. Treatment of HCV infection by targeting microRNA. *N. Engl. J. Med.* **2013**, *368*, 1685–1694. [CrossRef] [PubMed]

29. Young, D.D.; Connelly, C.M.; Grohmann, C.; Deiters, A. Small molecule modifiers of microRNA miR-122 function for the treatment of hepatitis C virus infection and hepatocellular carcinoma. *J. Am. Chem. Soc.* **2010**, *132*, 7976–7981. [CrossRef] [PubMed]

30. Shi, C.; Yang, Y.; Xia, Y.; Okugawa, Y.; Yang, J.; Liang, Y.; Chen, H.; Zhang, P.; Wang, F.; Han, H.; et al. Novel evidence for an oncogenic role of microRNA-21 in colitis-associated colorectal cancer. *Gut* **2016**, *65*, 1470–1481. [CrossRef] [PubMed]

31. Jiang, L.; Hermeking, H. miR-34a and miR-34b/c Suppress Intestinal Tumorigenesis. *Cancer Res.* **2017**, *77*, 2746–2758. [CrossRef] [PubMed]

32. Rokavec, M.; Öner, M.G.; Li, H.; Jackstadt, R.; Jiang, L.; Lodygin, D.; Kaller, M.; Horst, D.; Ziegler, P.K.; Schwitalla, S.; et al. IL-6R/STAT3/miR-34a feedback loop promotes EMT-mediated colorectal cancer invasion and metastasis. *J. Clin. Investig.* **2014**, *124*, 1853–1867. [CrossRef] [PubMed]

33. Toiyama, Y.; Takahashi, M.; Hur, K.; Nagasaka, T.; Tanaka, K.; Inoue, Y.; Kusunoki, M.; Boland, C.R.; Goel, A. Serum miR-21 as a diagnostic and prognostic biomarker in colorectal cancer. *J. Natl. Cancer Inst.* **2013**, *105*, 849–859. [CrossRef] [PubMed]

34. Wu, Y.; Song, Y.; Xiong, Y.; Wang, X.; Xu, K.; Han, B.; Bai, Y.; Li, L.; Zhang, Y.; Zhou, L. MicroRNA-21 (Mir-21) Promotes Cell Growth and Invasion by Repressing Tumor Suppressor PTEN in Colorectal Cancer. *Cell. Physiol. Biochem.* **2017**, *43*, 945–958. [CrossRef] [PubMed]

35. Nana-Sinkam, S.P.; Fabbri, M.; Croce, C.M. MicroRNAs in cancer: Personalizing diagnosis and therapy. *Ann. N. Y. Acad. Sci.* **2010**, *1210*, 25–33. [CrossRef] [PubMed]

36. Yamamichi, N.; Shimomura, R.; Inada, K.; Sakurai, K.; Haraguchi, T.; Ozaki, Y.; Fujita, S.; Mizutani, T.; Furukawa, C.; Fujishiro, M.; et al. Locked nucleic acid in situ hybridization analysis of miR-21 expression during colorectal cancer development. *Clin. Cancer Res.* **2009**, *15*, 4009–4016. [CrossRef] [PubMed]

37. Bastaminejad, S.; Taherikalani, M.; Ghanbari, R.; Akbari, A.; Shabab, N.; Saidijam, M. Investigation of MicroRNA-21 Expression Levels in Serum and Stool as a Potential Non-Invasive Biomarker for Diagnosis of Colorectal Cancer. *Iran. Biomed. J.* **2017**, *21*, 106–113. [CrossRef] [PubMed]

38. Wu, C.W.; Ng, S.S.M.; Dong, Y.J.; Ng, S.C.; Leung, W.W.; Lee, C.W.; Wong, Y.N.; Chan, F.K.L.; Yu, J.; Sung, J.J.Y. Detection of miR-92a and miR-21 in stool samples as potential screening biomarkers for colorectal cancer and polyps. *Gut* **2012**, *61*, 739–745. [CrossRef] [PubMed]

39. Link, A.; Balaguer, F.; Shen, Y.; Nagasaka, T.; Lozano, J.J.; Boland, C.R.; Goel, A. Fecal MicroRNAs as novel biomarkers for colon cancer screening. *Cancer Epidemiol. Biomark. Prev.* **2010**, *19*, 1766–1774. [CrossRef] [PubMed]

40. Fukushima, Y.; Iinuma, H.; Tsukamoto, M.; Matsuda, K.; Hashiguchi, Y. Clinical significance of microRNA-21 as a biomarker in each Dukes' stage of colorectal cancer. *Oncol. Rep.* **2015**, *33*, 573–582. [CrossRef] [PubMed]

41. Kjaer-Frifeldt, S.; Hansen, T.F.; Nielsen, B.S.; Joergensen, S.; Lindebjerg, J.; Soerensen, F.B.; dePont Christensen, R.; Jakobsen, A.; Danish Colorectal Cancer Group. The prognostic importance of miR-21 in stage II colon cancer: A population-based study. *Br. J. Cancer* **2012**, *107*, 1169–1174. [CrossRef] [PubMed]

42. Nielsen, B.S.; Jørgensen, S.; Fog, J.U.; Søkilde, R.; Christensen, I.J.; Hansen, U.; Brünner, N.; Baker, A.; Møller, S.; Nielsen, H.J. High levels of microRNA-21 in the stroma of colorectal cancers predict short disease-free survival in stage II colon cancer patients. *Clin. Exp. Metastasis* **2011**, *28*, 27–38. [CrossRef] [PubMed]

43. Shibuya, H.; Iinuma, H.; Shimada, R.; Horiuchi, A.; Watanabe, T. Clinicopathological and prognostic value of microRNA-21 and microRNA-155 in colorectal cancer. *Oncology* **2010**, *79*, 313–320. [CrossRef] [PubMed]

44. Xia, X.; Yang, B.; Zhai, X.; Liu, X.; Shen, K.; Wu, Z.; Cai, J. Prognostic role of microRNA-21 in colorectal cancer: A meta-analysis. *PLoS ONE* **2013**, *8*, e80426. [CrossRef] [PubMed]

45. Chen, Z.; Liu, H.; Jin, W.; Ding, Z.; Zheng, S.; Yu, Y. Tissue microRNA-21 expression predicted recurrence and poor survival in patients with colorectal cancer—A meta-analysis. *Oncotargets Ther.* **2016**, *9*, 2615–2624. [CrossRef]

46. Peng, Q.; Zhang, X.; Min, M.; Zou, L.; Shen, P.; Zhu, Y. The clinical role of microRNA-21 as a promising biomarker in the diagnosis and prognosis of colorectal cancer: A systematic review and meta-analysis. *Oncotarget* **2017**, *8*, 44893–44909. [CrossRef] [PubMed]

47. Kang, W.K.; Lee, J.K.; Oh, S.T.; Lee, S.H.; Jung, C.K. Stromal expression of miR-21 in T3-4a colorectal cancer is an independent predictor of early tumor relapse. *BMC Gastroenterol.* **2015**, *15*, 2. [CrossRef] [PubMed]

48. Kanaan, Z.; Rai, S.N.; Eichenberger, M.R.; Roberts, H.; Keskey, B.; Pan, J.; Galandiuk, S. Plasma miR-21: A potential diagnostic marker of colorectal cancer. *Ann. Surg.* **2012**, *256*, 544–551. [CrossRef] [PubMed]

49. Tsukamoto, M.; Iinuma, H.; Yagi, T.; Matsuda, K.; Hashiguchi, Y. Circulating Exosomal MicroRNA-21 as a Biomarker in Each Tumor Stage of Colorectal Cancer. *Oncology* **2017**, *92*, 360–370. [CrossRef] [PubMed]

50. Ahmed, F.E.; Ahmed, N.C.; Vos, P.W.; Bonnerup, C.; Atkins, J.N.; Casey, M.; Nuovo, G.J.; Naziri, W.; Wiley, J.E.; Mota, H.; et al. Diagnostic microRNA markers to screen for sporadic human colon cancer in stool: I. Proof of principle. *Cancer Genom. Proteom.* **2013**, *10*, 93–113.

51. Wang, Y.; Zhang, X.; Li, H.; Yu, J.; Ren, X. The role of miRNA-29 family in cancer. *Eur. J. Cell Biol.* **2013**, *92*, 123–128. [CrossRef] [PubMed]

52. Wang, L.-G.; Gu, J. Serum microRNA-29a is a promising novel marker for early detection of colorectal liver metastasis. *Cancer Epidemiol.* **2012**, *36*. [CrossRef] [PubMed]

53. Huang, Z.; Huang, D.; Ni, S.; Peng, Z.; Sheng, W.; Du, X. Plasma microRNAs are promising novel biomarkers for early detection of colorectal cancer. *Int. J. Cancer* **2010**, *127*, 118–126. [CrossRef] [PubMed]

54. Al-Sheikh, Y.A.; Ghneim, H.K.; Softa, K.I.; Al-Jobran, A.A.; Al-Obeed, O.; Mohamed, M.A.V.; Abdulla, M.; Aboul-Soud, M.A.M. Expression profiling of selected microRNA signatures in plasma and tissues of Saudi colorectal cancer patients by qPCR. *Oncol. Lett.* **2016**, *11*, 1406–1412. [CrossRef] [PubMed]

55. Li, L.; Guo, Y.; Chen, Y.; Wang, J.; Zhen, L.; Guo, X.; Liu, J.; Jing, C. The Diagnostic Efficacy and Biological Effects of microRNA-29b for Colon Cancer. *Technol. Cancer Res. Treat.* **2016**, *15*, 772–779. [CrossRef] [PubMed]

56. Basati, G.; Razavi, A.E.; Pakzad, I.; Malayeri, F.A. Circulating levels of the miRNAs, miR-194, and miR-29b, as clinically useful biomarkers for colorectal cancer. *Tumour Biol.* **2016**, *37*, 1781–1788. [CrossRef] [PubMed]

57. Tang, W.; Zhu, Y.; Gao, J.; Fu, J.; Liu, C.; Liu, Y.; Song, C.; Zhu, S.; Leng, Y.; Wang, G.; et al. MicroRNA-29a promotes colorectal cancer metastasis by regulating matrix metalloproteinase 2 and E-cadherin via KLF4. *Br. J. Cancer* **2014**, *110*, 450–458. [CrossRef] [PubMed]

58. Weissmann-Brenner, A.; Kushnir, M.; Lithwick Yanai, G.; Aharonov, R.; Gibori, H.; Purim, O.; Kundel, Y.; Morgenstern, S.; Halperin, M.; Niv, Y.; et al. Tumor microRNA-29a expression and the risk of recurrence in stage II colon cancer. *Int. J. Oncol.* **2012**, *40*, 2097–2103. [CrossRef] [PubMed]

59. Kuo, T.-Y.; Hsi, E.; Yang, I.-P.; Tsai, P.-C.; Wang, J.-Y.; Juo, S.-H.H. Computational analysis of mRNA expression profiles identifies microRNA-29a/c as predictor of colorectal cancer early recurrence. *PLoS ONE* **2012**, *7*, e31587. [CrossRef] [PubMed]

60. Inoue, A.; Yamamoto, H.; Uemura, M.; Nishimura, J.; Hata, T.; Takemasa, I.; Ikenaga, M.; Ikeda, M.; Murata, K.; Mizushima, T.; et al. MicroRNA-29b is a Novel Prognostic Marker in Colorectal Cancer. *Ann. Surg. Oncol.* **2015**, *22* (Suppl. 3), S1410–S1418. [CrossRef] [PubMed]

61. Yuan, L.; Zhou, C.; Lu, Y.; Hong, M.; Zhang, Z.; Zhang, Z.; Chang, Y.; Zhang, C.; Li, X. IFN-γ-mediated IRF1/miR-29b feedback loop suppresses colorectal cancer cell growth and metastasis by repressing IGF1. *Cancer Lett.* **2015**, *359*, 136–147. [CrossRef] [PubMed]

62. Ulivi, P.; Canale, M.; Passardi, A.; Marisi, G.; Valgiusti, M.; Frassineti, G.L.; Calistri, D.; Amadori, D.; Scarpi, E. Circulating Plasma Levels of miR-20b, miR-29b and miR-155 as Predictors of Bevacizumab Efficacy in Patients with Metastatic Colorectal Cancer. *Int. J. Mol. Sci.* **2018**, *19*, 307. [CrossRef] [PubMed]

63. Misso, G.; Di Martino, M.T.; De Rosa, G.; Farooqi, A.A.; Lombardi, A.; Campani, V.; Zarone, M.R.; Gullà, A.; Tagliaferri, P.; Tassone, P.; et al. Mir-34: A new weapon against cancer? *Mol. Ther. Nucleic Acids* **2014**, *3*, e194. [CrossRef] [PubMed]

64. Bu, P.; Chen, K.-Y.; Chen, J.H.; Wang, L.; Walters, J.; Shin, Y.J.; Goerger, J.P.; Sun, J.; Witherspoon, M.; Rakhilin, N.; et al. A microRNA miR-34a-regulated bimodal switch targets Notch in colon cancer stem cells. *Cell Stem Cell* **2013**, *12*, 602–615. [CrossRef] [PubMed]

65. Wu, X.; Song, Y.-C.; Cao, P.-L.; Zhang, H.; Guo, Q.; Yan, R.; Diao, D.-M.; Cheng, Y.; Dang, C.-X. Detection of miR-34a and miR-34b/c in stool sample as potential screening biomarkers for noninvasive diagnosis of colorectal cancer. *Med. Oncol.* **2014**, *31*. [CrossRef] [PubMed]

66. Aherne, S.T.; Madden, S.F.; Hughes, D.J.; Pardini, B.; Naccarati, A.; Levy, M.; Vodicka, P.; Neary, P.; Dowling, P.; Clynes, M. Circulating miRNAs miR-34a and miR-150 associated with colorectal cancer progression. *BMC Cancer* **2015**, *15*, 329. [CrossRef] [PubMed]

67. Gao, J.; Li, N.; Dong, Y.; Li, S.; Xu, L.; Li, X.; Li, Y.; Li, Z.; Ng, S.S.; Sung, J.J.; et al. miR-34a-5p suppresses colorectal cancer metastasis and predicts recurrence in patients with stage II/III colorectal cancer. *Oncogene* **2015**, *34*, 4142–4152. [CrossRef] [PubMed]

68. Ma, Y.; Bao-Han, W.; Lv, X.; Su, Y.; Zhao, X.; Yin, Y.; Zhang, X.; Zhou, Z.; MacNaughton, W.K.; Wang, H. MicroRNA-34a mediates the autocrine signaling of PAR2-activating proteinase and its role in colonic cancer cell proliferation. *PLoS ONE* **2013**, *8*, e72383. [CrossRef] [PubMed]

69. Siemens, H.; Neumann, J.; Jackstadt, R.; Mansmann, U.; Horst, D.; Kirchner, T.; Hermeking, H. Detection of miR-34a promoter methylation in combination with elevated expression of c-Met and β-catenin predicts distant metastasis of colon cancer. *Clin. Cancer Res.* **2013**, *19*, 710–720. [CrossRef] [PubMed]

70. Tazawa, H.; Tsuchiya, N.; Izumiya, M.; Nakagama, H. Tumor-suppressive miR-34a induces senescence-like growth arrest through modulation of the E2F pathway in human colon cancer cells. *Proc. Natl. Acad. Sci. USA* **2007**, *104*, 15472–15477. [CrossRef] [PubMed]

71. de Almeida, A.L.N.R.; Bernardes, M.V.A.A.; Feitosa, M.R.; Peria, F.M.; da Tirapelli, D.P.C.; da Rocha, J.J.R.; Feres, O. Serological under expression of microRNA-21, microRNA-34a and microRNA-126 in colorectal cancer. *Acta Cir. Bras.* **2016**, *31* (Suppl. 1), 13–18. [CrossRef] [PubMed]

72. Li, C.; Wang, Y.; Lu, S.; Zhang, Z.; Meng, H.; Liang, L.; Zhang, Y.; Song, B. MiR-34a inhibits colon cancer proliferation and metastasis by inhibiting platelet-derived growth factor receptor α. *Mol. Med. Rep.* **2015**, *12*, 7072–7078. [CrossRef] [PubMed]

73. Zhang, X.; Ai, F.; Li, X.; Tian, L.; Wang, X.; Shen, S.; Liu, F. MicroRNA-34a suppresses colorectal cancer metastasis by regulating Notch signaling. *Oncol. Lett.* **2017**, *14*, 2325–2333. [CrossRef] [PubMed]

74. Hiyoshi, Y.; Schetter, A.J.; Okayama, H.; Inamura, K.; Anami, K.; Nguyen, G.H.; Horikawa, I.; Hawkes, J.E.; Bowman, E.D.; Leung, S.Y.; et al. Increased microRNA-34b and -34c predominantly expressed in stromal tissues is associated with poor prognosis in human colon cancer. *PLoS ONE* **2015**, *10*, e0124899. [CrossRef] [PubMed]

75. Akao, Y.; Noguchi, S.; Iio, A.; Kojima, K.; Takagi, T.; Naoe, T. Dysregulation of microRNA-34a expression causes drug-resistance to 5-FU in human colon cancer DLD-1 cells. *Cancer Lett.* **2011**, *300*, 197–204. [CrossRef] [PubMed]

76. Sun, C.; Wang, F.-J.; Zhang, H.-G.; Xu, X.-Z.; Jia, R.-C.; Yao, L.; Qiao, P.-F. miR-34a mediates oxaliplatin resistance of colorectal cancer cells by inhibiting macroautophagy via transforming growth factor-β/Smad4 pathway. *World J. Gastroenterol.* **2017**, *23*, 1816–1827. [CrossRef] [PubMed]

77. Xi, Z.-W.; Xin, S.-Y.; Zhou, L.-Q.; Yuan, H.-X.; Wang, Q.; Chen, K.-X. Downregulation of rho-associated protein kinase 1 by miR-124 in colorectal cancer. *World J. Gastroenterol.* **2015**, *21*, 5454–5464. [CrossRef] [PubMed]

78. Feng, T.; Xu, D.; Tu, C.; Li, W.; Ning, Y.; Ding, J.; Wang, S.; Yuan, L.; Xu, N.; Qian, K.; et al. MiR-124 inhibits cell proliferation in breast cancer through downregulation of CDK4. *Tumour Biol.* **2015**, *36*, 5987–5997. [CrossRef] [PubMed]

79. Harada, T.; Yamamoto, E.; Yamano, H.; Nojima, M.; Maruyama, R.; Kumegawa, K.; Ashida, M.; Yoshikawa, K.; Kimura, T.; Harada, E.; et al. Analysis of DNA methylation in bowel lavage fluid for detection of colorectal cancer. *Cancer Prev. Res.* **2014**, *7*, 1002–1010. [CrossRef] [PubMed]

80. Faber, C.; Kirchner, T.; Hlubek, F. The impact of microRNAs on colorectal cancer. *Virchows Arch. Int. J. Pathol.* **2009**, *454*, 359–367. [CrossRef] [PubMed]

81. Ueda, Y.; Ando, T.; Nanjo, S.; Ushijima, T.; Sugiyama, T. DNA methylation of microRNA-124a is a potential risk marker of colitis-associated cancer in patients with ulcerative colitis. *Dig. Dis. Sci.* **2014**, *59*, 2444–2451. [CrossRef] [PubMed]

82. Wang, M.-J.; Li, Y.; Wang, R.; Wang, C.; Yu, Y.-Y.; Yang, L.; Zhang, Y.; Zhou, B.; Zhou, Z.-G.; Sun, X.-F. Downregulation of microRNA-124 is an independent prognostic factor in patients with colorectal cancer. *Int. J. Colorectal Dis.* **2013**, *28*, 183–189. [CrossRef] [PubMed]

83. Jinushi, T.; Shibayama, Y.; Kinoshita, I.; Oizumi, S.; Jinushi, M.; Aota, T.; Takahashi, T.; Horita, S.; Dosaka-Akita, H.; Iseki, K. Low expression levels of microRNA-124-5p correlated with poor prognosis in colorectal cancer via targeting of SMC4. *Cancer Med.* **2014**, *3*, 1544–1552. [CrossRef] [PubMed]

84. Slattery, M.L.; Pellatt, A.J.; Lee, F.Y.; Herrick, J.S.; Samowitz, W.S.; Stevens, J.R.; Wolff, R.K.; Mullany, L.E. Infrequently expressed miRNAs influence survival after diagnosis with colorectal cancer. *Oncotarget* **2017**, *8*, 83845–83859. [CrossRef] [PubMed]

85. Colangelo, T.; Fucci, A.; Votino, C.; Sabatino, L.; Pancione, M.; Laudanna, C.; Binaschi, M.; Bigioni, M.; Maggi, C.A.; Parente, D.; et al. MicroRNA-130b promotes tumor development and is associated with poor prognosis in colorectal cancer. *Neoplasia* **2013**, *15*, 1086–1099. [CrossRef] [PubMed]

86. Chen, W.; Tong, K.; Yu, J. MicroRNA-130a is upregulated in colorectal cancer and promotes cell growth and motility by directly targeting forkhead box F2. *Mol. Med. Rep.* **2017**, *16*, 5241–5248. [CrossRef] [PubMed]

87. Sempere, L.F.; Preis, M.; Yezefski, T.; Ouyang, H.; Suriawinata, A.A.; Silahtaroglu, A.; Conejo-Garcia, J.R.; Kauppinen, S.; Wells, W.; Korc, M. Fluorescence-based codetection with protein markers reveals distinct cellular compartments for altered MicroRNA expression in solid tumors. *Clin. Cancer Res.* **2010**, *16*, 4246–4255. [CrossRef] [PubMed]

88. Lv, Z.; Fan, Y.; Chen, H.; Zhao, D. Investigation of microRNA-155 as a serum diagnostic and prognostic biomarker for colorectal cancer. *Tumour Biol.* **2015**, *36*, 1619–1625. [CrossRef] [PubMed]

89. Hongliang, C.; Shaojun, H.; Aihua, L.; Hua, J. Correlation between expression of miR-155 in colon cancer and serum carcinoembryonic antigen level and its contribution to recurrence and metastasis forecast. *Saudi Med. J.* **2014**, *35*, 547–553. [PubMed]

90. Zhang, G.-J.; Xiao, H.-X.; Tian, H.-P.; Liu, Z.-L.; Xia, S.-S.; Zhou, T. Upregulation of microRNA-155 promotes the migration and invasion of colorectal cancer cells through the regulation of claudin-1 expression. *Int. J. Mol. Med.* **2013**, *31*, 1375–1380. [CrossRef] [PubMed]

91. Qu, Y.-L.; Wang, H.-F.; Sun, Z.-Q.; Tang, Y.; Han, X.-N.; Yu, X.-B.; Liu, K. Up-regulated miR-155-5p promotes cell proliferation, invasion and metastasis in colorectal carcinoma. *Int. J. Clin. Exp. Pathol.* **2015**, *8*, 6988–6994. [PubMed]

92. Wang, Y.; Lee, C.G.L. Role of miR-224 in hepatocellular carcinoma: A tool for possible therapeutic intervention? *Epigenomics* **2011**, *3*, 235–243. [CrossRef] [PubMed]

93. Zhu, Y.; Xu, A.; Li, J.; Fu, J.; Wang, G.; Yang, Y.; Cui, L.; Sun, J. Fecal miR-29a and miR-224 as the noninvasive biomarkers for colorectal cancer. *Cancer Biomark. Sect. Dis. Markers* **2016**, *16*, 259–264. [CrossRef] [PubMed]

94. Zhang, G.-J.; Zhou, H.; Xiao, H.-X.; Li, Y.; Zhou, T. Up-regulation of miR-224 promotes cancer cell proliferation and invasion and predicts relapse of colorectal cancer. *Cancer Cell Int.* **2013**, *13*, 104. [CrossRef] [PubMed]

95. Ling, H.; Pickard, K.; Ivan, C.; Isella, C.; Ikuo, M.; Mitter, R.; Spizzo, R.; Bullock, M.; Braicu, C.; Pileczki, V.; et al. The clinical and biological significance of MIR-224 expression in colorectal cancer metastasis. *Gut* **2016**, *65*, 977–989. [CrossRef] [PubMed]

96. Wang, Z.; Yang, J.; Di, J.; Cui, M.; Xing, J.; Wu, F.; Wu, W.; Yang, H.; Zhang, C.; Yao, Z.; et al. Downregulated USP3 mRNA functions as a competitive endogenous RNA of SMAD4 by sponging miR-224 and promotes metastasis in colorectal cancer. *Sci. Rep.* **2017**, *7*, 4281. [CrossRef] [PubMed]

97. Liao, W.-T.; Li, T.-T.; Wang, Z.-G.; Wang, S.-Y.; He, M.-R.; Ye, Y.-P.; Qi, L.; Cui, Y.-M.; Wu, P.; Jiao, H.-L.; et al. microRNA-224 promotes cell proliferation and tumor growth in human colorectal cancer by repressing PHLPP1 and PHLPP2. *Clin. Cancer Res.* **2013**, *19*, 4662–4672. [CrossRef] [PubMed]

98. Ke, T.-W.; Hsu, H.-L.; Wu, Y.-H.; Chen, W.T.-L.; Cheng, Y.-W.; Cheng, C.-W. MicroRNA-224 suppresses colorectal cancer cell migration by targeting Cdc42. *Dis. Markers* **2014**. [CrossRef] [PubMed]

99. Zhang, L.; Huang, L.-S.; Chen, G.; Feng, Z.-B. Potential Targets and Clinical Value of MiR-224-5p in Cancers of the Digestive Tract. *Cell. Physiol. Biochem.* **2017**, *44*, 682–700. [CrossRef] [PubMed]

100. Mosakhani, N.; Sarhadi, V.K.; Borze, I.; Karjalainen-Lindsberg, M.-L.; Sundström, J.; Ristamäki, R.; Osterlund, P.; Knuutila, S. MicroRNA profiling differentiates colorectal cancer according to KRAS status. *Genes Chromosomes Cancer* **2012**, *51*. [CrossRef] [PubMed]

101. Wang, Y.X.; Zhang, X.Y.; Zhang, B.F.; Yang, C.Q.; Chen, X.M.; Gao, H.J. Initial study of microRNA expression profiles of colonic cancer without lymph node metastasis. *J. Dig. Dis.* **2010**, *11*, 50–54. [CrossRef] [PubMed]

102. Chen, L.; Xu, S.; Xu, H.; Zhang, J.; Ning, J.; Wang, S. MicroRNA-378 is associated with non-small cell lung cancer brain metastasis by promoting cell migration, invasion and tumor angiogenesis. *Med. Oncol.* **2012**, *29*, 1673–1680. [CrossRef] [PubMed]

103. Zanutto, S.; Pizzamiglio, S.; Ghilotti, M.; Bertan, C.; Ravagnani, F.; Perrone, F.; Leo, E.; Pilotti, S.; Verderio, P.; Gariboldi, M.; et al. Circulating miR-378 in plasma: A reliable, haemolysis-independent biomarker for colorectal cancer. *Br. J. Cancer* **2014**, *110*, 1001–1007. [CrossRef] [PubMed]

104. Wang, Z.; Ma, B.; Ji, X.; Deng, Y.; Zhang, T.; Zhang, X.; Gao, H.; Sun, H.; Wu, H.; Chen, X.; et al. MicroRNA-378-5p suppresses cell proliferation and induces apoptosis in colorectal cancer cells by targeting BRAF. *Cancer Cell Int.* **2015**, *15*. [CrossRef] [PubMed]

105. Zhang, G.-J.; Zhou, H.; Xiao, H.-X.; Li, Y.; Zhou, T. MiR-378 is an independent prognostic factor and inhibits cell growth and invasion in colorectal cancer. *BMC Cancer* **2014**, *14*, 109. [CrossRef] [PubMed]

106. Zeng, M.; Zhu, L.; Li, L.; Kang, C. miR-378 suppresses the proliferation, migration and invasion of colon cancer cells by inhibiting SDAD1. *Cell. Mol. Biol. Lett.* **2017**, *22*. [CrossRef] [PubMed]

107. Li, H.; Dai, S.; Zhen, T.; Shi, H.; Zhang, F.; Yang, Y.; Kang, L.; Liang, Y.; Han, A. Clinical and biological significance of miR-378a-3p and miR-378a-5p in colorectal cancer. *Eur. J. Cancer* **2014**, *50*, 1207–1221. [CrossRef] [PubMed]

108. Zekri, A.-R.N.; Youssef, A.S.E.-D.; Lotfy, M.M.; Gabr, R.; Ahmed, O.S.; Nassar, A.; Hussein, N.; Omran, D.; Medhat, E.; Eid, S.; et al. Circulating Serum miRNAs as Diagnostic Markers for Colorectal Cancer. *PLoS ONE* **2016**, *11*, e0154130. [CrossRef] [PubMed]

109. Li, B.; Li, B.; Sun, H.; Zhang, H. The predicted target gene validation, function, and prognosis studies of miRNA-22 in colorectal cancer tissue. *Tumour Biol.* **2017**, *39*, 1010428317692257. [CrossRef] [PubMed]

110. Kerimis, D.; Kontos, C.K.; Christodoulou, S.; Papadopoulos, I.N.; Scorilas, A. Elevated expression of miR-24-3p is a potentially adverse prognostic factor in colorectal adenocarcinoma. *Clin. Biochem.* **2017**, *50*, 285–292. [CrossRef] [PubMed]

111. Li, Y.; Sun, Z.; Liu, B.; Shan, Y.; Zhao, L.; Jia, L. Tumor-suppressive miR-26a and miR-26b inhibit cell aggressiveness by regulating FUT4 in colorectal cancer. *Cell Death Dis.* **2017**, *8*, e2892. [CrossRef] [PubMed]

112. Elshafei, A.; Shaker, O.; Abd El-Motaal, O.; Salman, T. The expression profiling of serum miR-92a, miR-375, and miR-760 in colorectal cancer: An Egyptian study. *Tumour Biol.* **2017**, *39*, 1010428317705765. [CrossRef] [PubMed]

113. Chang, P.-Y.; Chen, C.-C.; Chang, Y.-S.; Tsai, W.-S.; You, J.-F.; Lin, G.-P.; Chen, T.-W.; Chen, J.-S.; Chan, E.-C. MicroRNA-223 and microRNA-92a in stool and plasma samples act as complementary biomarkers to increase colorectal cancer detection. *Oncotarget* **2016**, *7*, 10663–10675. [CrossRef] [PubMed]

114. Fu, F.; Jiang, W.; Zhou, L.; Chen, Z. Circulating Exosomal miR-17-5p and miR-92a-3p Predict Pathologic Stage and Grade of Colorectal Cancer. *Transl. Oncol.* **2018**, *11*, 221–232. [CrossRef] [PubMed]

115. He, Y.; Wang, G.; Zhang, L.; Zhai, C.; Zhang, J.; Zhao, X.; Jiang, X.; Zhao, Z. Biological effects and clinical characteristics of microRNA-106a in human colorectal cancer. *Oncol. Lett.* **2017**, *14*, 830–836. [CrossRef] [PubMed]

116. Maierthaler, M.; Benner, A.; Hoffmeister, M.; Surowy, H.; Jansen, L.; Knebel, P.; Chang-Claude, J.; Brenner, H.; Burwinkel, B. Plasma miR-122 and miR-200 family are prognostic markers in colorectal cancer. *Int. J. Cancer* **2017**, *140*, 176–187. [CrossRef] [PubMed]

117. Liang, L.; Gao, C.; Li, Y.; Sun, M.; Xu, J.; Li, H.; Jia, L.; Zhao, Y. miR-125a-3p/FUT5-FUT6 axis mediates colorectal cancer cell proliferation, migration, invasion and pathological angiogenesis via PI3K-Akt pathway. *Cell Death Dis.* **2017**, *8*, e2968. [CrossRef] [PubMed]

118. Fiala, O.; Pitule, P.; Hosek, P.; Liska, V.; Sorejs, O.; Bruha, J.; Vycital, O.; Buchler, T.; Poprach, A.; Topolcan, O.; et al. The association of miR-126-3p, miR-126-5p and miR-664-3p expression profiles with outcomes of patients with metastatic colorectal cancer treated with bevacizumab. *Tumour Biol.* **2017**, *39*, 1010428317709283. [CrossRef] [PubMed]

119. Ng, L.; Wan, T.M.-H.; Man, J.H.-W.; Chow, A.K.-M.; Iyer, D.; Chen, G.; Yau, T.C.-C.; Lo, O.S.-H.; Foo, D.C.-C.; Poon, J.T.-C.; et al. Identification of serum miR-139-3p as a non-invasive biomarker for colorectal cancer. *Oncotarget* **2017**, *8*, 27393–27400. [CrossRef] [PubMed]

120. Miyoshi, J.; Toden, S.; Yoshida, K.; Toiyama, Y.; Alberts, S.R.; Kusunoki, M.; Sinicrope, F.A.; Goel, A. MiR-139-5p as a novel serum biomarker for recurrence and metastasis in colorectal cancer. *Sci. Rep.* **2017**, *7*, 43393. [CrossRef] [PubMed]

121. Baltruskeviciene, E.; Schveigert, D.; Stankevicius, V.; Mickys, U.; Zvirblis, T.; Bublevic, J.; Suziedelis, K.; Aleknavicius, E. Down-regulation of miRNA-148a and miRNA-625-3p in colorectal cancer is associated with tumor budding. *BMC Cancer* **2017**, *17*, 607. [CrossRef] [PubMed]

122. Gu, X.; Jin, R.; Mao, X.; Wang, J.; Yuan, J.; Zhao, G. Prognostic value of miRNA-181a/b in colorectal cancer: A meta-analysis. *Biomark. Med.* **2018**, *12*, 299–308. [CrossRef] [PubMed]

123. Yamazaki, N.; Koga, Y.; Taniguchi, H.; Kojima, M.; Kanemitsu, Y.; Saito, N.; Matsumura, Y. High expression of miR-181c as a predictive marker of recurrence in stage II colorectal cancer. *Oncotarget* **2017**, *8*, 6970–6983. [CrossRef] [PubMed]

124. Guo, X.; Zhu, Y.; Hong, X.; Zhang, M.; Qiu, X.; Wang, Z.; Qi, Z.; Hong, X. miR-181d and c-myc-mediated inhibition of CRY2 and FBXL3 reprograms metabolism in colorectal cancer. *Cell Death Dis.* **2017**, *8*, e2958. [CrossRef] [PubMed]

125. Lin, M.; Duan, B.; Hu, J.; Yu, H.; Sheng, H.; Gao, H.; Huang, J. Decreased expression of miR-193a-3p is associated with poor prognosis in colorectal cancer. *Oncol. Lett.* **2017**, *14*, 1061–1067. [CrossRef] [PubMed]

126. Li, W.; Chang, J.; Tong, D.; Peng, J.; Huang, D.; Guo, W.; Zhang, W.; Li, J. Differential microRNA expression profiling in primary tumors and matched liver metastasis of patients with colorectal cancer. *Oncotarget* **2017**, *8*, 35783–35791. [CrossRef] [PubMed]

127. Tayel, S.I.; Fouda, E.A.M.; Gohar, S.F.; Elshayeb, E.I.; El-Sayed, E.H.; El-Kousy, S.M. Potential role of MicroRNA 200c gene expression in assessment of colorectal cancer. *Arch. Biochem. Biophys.* **2018**, *647*, 41–46. [CrossRef] [PubMed]

128. Krawczyk, P.; Powrózek, T.; Olesiński, T.; Dmitruk, A.; Dziwota, J.; Kowalski, D.; Milanowski, J. Evaluation of miR-506 and miR-4316 expression in early and non-invasive diagnosis of colorectal cancer. *Int. J. Colorectal Dis.* **2017**, *32*, 1057–1060. [CrossRef] [PubMed]

129. Ye, L.; Jiang, T.; Shao, H.; Zhong, L.; Wang, Z.; Liu, Y.; Tang, H.; Qin, B.; Zhang, X.; Fan, J. miR-1290 Is a Biomarker in DNA-Mismatch-Repair-Deficient Colon Cancer and Promotes Resistance to 5-Fluorouracil by Directly Targeting hMSH2. *Mol. Ther. Nucleic Acids* **2017**, *7*, 453–464. [CrossRef] [PubMed]

130. Kiss, I.; Mlčochová, J.; Součková, K.; Fabian, P.; Poprach, A.; Halamkova, J.; Svoboda, M.; Vyzula, R.; Slaby, O. MicroRNAs as outcome predictors in patients with metastatic colorectal cancer treated with bevacizumab in combination with FOLFOX. *Oncol. Lett.* **2017**, *14*, 743–750. [CrossRef] [PubMed]

131. Chang, J.; Huang, L.; Cao, Q.; Liu, F. Identification of colorectal cancer-restricted microRNAs and their target genes based on high-throughput sequencing data. *Oncotargets Ther.* **2016**, *9*, 1787–1794. [CrossRef]

132. Koduru, S.V.; Tiwari, A.K.; Hazard, S.W.; Mahajan, M.; Ravnic, D.J. Exploration of small RNA-seq data for small non-coding RNAs in Human Colorectal Cancer. *J. Genom.* **2017**, *5*, 16–31. [CrossRef] [PubMed]

133. Pan, C.; Yan, X.; Li, H.; Huang, L.; Yin, M.; Yang, Y.; Gao, R.; Hong, L.; Ma, Y.; Shi, C.; et al. Systematic literature review and clinical validation of circulating microRNAs as diagnostic biomarkers for colorectal cancer. *Oncotarget* **2017**, *8*, 68317–68328. [CrossRef] [PubMed]

134. Guo, S.; Zhang, J.; Wang, B.; Zhang, B.; Wang, X.; Huang, L.; Liu, H.; Jia, B. A 5-serum miRNA panel for the early detection of colorectal cancer. *Oncotargets Ther.* **2018**, *11*, 2603–2614. [CrossRef] [PubMed]

135. Zhu, M.; Huang, Z.; Zhu, D.; Zhou, X.; Shan, X.; Qi, L.-W.; Wu, L.; Cheng, W.; Zhu, J.; Zhang, L.; et al. A panel of microRNA signature in serum for colorectal cancer diagnosis. *Oncotarget* **2017**, *8*, 17081–17091. [CrossRef] [PubMed]

136. Wang, S.; Xiang, J.; Li, Z.; Lu, S.; Hu, J.; Gao, X.; Yu, L.; Wang, L.; Wang, J.; Wu, Y.; et al. A plasma microRNA panel for early detection of colorectal cancer. *Int. J. Cancer* **2015**, *136*, 152–161. [CrossRef] [PubMed]

137. Kanaan, Z.; Roberts, H.; Eichenberger, M.R.; Billeter, A.; Ocheretner, G.; Pan, J.; Rai, S.N.; Jorden, J.; Williford, A.; Galandiuk, S. A plasma microRNA panel for detection of colorectal adenomas: A step toward more precise screening for colorectal cancer. *Ann. Surg.* **2013**, *258*, 400–408. [CrossRef] [PubMed]

138. Wikberg, M.L.; Myte, R.; Palmqvist, R.; van Guelpen, B.; Ljuslinder, I. Plasma miRNA can detect colorectal cancer, but how early? *Cancer Med.* **2018**, *7*, 1697–1705. [CrossRef] [PubMed]

139. Yang, I.-P.; Tsai, H.-L.; Miao, Z.-F.; Huang, C.-W.; Kuo, C.-H.; Wu, J.-Y.; Wang, W.-M.; Juo, S.-H.H.; Wang, J.-Y. Development of a deregulating microRNA panel for the detection of early relapse in postoperative colorectal cancer patients. *J. Transl. Med.* **2016**, *14*. [CrossRef] [PubMed]

140. Liu, H.-N.; Liu, T.-T.; Wu, H.; Chen, Y.-J.; Tseng, Y.-J.; Yao, C.; Weng, S.-Q.; Dong, L.; Shen, X.-Z. Serum microRNA signatures and metabolomics have high diagnostic value in colorectal cancer using two novel methods. *Cancer Sci.* **2018**, *109*, 1185–1194. [CrossRef] [PubMed]

141. Wong, C.K.W.; Fedorak, R.N.; Prosser, C.I.; Stewart, M.E.; van Zanten, S.V.; Sadowski, D.C. The sensitivity and specificity of guaiac and immunochemical fecal occult blood tests for the detection of advanced colonic adenomas and cancer. *Int. J. Colorectal Dis.* **2012**, *27*, 1657–1664. [CrossRef] [PubMed]

142. Barzi, A.; Lenz, H.-J.; Quinn, D.I.; Sadeghi, S. Comparative effectiveness of screening strategies for colorectal cancer. *Cancer* **2017**, *123*, 1516–1527. [CrossRef] [PubMed]

International Journal of
Molecular Sciences

MDPI

Article

Circulating Plasma Levels of miR-20b, miR-29b and miR-155 as Predictors of Bevacizumab Efficacy in Patients with Metastatic Colorectal Cancer

Paola Ulivi [1,*], **Matteo Canale** [1], **Alessandro Passardi** [2], **Giorgia Marisi** [1], **Martina Valgiusti** [2], **Giovanni Luca Frassineti** [2], **Daniele Calistri** [1], **Dino Amadori** [2] and **Emanuela Scarpi** [3]

[1] Biosciences Laboratory, Istituto Scientifico Romagnolo per lo Studio e la Cura dei Tumori (IRST) IRCCS, 47014 Meldola, Italy; matteo.canale@irst.emr.it (M.C.); giorgia.marisi@irst.emr.it (G.M.); daniele.calistri@irst.emr.it (D.C.)

[2] Department of Medical Oncology, Istituto Scientifico Romagnolo per lo Studio e la Cura dei Tumori (IRST) IRCCS, 47014 Meldola, Italy; alessandro.passardi@irst.emr.it (A.P.); martina.valgiusti@irst.emr.it (M.V.); luca.frassineti@irst.emr.it (G.L.F.); dino.amadori@irst.emr.it (D.A.)

[3] Unit of Biostatistics and Clinical Trials, Istituto Scientifico Romagnolo per lo Studio e la Cura dei Tumori (IRST) IRCCS, 47014 Meldola, Italy; emanuela.scarpi@irst.emr.it

* Correspondence: paola.ulivi@irst.emr.it; Tel.: +39-0543-739277; Fax: +39-0543-739221

Received: 3 January 2018; Accepted: 18 January 2018; Published: 20 January 2018

Abstract: Targeting angiogenesis in the treatment of colorectal cancer (CRC) is a common strategy, for which potential predictive biomarkers have been studied. miRNAs are small non-coding RNAs involved in several processes including the angiogenic pathway. They are very stable in biological fluids, which turns them into potential circulating biomarkers. In this study, we considered a case series of patients with metastatic (m) CRC treated with a bevacizumab (B)-based treatment, enrolled in the prospective multicentric Italian Trial in Advanced Colorectal Cancer (ITACa). We then analyzed a panel of circulating miRNAs in relation to the patient outcome. In multivariate analysis, circulating basal levels of hsa-miR-20b-5p, hsa-miR-29b-3p and hsa-miR-155-5p resulted in being significantly associated with progression-free survival (PFS) ($p = 0.027$, $p = 0.034$ and $p = 0.039$, respectively) and overall survival (OS) ($p = 0.044$, $p = 0.024$ and $p = 0.032$, respectively). We also observed that an increase in hsa-miR-155-5p at the first clinical evaluation was significantly associated with shorter PFS (HR 3.03 (95% CI 1.06–9.09), $p = 0.040$) and OS (HR 3.45 (95% CI 1.18–10.00), $p = 0.024$), with PFS and OS of 9.5 (95% CI 6.8–18.7) and 15.9 (95% CI 8.4–not reached), respectively, in patients with an increase $\geq 30\%$ of hsa-miR-155-5p and 22.3 (95% CI 10.2–25.5) and 42.9 (24.8–not reached) months, respectively, in patients without such increase. In conclusion, our results highlight the potential usefulness of circulating basal levels of hsa-miR-20b-5p, hsa-miR-29b-3p and hsa-miR-155-5p in predicting the outcome of patients with mCRC treated with B. In addition, the variation of circulating hsa-miR-155-5p could also be indicative of the patient survival.

Keywords: plasma; miRNAs; colorectal cancer; bevacizumab

---

1. Introduction

Targeting angiogenesis has been the standard of treatment in metastatic colorectal cancer (mCRC) for more than a decade, and novel anti-angiogenic agents are emerging each year. However, despite improvements in our understanding of the molecular biology of colorectal cancer (CRC), there are still no validated biomarkers for anti-angiogenic treatment. According to randomized clinical trials [1–3], bevacizumab (B), a monoclonal antibody against the vascular endothelial growth factor A, is widely used in combination with the chemotherapeutic regimen in a number of countries. Although several biomarkers have been studied and hypothesized to be useful for patient selection, none of these have yet been validated for use in clinical practice [4–6].

MicroRNAs (miRNAs) are a class of small non-coding RNAs approximately 18–25 nucleotides long with an important role in regulating gene expression. Expression patterns of miRNAs correlate with specific clinical pathological parameters in different cancer subtypes, suggesting that miRNAs could be potential biomarkers on the basis of tumor origin, histology, aggressiveness or chemosensitivity [7]. It has been reported that miRNAs may regulate the angiogenic process by exerting pro-angiogenic or anti-angiogenic effects [8–14]. Specific tumor tissue miRNAs have been shown to be predictive of the effectiveness of B in CRC patients [15]. The nature of miRNAs renders them particularly stable in biological fluids such as serum and plasma, making them potentially ideal circulating biomarkers for diagnosis, prognosis and as predictors of response to treatment [7,16]. With regard to this last characteristic, a miRNA signature composed of eight circulating miRNAs has been found to significantly correlate with overall survival (OS) in patients with glioblastoma treated with B [17], whereas a six-circulating miRNA signature has proven prognostic in patients with advanced non-small cell lung cancer treated with B plus erlotinib followed by platinum-based chemotherapy (CT) [18].

A previous study analyzed the role of circulating miR-126 in relation to outcome in patients treated with CT plus B. The authors demonstrated that an increased level of miR-126 from baseline to the first clinical evaluation was associated with a lack of benefit to treatment, concluding that it could represent a resistance mechanism to B [19].

In this study, we evaluated a panel of circulating miRNAs, including miR-126, selected on the basis of their role in the angiogenic process, as bio-markers of the treatment with bevacizumab. To this aim, we determined miRNA plasma levels in a case series of patients treated with a B-based CT, enrolled into the prospective multicentric randomized phase III study "Italian Trial in Advanced Colorectal Cancer" (ITACa) [20].

2. Results

2.1. Case Series

The clinical pathological characteristics of patients are shown in Table 1.

Median age was 65 years (range 37–83 years), and about two thirds (35 patients, 67.3%) were male, in a good performance status and in an advanced stage of the disease. The tumor localization was mainly the colon (71.1%), compared to the rectum (28.9%), and equally distributed as left- and right-sided tumors. A total of 48.1% had a *RAS* mutation (21 patients were *KRAS* mutated and four were *NRAS* mutated), and 11.5% had a *BRAF* mutation.

Table 1. Patient characteristics (*n* = 52).

Patient Characteristics	No. (%)
Median age, years (range)	65 (37–83)
Gender	
Male	35 (67.3)
Female	17 (32.7)
Performance Status (ECOG)	
0	44 (84.6)
1–2	8 (15.4)
Stage at Diagnosis	
I–III	12 (23.1)
IV	40 (76.9)
Tumor Localization	
Colon	37 (71.1)
Rectum	15 (28.9)
Left-sided	27 (55.1)
Right-sided	22 (44.9)

Table 1. *Cont.*

Patient Characteristics	No. (%)
Grading	
1–2	25 (59.5)
3	17 (40.5)
Missing	10 (19.0)
CT Regimen	
FOLFOX4	27 (51.9)
FOLFIRI	25 (48.1)
Prior Cancer Therapy	
Surgery	40 (76.9)
Radiotherapy	4 (7.7)
Adjuvant CT	9 (17.3)
RAS Status	
Wild type	27 (51.9)
Mutated	25 (48.1)
BRAF Status	
Wild type	46 (88.5)
Mutated	6 (11.5)

ECOG, Eastern Cooperative Oncology Group; CT: chemotherapy; FOLFOX4, folinic acid, 5-fluorouracil and oxaliplatin; FOLFIRI, folinic acid, 5-fluorouracil and irinotecan.

2.2. Baseline Circulating miRNAs in Relation to Clinical Pathological Characteristics of Patients

By GeNorm analysis, two miRNAs (hsa-miR-484 and hsa-miR-223-3p) resulted in being the more stable and were used for the normalization analysis together with the spike in cel-miR-39-3p.

Baseline circulating levels of some miRNAs were significantly associated with the clinical pathological characteristics of patients. Of the analyzed miRNAs, three (hsa-miR-199a-5p, hsa-miR-335-5p and hsa-miR-520d-3p) were significantly upregulated in left-sided compared to right-sided lesions (Table 2) and two were significantly correlated with *RAS* status.

Table 2. miRNAs significantly correlated with tumor localization.

miRNA	Left-Sided	Right-Sided	*p*
	Median Value (Range)		
hsa-miR-199a-5p	3188 (0.5–149,395)	1960.5 (0.47–48,761)	0.034
hsa-miR-335-5p	6574.5 (1493–1,332,286)	3214 (2.14–40,038)	0.006
hsa-miR-520d-3p	5087 (3.09–2,831,724)	1505 (0.49–48,452)	0.008

In particular, hsa-miR-21-5p was significantly downregulated in both *KRAS* and *NRAS* mutated patients. Conversely, hsa-miR-221-3p was significantly upregulated in *RAS* mutated patients (Table 3).

Table 3. miRNAs significantly correlated with *RAS* status.

miRNA	*KRAS*		*p*	*NRAS*		*p*
	Median Value (Range)			Median Value (Range)		
	Wild Type	Mutated		Wild Type	Mutated	
hsa-miR-21-5p	1424 (0.57–4627)	1.71 (0.53–3594)	0.019	1558 (0.57–4627)	1011 (0.53–3594)	0.008
hsa-miR-221-3p	1163 (0.03–5499)	1878 (0.58–34,375)	0.050	1122 (0.03–5499)	1866 (0.58–34,375)	0.010

2.3. Response to Therapy and Prognosis in Relation to Clinical Pathological Characteristics of Patients

Overall, an objective response rate (ORR) of 62.7% was observed. Progression-free (PFS) and overall survival (OS) were 9.7 months (95% confidence interval (CI) 8.1–14.1) and 22.7 months (95% CI

13.1–28.8), respectively. No correlation was found between response to therapy and clinical pathological characteristics of patients. Conversely, performance and *BRAF* statuses were significantly associated with both PFS and OS. In particular, a hazard ratio (HR) of 2.32 (95% CI 1.06–5.08), $p = 0.036$, and HR of 3.27 (95% CI 1.45–7.41), $p = 0.004$, were observed for performance status in relation to PFS and OS, respectively, whereas HR of 3.41 (95% CI 1.35–8.59), $p = 0.009$, and HR of 3.62 (95% CI 1.45–9.07), $p = 0.006$, were observed for *BRAF* status in relation to PFS and OS, respectively. Age, dichotomized on the basis of the median value, was significantly associated with PFS (HR 2.11 (95% CI 1.15–3.89), $p = 0.016$) but not with OS. Moreover, CT regimen was associated with OS: HR 1.99 (95% CI 1.05–3.79), $p = 0.035$ (Table S1).

2.4. Baseline Circulating miRNAs in Relation to Response to Therapy and Patient Prognosis

With regard to response to therapy, only hsa-miR-17-5p resulted in being significantly correlated, with an odds ratio (OR) of 0.87 (95% CI 0.77–0.99).

In univariate analysis, two miRNAs, hsa-miR-20b-5p and hsa-miR424-5p, were significantly associated with PFS and OS. In particular, HR of 0.931 (95% CI 0.880–0.986, $p = 0.014$) and of 0.932 (95% CI 0.869–0.999, $p = 0.048$) were observed for PFS. With regard to OS, HR of 0.922 (95% CI 0.869–0.978, $p = 0.007$) and of 0.891 (95% CI 0.827–0.960, $p = 0.002$) were observed. miRNA hsa-miR-29b-3p resulted in being significantly correlated with PFS (HR 0.868 (95% CI 0.796–0.948), $p = 0.002$), but not with OS ($p = 0.070$). In addition, hsa-miR-155-5p was borderline associated with PFS and OS ($p = 0.078$ and 0.065, respectively) (Table S2).

In multivariate analysis, considering miRNAs levels as continuous variables, hsa-miR-20b-5p, hsa-miR-29b-3p and hsa-miR-155-5p resulted in being significantly associated with PFS ($p = 0.027$, $p = 0.034$ and $p = 0.039$, respectively) and OS ($p = 0.044$, $p = 0.024$ and $p = 0.032$, respectively) (Table 4).

Table 4. Multivariate analysis of PFS and OS.

Baseline	PFS		OS	
	HR (95% CI)	p	HR (95% CI)	p
has-miR-20b-5p	0.922 (0.847–0.989)	0.035	0.930 (0.850–0.995)	0.046
has-miR-29b-3p	0.854 (0.728–0.997)	0.045	0.872 (0.753–0.991)	0.039
has-miR-424-5p	0.968 (0.877–1.069)	0.517	0.936 (0.838–1.046)	0.242
has-miR-155-5p	0.927 (0.863–0.997)	0.040	0.917 (0.850–0.990)	0.026
ECOG PS (1–2 vs. 0)	1.206 (0.424–3.433)	0.725	1.838 (0.667–5.060)	0.239
BRAF (mutated vs. wild type)	3.574 (1.075–11.882)	0.038	3.628 (1.063–12.378)	0.040
Age, years (≥65 vs. <65)	2.207 (0.987–4.935)	0.054	1.478 (0.650–3.364)	0.351

Setting the median value as the cutoff, statistically-significant differences were seen in terms of PFS and OS for the three miRNAs. In particular, significantly longer PFS and OS were observed for patients with circulating miRNA values over the cutoff (Table 5 and Figure 1).

Table 5. Univariate analysis of PFS and OS in relation to miRNA cutoff values.

PFS	No. Patients	No. Events	Median PFS (Months) (95% CI)	p	HR (95% CI)	p
hsa-miR-20b-5p						
<1293	26	24	8.1 (5.0–12.5)		1.00	
≥1293	26	20	14.0 (9.4–21.3)	0.008	0.44 (0.24–0.82)	0.010
hsa-miR-29b-3p						
<3138	25	23	8.2 (5.0–12.4)		1.00	
≥3138	27	21	14.9 (9.1–21.3)	0.021	0.50 (0.27–0.91)	0.024
hsa-miR-155-5p						
<0.73	32	30	8.3 (6.1–9.7)		1.00	
≥0.73	20	14	16.0 (10.2–23.0)	0.007	0.42 (0.22–0.81)	0.009

Table 5. *Cont.*

OS	No. Patients	No. Events	Median OS (Months) (95% CI)	p	HR (95% CI)	p
hsa-miR-20b-5p						
<1293	26	23	11.6 (8.2–23.4)		1.00	
≥1293	26	17	28.8 (19.3–42.9)	0.004	0.40 (0.21–0.77)	0.005
hsa-miR-29b-3p						
<3138	25	22	15.5 (6.8–24.8)		1.00	
≥3138	27	18	31.7 (13.9–47.1)	0.005	0.40 (0.21–0.78)	0.007
hsa-miR-155-5p						
<0.73	32	27	13.5 (8.2–23.4)		1.00	
≥0.73	20	13	31.6 (21.8–42.9)	0.024	0.47 (0.24–0.92)	0.028

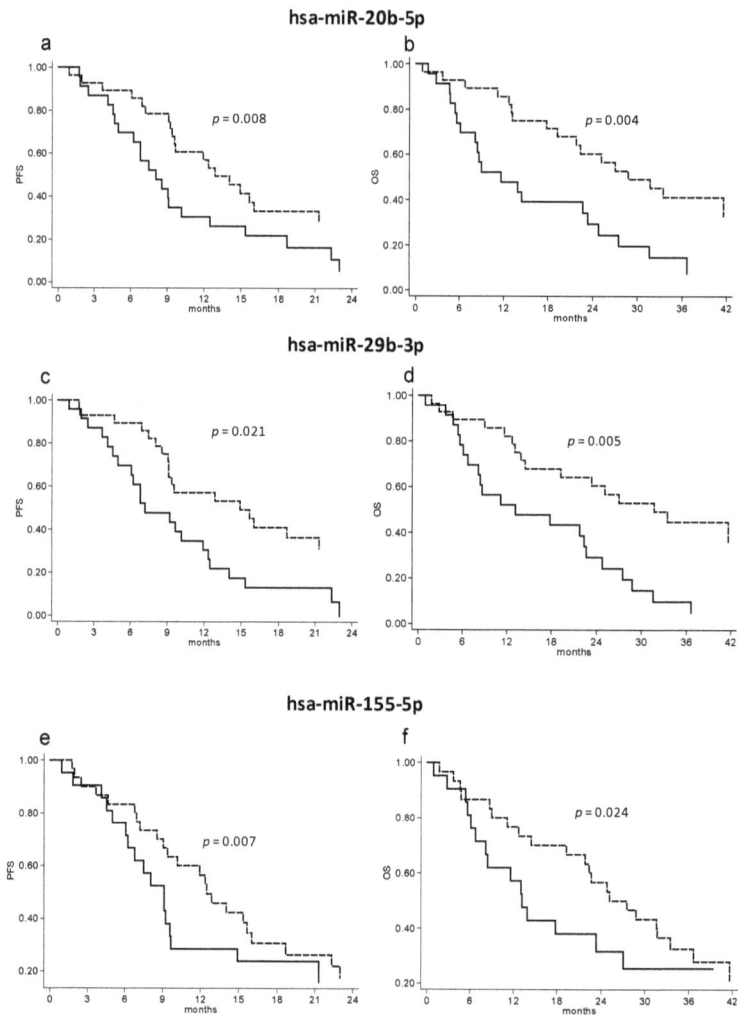

Figure 1. PFS and OS of basal circulating levels of hsa-miR-20b-5p (**a,b**), hsa-miR-29b-5p (**c,d**) and hsa-miR-155-5p (**e,f**). Dashed lines represent patients with miRNA values greater than the median value, whereas continuous lines represent patients with miRNA values lower than the median value.

2.5. Circulating miRNAs' Variations during Treatment in Relation to Patient Outcome

Variations of miRNAs at the first clinical evaluation with respect to baseline were analyzed. Differences in miRNAs variation were observed in relation to the clinical pathological characteristics of patients. In particular, different variations were observed in relation to left- or right-sided tumors for hsa-miR-16-5p (p = 0.049), hsa-miR-221-3p (p = 0.011), hsa-miR-29b-3p (p = 0.015) and hsa-miR-335-5p (p = 0.026). Variations of hsa-miR-221-3p were also associated with the *BRAF* status (p = 0.049). Moreover, hsa-miR-194-5p variations were significantly associated with the *KRAS* status (p = 0.040).

We analyzed the variation of circulating miRNA expression at the first clinical evaluation in relation to response to treatment and PFS and OS. No significant associations were observed between miRNA variations and response to therapy. Conversely, we observed that an increase of hsa-miR-155-5p was significantly associated with shorter PFS (HR 3.03 (95% CI 1.06–9.09), p = 0.040) and OS (HR 3.45 (95% CI 1.18–10.00), p = 0.024), with PFS and OS of 9.5 (95% CI 6.8–18.7) and 15.9 (95% CI 8.4– not reached), respectively, in patients with an increase $\geq$30% of hsa-miR-155-5p and 22.3 (95% CI 10.2–25.5) and 42.9 (24.8–not reached) months, respectively, in patients without such increase (Figure 2). An increase of hsa-miR-24-3p was also associated with a significantly shorter PFS (HR 2.22 (95% CI 0.99–5.00), p = 0.053), and OS (HR 2.13 (95% CI 0.89–5.00), p = 0.087).

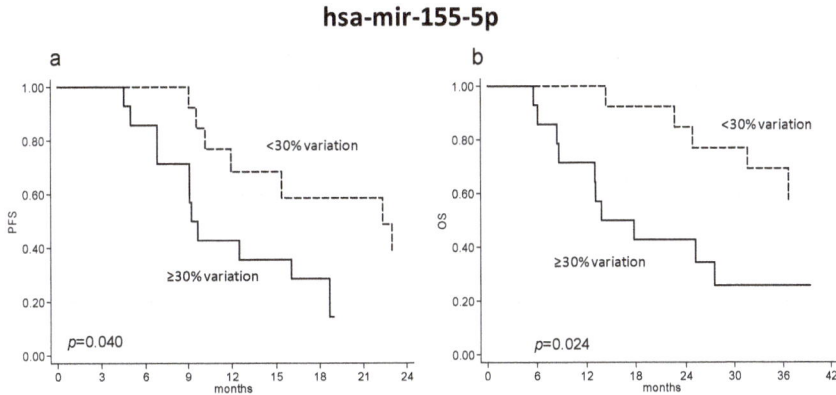

Figure 2. PFS (**a**) and OS (**b**) of patients with an increase $\geq$ or <30% of circulating hsa-miR-155-5p at the first clinical evaluation. Dashed lines represent patients with miRNA values greater than the median value, whereas continuous lines represent patients with miRNA values lower than the median value.

3. Discussion

In this study, we found that specific circulating miRNAs are associated with prognosis in mCRC patients treated with CT plus B. Baseline circulating levels of hsa-miR-20b-5p, hsa-miR-29b-3p and hsa-miR-155-5p were significantly correlated with PFS and OS. Patients with higher baseline levels of the three miRNAs showed longer PFS and OS, suggesting that they could be involved in pathways potentially correlated with the angiogenic pathway and, as a consequence, with B efficacy. It has been demonstrated that hsa-miR-29b is capable of repressing tumor angiogenesis, invasion and metastasis, by targeting metalloproteinase-2 (MMP2) [21]. Similarly, another study has demonstrated that miR-29b in non-small cell lung cancer models could suppress cells proliferation, migration and invasion by targeting the 3′-UTR of MMP2 and PTEN mRNA [22]. More recently, it has been reported that hsa-miR-29b suppresses tumor growth through simultaneously inhibiting angiogenesis and tumorigenesis by targeting Akt3 [23]. These findings are in agreement with our results, suggesting that patients with higher levels of hsa-miR-29b could have a greater benefit from B as both exert an anti-angiogenic effect. Although little evidence is available on the correlation between miR-20b and the angiogenic process, a recent study reported the role of hsa-miR-20b in regulating proliferation and senescence of endothelial cells, through the involvement of RBL1 [24]. We also observed that patients with a high basal level of hsa-miR-155-5p had a better outcome and that patients with a rise in the level of this type of miRNA at the first clinical evaluation (i.e., after one month of treatment) had a considerable shorter PFS and OS. Of the many studies on the role of hsa-miR-155-5p in the angiogenic process [25–27], some have reported a role in the process of hypoxia [25,28], showing that hsa-miR-155 contributes to controlling hypoxia-inducible factor 1-alpha (HIF-1α) expression and promotes angiogenesis under hypoxia condition. The association between high basal levels of hsa-miR-155-5p and better outcome of patients treated with B is consistent with the link between this miRNA and angiogenesis and inflammation processes [27,29], both targets of the drug. On the other hand, the induction of circulating hsa-miR-155 after treatment with B could indicate a process of drug resistance due to the stimulation of angiogenesis that could contrast with the B activity and that could be in line with the poor prognosis of such patients. We also observed a less evident association between the induction of circulating hsa-miR-24-3p and a worse outcome. It has been shown that the endothelial nitric oxide synthase (eNOS) gene is one of the targets of hsa-miR-24-3p and that hsa-miR-24-3p inhibition increases eNOS protein expression [30,31] with a consequent role in the angiogenic process. In contrast with previous findings [19], our study did not reveal any correlation between miR-126 circulating levels and patient outcome.

We also showed that baseline circulating miRNA levels differed with respect to patient clinical pathological characteristics, in particular tumor location and *RAS* status. As defined previously [32], left-sided tumors (originating in the splenic flexure, descending colon, sigmoid colon, rectum or one-third of the transverse colon) derive from the embryonic hindgut, whereas right-sided tumors (originating in the appendix, cecum, ascending colon, hepatic flexure or two-thirds of the transverse colon) derive from the embryonic midgut. The hsa-miR-199a-5p, hsa-miR-335-5p and hsa-miR-520d-3p miRNAs were significantly more upregulated in patients with left-sided than right-sided lesions, reflecting the different tumor biology. As all three miRNAs act as tumor suppressors [33–36], their overexpression in left-sided tumors could partially explain the better outcome of this group of patients. These differences in circulating miRNA expression with respect to tumor side agree with our recent report indicating different gene expression profiles, inflammatory indexes and responses to B in patients with left- and right-sided tumors [37]. Furthermore, miRNAs has-miR-21-5p and hsa-miR-221-3p were found to be significantly correlated with RAS status. Decreased hsa-miR-21-5p and increased hsa-miR-221-3p were observed in RAS mutated patients with respect to *RAS* wt patients. Although it has been demonstrated that these miRNAs are involved in the angiogenic process, we did not observe a correlation between hsa-miR-21-5p or hsa-miR-221-3p and response to bevacizumab.

This study has some limitations. First, the sample size was small, making it necessary to confirm results in a larger case series. Moreover, the lack of a control group treated with chemotherapy alone did not permit us to understand whether miRNAs were of prognostic value or predictive of response to B. Finally, we restricted our analysis to a specific panel of miRNAs on the basis of literature results, but cannot exclude that other miRNAs may play a role in response to B.

4. Materials and Methods

4.1. Case Series

This study included patients enrolled in the ITACa clinical trial [17], randomized to be treated with first-line CT (FOLFOX4 or FOLFIRI) only or CT plus B. Fifty-two patients in the CT + B arm, whose biological material was available, were analyzed for this study. All patients were characterized for RAS and BRAF status by MassARRAY (Sequenom, San Diego, CA, USA) using the Myriapod Colon status (Diatech Pharmacogenetics, Jesi, Italy) as the routine diagnostic procedure. Consenting patients underwent periodic blood sampling: peripheral blood samples were collected at baseline (before treatment began), at the first evaluation (after around 2 months) and when progressive disease

(PD) was documented. All patients were assessed for response, PFS and OS according to RECIST (Response Evaluation Criteria In Solid Tumors) criteria Version 1.1. Tumor response was evaluated every 2 months by CT scan. Responders included patients with a complete response (CR) and a partial response (PR). Non-responders included patients with stable disease (SD) or PD. The study protocol was approved by the Local Ethics Committee (Comitato Etico Area Vasta e Istituto Scientifico Romagnolo per lo Studio e la Cura dei Tumori, no. 674) on 19 September 2007. All patients gave informed consent before blood sample collection.

4.2. miRNA Selection

A review of the literature was made for selecting the panel of miRNAs for analysis in plasma. The miRNAs were selected on the basis of their role in the angiogenic process, especially in CRC, and evidence of their possible determination in plasma or serum [38–40]. Twenty-three miRNAs were selected for analysis as follows: has-miR-107, hsa-miR-126-3hashsa-miR-145-5p, hsa-miR-194-5p, hsa-miR-199a-5p, hsa-miR-200b-3p, hsa-miR-20b-5p, hsa-miR-21-5p, hsa-miR-210-3p, hsa-miR-221-3p, hsa-miR-223-3p, hsa-miR-24-3p, hsa-miR-27a-3p, hsa-miR-29b-3p, hsa-miR-335-5p, hsa-miR-424-5p, hsa-miR-484, hsa-miR-497-5p, hsa-miR-520d-3p, hsa-miR-92a-3p, hsa-miR-17-5p, hsa-miR-155-5p, hsa-miR-16-5p. Moreover, cel-miR-39-3p was used as a spike-in for exogenous normalization.

4.3. Circulating miRNA Expression Analysis

Plasma was obtained from peripheral blood collected in EDTA-tubes, after centrifugation at 3000 rpm for 15 min. Plasma samples were stored at $-80\,^\circ$C until miRNA extraction. miRNAs were extracted from 400 µL of plasma using the miRVANA PARIS kit (Thermofisher, Monza, Italy). The 24 selected miRNAs were then spotted in array custom plates. For the selection of housekeeping (HSK) miRNAs, results were analyzed by GeNorm software (v. 3.2) to evaluate the stablest miRNAs. Assays were run on a 7500 Real-Time PCR System (Thermofisher). The reactions were initiated at 95 $^\circ$C for 5 min followed by 40 cycles of 95 $^\circ$C for 15 s and 60 $^\circ$C for 1 min. All reactions, including the no template controls, were run in duplicate. Data were analyzed using Expression suite software v1.1 (Thermofisher) according to the $\Delta\Delta$Ct method.

4.4. Statistical Analysis

The aim of this analysis was to examine the association between baseline circulating miRNA expression levels and PFS, OS and ORR in the ITACa case series and to evaluate their modification during CT + B therapy. The primary objective of the ITACa trial was PFS. Secondary efficacy endpoints were ORR and OS. PFS was calculated as the time from the date of randomization to the date of the first documented evidence of PD (per investigator assessment), last tumor assessment or death in the absence of disease progression. Patients submitted to curative metastasectomy were censored at the time of surgery. OS was calculated as the time from the date of randomization to the date of death from any cause or last follow-up. Descriptive statistics were used to describe patients. The relationship between baseline miRNA expression and clinical pathological factors was evaluated using a nonparametric ranking statistic test (median test). The median value of variation in the case series (30%) was set as the cutoff point. Time-to-event data (PFS, OS) were described using the Kaplan–Meier method and compared using the log rank test (significance level of 5%). Ninety-five percent confidence intervals (95% CI) were calculated using nonparametric methods. Estimated HR and their 95% CI were calculated by the Cox regression model. The multivariate Cox regression model was used to select the most useful prognostic markers of all the miRNAs used (considered as continuous variables) adjusting for clinical pathological characteristics of patients statistically significant at univariate analysis. Circulating basal levels of miRNAs were dichotomized into "high" or "low" according to median values [41]. We also conducted landmark analyses to reduce possible confounding by time on treatment by assessing the impact of miRNA level change from baseline to first tumor evaluation (about 2 months of onset of the treatment protocol) of survival outcomes. Patients who were still alive and not lost to follow-up at the landmark time were divided into two categories, i.e., patients who had progressed and patients who had not progressed by that time. PFS and OS after the landmark time were computed using the Kaplan–Meier curves. Logistic regression models were used to assess OR and their 95% CI in order to evaluate the association between miRNA baseline levels and ORR (CR + PR). All *p*-values were based on two-sided testing, and statistical analyses were performed using SAS statistical software Version 9.4 (SAS Institute, Cary, NC, USA).

5. Conclusions

This study showed that circulating higher levels of hsa-miR-20b-5p, hsa-miR-29b-3p and hsa-miR-155-5p at baseline are associated with a better prognosis in mCRC patients treated with B-based CT. Measuring these miRNAs before treatment could be helpful in selecting the patients who are more likely to benefit from the drug. The increase in circulating hsa-miR-155-5p after one month of treatment is associated with much shorter PFS and OS, suggesting that the determination of this miRNA during treatment could give important information for monitoring the drug response. These results should be confirmed and verified in an independent case series before being translated into the clinical setting.

Supplementary Materials: Supplementary materials can be found at www.mdpi.com/1422-0067/19/1/307/s1.

Acknowledgments: The authors would like to thank Veronica Zanoni for editorial assistance.

Author Contributions: Paola Ulivi conceived of and designed the experiments. Alessandro Passardi, Martina Valgiusti and Giovanni Luca Frassineti were responsible for data collection. Matteo Canale and Giorgia Marisi performed the experiments. Paola Ulivi and Emanuela Scarpi analyzed the data and wrote the manuscript. Dino Amadori and Daniele Calistri critically revised the manuscript for important intellectual content. All authors read and approved the final version for submission.

Conflicts of Interest: The authors declare no conflict of interest.

Abbreviations

B	bevacizumab
CT	chemotherapy
eNOS	nitric oxide synthase
mCRC	metastatic colorectal cancer
miRNA	microRNA
MMP2	metalloproteinase-2
ORR	objective response rate
OS	overall survival
PFS	progression-free survival

References

1. Hochster, H.S.; Hart, L.L.; Ramanathan, R.K.; Childs, B.H.; Hainsworth, J.D.; Cohn, A.L.; Wong, L.; Fehrenbacher, L.; Abubakr, Y.; Saif, M.W.; et al. Safety and Efficacy of Oxaliplatin and Fluoropyrimidine Regimens with or without Bevacizumab as First-Line Treatment of Metastatic Colorectal Cancer: Results of the TREE Study. *J. Clin. Oncol.* **2008**, *26*, 3523–3529. [CrossRef] [PubMed]
2. Saltz, L.B.; Clarke, S.; Diaz-Rubio, E.; Scheithauer, W.; Figer, A.; Wong, R.; Koski, S.; Lichinitser, M.; Yang, T.S.; Rivera, F.; et al. Bevacizumab in Combination with Oxaliplatin-Based Chemotherapy as First-Line Therapy in Metastatic Colorectal Cancer: A Randomized Phase III Study. *J. Clin. Oncol.* **2008**, *26*, 2013–2019. [CrossRef] [PubMed]
3. Hurwitz, H.; Fehrenbacher, L.; Novotny, W.; Cartwright, T.; Hainsworth, J.; Heim, W.; Berlin, J.; Baron, A.; Griffing, S.; Holmgren, E.; et al. Bevacizumab Plus Irinotecan, Fluorouracil, and Leucovorin for Metastatic Colorectal Cancer. *N. Engl. J. Med.* **2004**, *350*, 2335–2342. [CrossRef] [PubMed]

4. Gonzalez-Vacarezza, N.; Alonso, I.; Arroyo, G.; Martinez, J.; De Andres, F.; LLerena, A.; Estevez-Carrizo, F. Predictive Biomarkers Candidates for Patients with Metastatic Colorectal Cancer Treated with Bevacizumab-Containing Regimen. *Drug Metab. Pers. Ther.* **2016**, *31*, 83–90. [CrossRef] [PubMed]

5. Marien, K.M.; Croons, V.; Martinet, W.; De Loof, H.; Ung, C.; Waelput, W.; Scherer, S.J.; Kockx, M.M.; De Meyer, G.R. Predictive Tissue Biomarkers for Bevacizumab-Containing Therapy in Metastatic Colorectal Cancer: An Update. *Expert Rev. Mol. Diagn.* **2015**, *15*, 399–414. [CrossRef] [PubMed]

6. Cidon, E.U.; Alonso, P.; Masters, B. Markers of Response to Antiangiogenic Therapies in Colorectal Cancer: Where are we Now and what should be Next? *Clin. Med. Insights Oncol.* **2016**, *10*, 41–55. [CrossRef] [PubMed]

7. Inamura, K.; Ishikawa, Y. MicroRNA in Lung Cancer: Novel Biomarkers and Potential Tools for Treatment. *J. Clin. Med.* **2016**, *5*. [CrossRef] [PubMed]

8. Fish, J.E.; Santoro, M.M.; Morton, S.U.; Yu, S.; Yeh, R.F.; Wythe, J.D.; Ivey, K.N.; Bruneau, B.G.; Stainier, D.Y.; Srivastava, D. MiR-126 Regulates Angiogenic Signaling and Vascular Integrity. *Dev. Cell.* **2008**, *15*, 272–284. [CrossRef] [PubMed]

9. Ota, T.; Doi, K.; Fujimoto, T.; Tanaka, Y.; Ogawa, M.; Matsuzaki, H.; Kuroki, M.; Miyamoto, S.; Shirasawa, S.; Tsunoda, T. KRAS Up-Regulates the Expression of miR-181a, miR-200c and miR-210 in a Three-Dimensional-Specific Manner in DLD-1 Colorectal Cancer Cells. *Anticancer Res.* **2012**, *32*, 2271–2275. [PubMed]

10. Wang, X.; Wang, J.; Ma, H.; Zhang, J.; Zhou, X. Downregulation of miR-195 Correlates with Lymph Node Metastasis and Poor Prognosis in Colorectal Cancer. *Med. Oncol.* **2012**, *29*, 919–927. [CrossRef] [PubMed]

11. Wang, C.J.; Stratmann, J.; Zhou, Z.G.; Sun, X.F. Suppression of microRNA-31 Increases Sensitivity to 5-FU at an Early Stage, and Affects Cell Migration and Invasion in HCT-116 Colon Cancer Cells. *BMC Cancer* **2010**, *10*, 616. [CrossRef] [PubMed]

12. Kanaan, Z.; Rai, S.N.; Eichenberger, M.R.; Barnes, C.; Dworkin, A.M.; Weller, C.; Cohen, E.; Roberts, H.; Keskey, B.; Petras, R.E.; et al. Differential microRNA Expression Tracks Neoplastic Progression in Inflammatory Bowel Disease-Associated Colorectal Cancer. *Hum. Mutat.* **2012**, *33*, 551–560. [CrossRef] [PubMed]

13. Hur, K.; Toiyama, Y.; Takahashi, M.; Balaguer, F.; Nagasaka, T.; Koike, J.; Hemmi, H.; Koi, M.; Boland, C.R.; Goel, A. MicroRNA-200c Modulates Epithelial-to-Mesenchymal Transition (EMT) in Human Colorectal Cancer Metastasis. *Gut* **2013**, *62*, 1315–1326. [CrossRef] [PubMed]

14. Xu, K.; Liang, X.; Shen, K.; Sun, L.; Cui, D.; Zhao, Y.; Tian, J.; Ni, L.; Liu, J. miR-222 Modulates Multidrug Resistance in Human Colorectal Carcinoma by Down-Regulating ADAM-17. *Exp. Cell Res.* **2012**, *318*, 2168–2177. [CrossRef] [PubMed]

15. Boisen, M.K.; Dehlendorff, C.; Linnemann, D.; Nielsen, B.S.; Larsen, J.S.; Osterlind, K.; Nielsen, S.E.; Tarpgaard, L.S.; Qvortrup, C.; Pfeiffer, P.; et al. Tissue microRNAs as Predictors of Outcome in Patients with Metastatic Colorectal Cancer Treated with First Line Capecitabine and Oxaliplatin with or without Bevacizumab. *PLoS ONE* **2014**, *9*, e109430. [CrossRef] [PubMed]

16. Inamura, K. Diagnostic and Therapeutic Potential of MicroRNAs in Lung Cancer. *Cancers (Basel)* **2017**, *9*. [CrossRef] [PubMed]

17. Hayes, J.; Thygesen, H.; Gregory, W.; Westhead, D.R.; French, P.J.; Van Den Bent, M.J.; Lawler, S.E.; Short, S.C. A Validated microRNA Profile with Predictive Potential in Glioblastoma Patients Treated with Bevacizumab. *Mol. Oncol.* **2016**, *10*, 1296–1304. [CrossRef] [PubMed]

18. Joerger, M.; Baty, F.; Fruh, M.; Droege, C.; Stahel, R.A.; Betticher, D.C.; von Moos, R.; Ochsenbein, A.; Pless, M.; Gautschi, O.; et al. Circulating microRNA Profiling in Patients with Advanced Non-Squamous NSCLC Receiving bevacizumab/erlotinib Followed by Platinum-Based Chemotherapy at Progression (SAKK 19/05). *Lung Cancer* **2014**, *85*, 306–313. [CrossRef] [PubMed]

19. Hansen, T.F.; Carlsen, A.L.; Heegaard, N.H.; Sorensen, F.B.; Jakobsen, A. Changes in Circulating microRNA-126 during Treatment with Chemotherapy and Bevacizumab Predicts Treatment Response in Patients with Metastatic Colorectal Cancer. *Br. J. Cancer* **2015**, *112*, 624–629. [CrossRef] [PubMed]

20. Passardi, A.; Nanni, O.; Tassinari, D.; Turci, D.; Cavanna, L.; Fontana, A.; Ruscelli, S.; Mucciarini, C.; Lorusso, V.; Ragazzini, A.; et al. Effectiveness of Bevacizumab Added to Standard Chemotherapy in Metastatic Colorectal Cancer: Final Results for First-Line Treatment from the ITACa Randomized Clinical Trial. *Ann. Oncol.* **2015**, *26*, 1201–1207. [CrossRef] [PubMed]

21. Fang, J.H.; Zhou, H.C.; Zeng, C.; Yang, J.; Liu, Y.; Huang, X.; Zhang, J.P.; Guan, X.Y.; Zhuang, S.M. MicroRNA-29b Suppresses Tumor Angiogenesis, Invasion, and Metastasis by Regulating Matrix Metalloproteinase 2 Expression. *Hepatology* **2011**, *54*, 1729–1740. [CrossRef] [PubMed]

22. Wang, H.; Guan, X.; Tu, Y.; Zheng, S.; Long, J.; Li, S.; Qi, C.; Xie, X.; Zhang, H.; Zhang, Y. MicroRNA-29b Attenuates Non-Small Cell Lung Cancer Metastasis by Targeting Matrix Metalloproteinase 2 and PTEN. *J. Exp. Clin. Cancer Res.* **2015**, *34*. [CrossRef] [PubMed]

23. Li, Y.; Cai, B.; Shen, L.; Dong, Y.; Lu, Q.; Sun, S.; Liu, S.; Ma, S.; Ma, P.X.; Chen, J. MiRNA-29b Suppresses Tumor Growth through Simultaneously Inhibiting Angiogenesis and Tumorigenesis by Targeting Akt3. *Cancer Lett.* **2017**, *397*, 111–119. [CrossRef] [PubMed]

24. Wong, P.F.; Jamal, J.; Tong, K.L.; Khor, E.S.; Yeap, C.E.; Jong, H.L.; Lee, S.T.; Mustafa, M.R.; Abubakar, S. Deregulation of hsa-miR-20b Expression in TNF-α-Induced Premature Senescence of Human Pulmonary Microvascular Endothelial Cells. *Microvasc. Res.* **2017**, *114*, 26–33. [CrossRef] [PubMed]

25. Yang, D.; Wang, J.; Xiao, M.; Zhou, T.; Shi, X. Role of Mir-155 in Controlling HIF-1α Level and Promoting Endothelial Cell Maturation. *Sci. Rep.* **2016**, *6*, 35316. [CrossRef] [PubMed]

26. Pankratz, F.; Bemtgen, X.; Zeiser, R.; Leonhardt, F.; Kreuzaler, S.; Hilgendorf, I.; Smolka, C.; Helbing, T.; Hoefer, I.; Esser, J.S.; et al. MicroRNA-155 Exerts Cell-Specific Antiangiogenic but Proarteriogenic Effects during Adaptive Neovascularization. *Circulation* **2015**, *131*, 1575–1589. [CrossRef] [PubMed]

27. Kong, W.; He, L.; Richards, E.J.; Challa, S.; Xu, C.X.; Permuth-Wey, J.; Lancaster, J.M.; Coppola, D.; Sellers, T.A.; Djeu, J.Y.; et al. Upregulation of miRNA-155 Promotes Tumour Angiogenesis by Targeting VHL and is Associated with Poor Prognosis and Triple-Negative Breast Cancer. *Oncogene* **2014**, *33*, 679–689. [CrossRef] [PubMed]

28. Robertson, E.D.; Wasylyk, C.; Ye, T.; Jung, A.C.; Wasylyk, B. The Oncogenic MicroRNA hsa-miR-155–5p Targets the Transcription Factor ELK3 and Links it to the Hypoxia Response. *PLoS ONE* **2014**, *9*, e113050. [CrossRef] [PubMed]

29. Wan, J.; Xia, L.; Xu, W.; Lu, N. Expression and Function of miR-155 in Diseases of the Gastrointestinal Tract. *Int. J. Mol. Sci.* **2016**, *17*. [CrossRef] [PubMed]

30. Meloni, M.; Marchetti, M.; Garner, K.; Littlejohns, B.; Sala-Newby, G.; Xenophontos, N.; Floris, I.; Suleiman, M.S.; Madeddu, P.; Caporali, A.; et al. Local Inhibition of microRNA-24 Improves Reparative Angiogenesis and Left Ventricle Remodeling and Function in Mice with Myocardial Infarction. *Mol. Ther.* **2013**, *21*, 1390–1402. [CrossRef] [PubMed]

31. Marisi, G.; Scarpi, E.; Passardi, A.; Nanni, O.; Ragazzini, A.; Valgiusti, M.; Casadei Gardini, A.; Neri, L.; Frassineti, G.; Amadori, D.; et al. Circulating *VEGF* and *eNOS* Variations as Predictors of Outcome in Metastatic Colorectal Cancer Patients Receiving Bevacizumab. *Sci. Rep.* **2017**, *7*. [CrossRef] [PubMed]

32. Tejpar, S.; Stintzing, S.; Ciardiello, F.; Tabernero, J.; van Cutsem, E.; Beier, F.; Esser, R.; Lenz, H.J.; Heinemann, V. Prognostic and Predictive Relevance of Primary Tumor Location in Patients with RAS Wild-Type Metastatic Colorectal Cancer: Retrospective Analyses of the CRYSTAL and FIRE-3 Trials. *JAMA Oncol.* **2017**, *3*, 194–201. [CrossRef] [PubMed]

33. Sun, Z.; Zhang, Z.; Liu, Z.; Qiu, B.; Liu, K.; Dong, G. MicroRNA-335 Inhibits Invasion and Metastasis of Colorectal Cancer by Targeting ZEB2. *Med. Oncol.* **2014**, *31*. [CrossRef] [PubMed]

34. Sun, D.; Han, S.; Liu, C.; Zhou, R.; Sun, W.; Zhang, Z.; Qu, J. Microrna-199a-5p Functions as a Tumor Suppressor Via Suppressing Connective Tissue Growth Factor (CTGF) in Follicular Thyroid Carcinoma. *Med. Sci. Monit.* **2016**, *22*, 1210–1217. [CrossRef] [PubMed]

35. Yang, X.; Lei, S.; Long, J.; Liu, X.; Wu, Q. MicroRNA-199a-5p Inhibits Tumor Proliferation in Melanoma by Mediating HIF-1α. *Mol. Med. Rep.* **2016**, *13*, 5241–5247. [CrossRef] [PubMed]

36. Tsuno, S.; Wang, X.; Shomori, K.; Hasegawa, J.; Miura, N. Hsa-miR-520d Induces Hepatoma Cells to Form Normal Liver Tissues Via a Stemness-Mediated Process. *Sci. Rep.* **2014**, *4*, 3852. [CrossRef] [PubMed]

37. Ulivi, P.; Scarpi, E.; Chiadini, E.; Marisi, G.; Valgiusti, M.; Capelli, L.; Casadei Gardini, A.; Monti, M.; Ruscelli, S.; Frassineti, G.L.; et al. Right- vs. Left-Sided Metastatic Colorectal Cancer: Differences in Tumor Biology and Bevacizumab Efficacy. *Int. J. Mol. Sci.* **2017**, *18*. [CrossRef] [PubMed]

38. Hong, L.; Han, Y.; Zhou, Y.; Nita, A. Angiogenesis-Related microRNAs in Colon Cancer. *Expert Opin. Biol. Ther.* **2013**, *13*, 77–84. [CrossRef] [PubMed]

39. Muhammad, S.; Kaur, K.; Huang, R.; Zhang, Q.; Kaur, P.; Yazdani, H.O.; Bilal, M.U.; Zheng, J.; Zheng, L.; Wang, X.S. MicroRNAs in Colorectal Cancer: Role in Metastasis and Clinical Perspectives. *World J. Gastroenterol.* **2014**, *20*, 17011–17019. [CrossRef] [PubMed]

40. Gallach, S.; Calabuig-Farinas, S.; Jantus-Lewintre, E.; Camps, C. MicroRNAs: Promising New Antiangiogenic Targets in Cancer. Biomed. *Res. Int.* **2014**, *2014*, 878450. [CrossRef] [PubMed]

41. Altman, D.G.; McShane, L.M.; Sauerbrei, W.; Taube, S.E. Reporting Recommendations for Tumor Marker Prognostic Studies (REMARK): Explanation and Elaboration. *PLoS Med.* **2012**, *9*, e1001216. [CrossRef] [PubMed]

International Journal of
Molecular Sciences

MDPI

Article

Co-Detection of miR-21 and TNF-α mRNA in Budding Cancer Cells in Colorectal Cancer

Trine Møller [1], Jaslin P James [1], Kim Holmstrøm [1], Flemming B Sørensen [2,3], Jan Lindebjerg [2,4,5] and Boye S Nielsen [1,*]

[1] Bioneer A/S, Hørsholm, Kogle Allé 2, 2970 Hørsholm, Denmark; trm@bioneer.dk (T.M.); jaslin.pj@gmail.com (J.P.J.); kho@bioneer.dk (K.H.)

[2] Danish Colorectal Cancer Center South, Vejle Hospital, Part of Lillebaelt Hospital, Beriderbakken 4, 7100 Vejle, Denmark; flsoer@rm.dk (F.B.S.); Jan.Lindebjerg@rsyd.dk (J.L.)

[3] University Institute of Pathology, Aarhus University Hospital, Palle Juul-Jensens Boulevard 99, 8200 Aarhus N, Denmark

[4] Department of Pathology, Vejle Hospital, Part of Lillebaelt Hospital, Beriderbakken 4, 7100 Vejle, Denmark

[5] Institute of Regional Health Research, University of Southern Denmark, Winsløwparken 19,3, 5000 Odense C, Denmark

* Correspondence: bsn@bioneer.dk

Received: 31 January 2019; Accepted: 15 April 2019; Published: 17 April 2019

Abstract: MicroRNA-21 (miR-21) is upregulated in many cancers including colon cancers and is a prognostic indicator of recurrence and poor prognosis. In colon cancers, miR-21 is highly expressed in stromal fibroblastic cells and more weakly in a subset of cancer cells, particularly budding cancer cells. Exploration of the expression of inflammatory markers in colon cancers revealed tumor necrosis factor alpha (TNF-α) mRNA expression at the invasive front of colon cancers. Surprisingly, a majority of the TNF-α mRNA expressing cells were found to be cancer cells and not inflammatory cells. Because miR-21 is positively involved in cell survival and TNF-α promotes necrosis, we found it interesting to analyze the presence of miR-21 in areas of TNF-α mRNA expression at the invasive front of colon cancers. For this purpose, we developed an automated procedure for the co-staining of miR-21, TNF-α mRNA and the cancer cell marker cytokeratin based on analysis of frozen colon cancer tissue samples ($n = 4$) with evident cancer cell budding. In all four cases, TNF-α mRNA was seen in a small subset of cancer cells at the invasive front. Evaluation of miR-21 and TNF-α mRNA expression was performed on digital slides obtained by confocal slide scanning microscopy. Both co-expression and lack of co-expression with miR-21 in the budding cancer cells was noted, suggesting non-correlated expression. miR-21 was more often seen in cancer cells than TNF-α mRNA. In conclusion, we report that miR-21 is not linked to expression of the pro-inflammatory cytokine TNF-α mRNA, but that miR-21 and TNF-α both take part in the cancer expansion at the invasive front of colon cancers. We hypothesize that miR-21 may protect both fibroblasts and cancer cells from cell death directed by TNF-α paracrine and autocrine activity.

Keywords: colorectal cancer; confocal slide scanning microscopy; inflammation; interleukin-1β; microRNA; miR-21; TNF-α; tumor budding cells

1. Introduction

MicroRNAs are short regulatory RNAs that are formed as inactive precursors with hairpin-like structure from which a 3p and 5p strand containing 19–23 nucleotides are generated after cleavage by endonucleases [1–3]. The released 3p and 5p strand have different functions, but often only one of the strands associates with the RNA-induced silencing complex (RISC). The microRNA:RISC binds to unique complementary RNA sequences often located in the untranslated 3′ end of a mRNA and leads to de-stabilization or degradation of the individual target mRNA. microRNA-21 (miR-21) is one of most consistently upregulated microRNAs in cancer tissue including colorectal cancer [4–7]. miR-21 is expressed predominantly in stromal fibroblast-like cells, but is seen also in populations of cancer cells [7,8]. In colon cancer tissue, we recently described the presence of miR-21 in budding cancer cells [9]. Budding cancer cells are de-differentiated cancer cells that have detached from the cohesive, more differentiated, part of the colon cancer and are located as single cells or small clusters of cancer cells at the invasive front [10]. The prevalence of budding cancer cells in tumor microenvironment is associated with increased metastasis and poor prognosis [11,12]. The function of miR-21 in colon cancer fibroblasts and cancer cells is not clear, but its regulatory roles are likely different in different cell populations. miR-21 is involved in fibrosis as part of the transforming growth factor (TGF-β) induced fibrosis pathway [13], and targets programmed cell death-4 (PDCD4) to sustain cell survival, cancer cell invasion and metastasis [14]. miR-21 is also upregulated in inflammatory bowel disease [15–17], in which expression is seen in thus far uncharacterized inflammatory cells [16].

Inflammation is an inherent part of colorectal cancer progression. Several inflammatory cytokines are upregulated in cancer tissue, including tumor necrosis factor alpha (TNF-α), a potent cytokine that causes necrosis and inflammation [18–20] and promotes cancer [21]. Increased levels of TNF-α are associated with metastatic disease in several cancer types including colorectal cancer [22–24]. TNF-α is produced in cells as a type II transmembrane protein arranged in stable homo-trimers [25]. Soluble TNF-α is generated by proteolytic cleavage by TNF-α converting enzyme (TACE) that forms the active soluble homo-trimeric cytokine. TNF-α acts through TNF-α receptors (TNFR) to induce cell signaling. Interestingly, Cottonham et al. [26] showed that TNF-α in cooperation with TGF-β positively regulates the expression of miR-21 in colorectal cancer cells, and that the cells in an organoid model increase their motility and invasiveness. In addition, Qiu et al. [27] found that miR-21 expression was positively correlated with TNF-α in oral cancer cells and controlled proliferation and apoptosis, and Chen et al. [28] found that miR-21 depletion inhibited secretion of TNF-α in a rat model of diabetic nephropathy. Thus, several studies indicate an interplay between miR-21 and TNF-α putatively, of both paracrine and autocrine nature. If miR-21 regulates TNF-α protein expression, it would be expected to occur through an indirect pathway since TNF-α mRNA is not considered a direct target of miR-21 according to the microRNA target database (miRDB) (www.mirdb.org).

To study expression of individual molecules in clinical tissue samples, we have established in situ hybridization (ISH) technologies for microRNA and mRNA. Localization of the mRNA transcripts show the origin of protein synthesis in situ and thereby reveals the cellular expression in complex tissue structures. During recent years, new ISH technologies have been developed. For long RNAs, branched DNA (bDNA) probes (RNAscope) have allowed detection of mRNA expression with high specificity and high sensitivity [29]. The method allows detection of single molecules that are visible as single dots in cells and tissues and has been developed for both automated and manual procedures [30,31]. For microRNAs, the limited size of 19–23 nucleotides makes locked nucleic acid (LNA) probes advantageous [32]. LNA probes have been used for detection of multiple microRNAs both in manual and automated assays, on paraffin and frozen tissue sections, and in single-plex and du-plex applications [7,33–36].

This study was undertaken to explore expression of miR-21 and TNF-α mRNA at the invasive front of colon cancers to elucidate interplay between miR-21 and TNF-α. To do this, we developed a combined microRNA and mRNA ISH assay that allows the use of LNA probes and bDNA probes on the same section as well as enables immunohistochemical detection of cellular markers.

2. Results

In this study, we performed in situ expression analyses on sections from both formalin-fixed, paraffin embedded (FFPE) samples and frozen (cryo-embedded) samples. The FFPE samples were used for chromogenic single-plex analysis and the frozen samples were used for the subsequent combined multi-plex fluorescence analyses.

2.1. Detection of TNF-α mRNA in Colon Cancer Tissue

Three FFPE samples were stained with RNAscope probes to TNF-α and matrix metalloproteinase-9 (MMP9) mRNAs together with PPIB mRNA and bacterial dapB mRNA probes as positive and negative controls, respectively (Figure 1). TNF-α mRNA ISH signal was weak and restricted to a few cells in the invasive front, where both a subset of cancer cells and stromal cells were positive. The presence of TNF-α mRNA in cancer cells was more evident in the case with high cancer cell budding (Figure 1a), whereas the two cases with low budding contained very few positive cells, and then seen in the stromal compartment (example in Figure 1e) and as occasional focal accentuation in tumor cells bordering stroma. Expression of MMP9 mRNA, which is known to be stimulated by TNF-α [37], was seen in all three samples (example in Figure 1c,g) in stromal cells with macrophage-like morphology in the expected expression pattern in the invasive front [38]. PPIB mRNA ISH signal was prominent in all three cases and stained both cancer and stromal cells, and the dapB probe resulted in no ISH signal, with only some un-specific chromogen depositions in focal areas. In one of the three cases (Figure 1a–d), a small area with normal mucosa was seen. Here, TNF-α mRNA was seen as single dots (presumably representing individual molecules) in a subset of the epithelial cells (Figure 1i–l), suggesting a very low level of expression in normal mucosa.

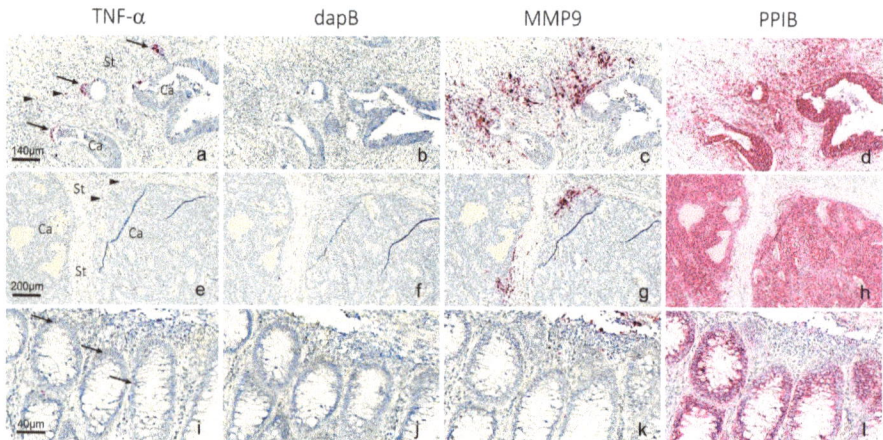

Figure 1. TNF-α and MMP9 mRNA in colorectal cancer samples. RNAscope probes to: TNF-α (**a,e**); dapB negative control (**b,f**); MMP9 (**c,g**); and PPIB positive control (**d,h**) mRNAs were used on serial sections from two FFPE colorectal cancer samples (**a–d**) with high budding and (**e–h**) with low budding, using automated RNAscope procedure and detection of the mRNAs with the AP and Fast Red substrate (red). In the sample shown in (**a–d**), an area with normal mucosa was present and expression of the RNA transcripts are shown in (**i–l**). Sections were counterstained with hematoxylin. TNF-α mRNA is seen as a distinct signal in a few cancer cells with branching characteristics (arrows) and a few stromal cells (arrowheads). MMP9 mRNA is very intense and seen in multiple stromal cells, typically macrophages. In normal mucosa, TNF-α mRNA is seen as single dots (presumably representing individual molecules) in a subset of the epithelial cells. The negative control probe dapB shows no signal, whereas the positive control probe to PPIB mRNA stains virtually all cells. Ca indicates cancer cell compartment; St indicates stromal cell compartment. Scale bars in (**a–d**): 140 μm; Scale bars in (**e–h**): 200 μm; Scale bars in (**i–l**): 40 μm.

2.2. Automation of microRNA ISH, mRNA ISH and IHC in a Combined Assay

To combine LNA-probe based microRNA ISH and RNAscope-probe based mRNA ISH on frozen sections, we first developed the microRNA ISH assay on the Ventana instrument on FFPE sections. As test probes, we included probes to microRNAs with previously reported consistent expression patterns [33,39], including miR-126 in endothelial cells, miR-21 in stromal cells and miR-17 in cancer cells (Figure 2). The microRNA ISH assay on the Ventana instrument was found to be equally or more sensitive than the Tecan assay [7] and was found suitable for unambiguous detection of microRNAs. The microRNA staining procedure with peroxidase and TSA substrate detection [34] was then combined with RNAscope procedure, using the AP (red kit) for mRNA detection and immunohistochemistry for cell marker detection on frozen sections from a colorectal cancer. Initially, miR-17 was combined with automated TNF-α and IL-1β mRNA RNAscope probe and cytokeratin immunofluorescence (Figure 3). The combined staining revealed localization of miR-17 in the cancer cells as expected, and TNF-α mRNA was, in this case, mostly seen in CK-positive cancer cells located at the invasive front (Figure 3). In contrast, IL-1β mRNA ISH signal was limited to stromal cells in focal areas close to cancer cell de-differentiation and budding (Figure 3). miR-17 expression was lost in the cancer cells at the outer invasive front including in the budding cancer cells (Figure 3) [39].

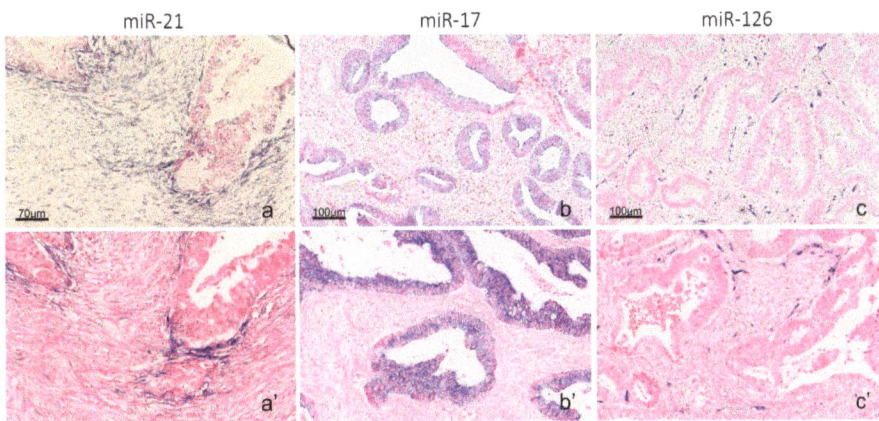

Figure 2. Automation of miR-21, miR-17 and miR-126 in a Ventana instrument. LNA probes to: miR-21 (**a**); miR-17 (**b**); and miR-126 (**c**) were used on FFPE colon cancer samples in a Tecan Genepaint instrument (**a**–**c**) and a Ventana Discovery Ultra instrument (**a′**–**c′**), using AP detection and NBT-BCIP substrate (blue). The sections were counterstained with nuclear fast red. At optimized experimental conditions, the Ventana and Tecan instruments showed identical staining patterns. miR-21 in stromal fibroblastic cells (**a,a′**); miR-17 in cancer cells (**b,b′**); and miR-126 in endothelial cells (**c,c′**). Scale bars in (**a**–**c**) are representative for (**a′**–**c′**).

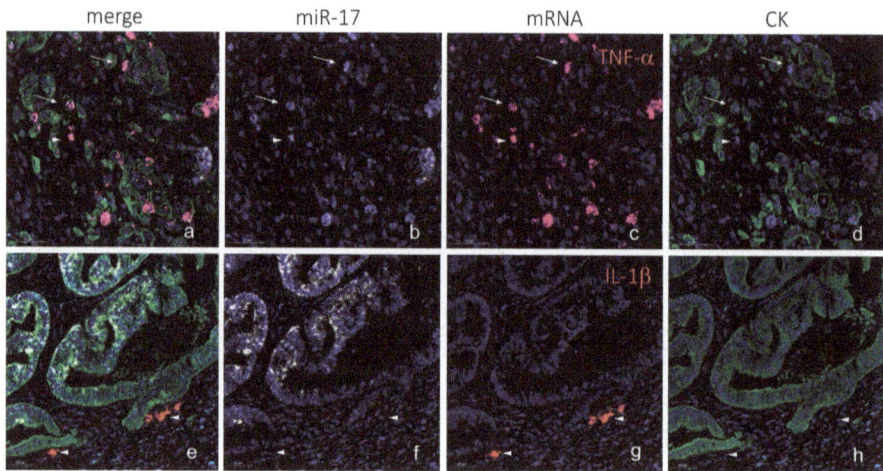

Figure 3. Combination of LNA ISH, RNAscope and IHC (AP procedure). Colorectal cancer sections from a cryo-embedded sample were stained for miR-17 using an LNA probe, and either TNF-α or IL-1β using RNAscope probes, and cytokeratin using immunohistochemistry. miR-17 (white in (**a,b,e,f**)) is seen in tumor cells in differentiated cancer cells (cohesive structures) and is absent in the de-differentiated cancer cells in areas with budding and/or branching. TNF-α mRNA (red in (**a,c**)) is seen in CK-positive cancer cells primarily located at the invasive front (arrows in (**a–d**)) and in CK-negative cells (arrowheads in (**a–d**)). IL-1β mRNA is seen CK-negative stromal cells (arrowheads in (**e–h**)). The cytokeratin staining (green in (**a,d,e,h**)) is seen in cancer cells including the differentiated cells of the cohesive structures and de-differentiated cells in branching and budding cancer cells. The CK staining was sub-optimal in this case—the miR-17 positive cancer cells are poorly stained with cytokeratin, and there is a dubious bleed-through of the red signal into the white channel. The blue fluorescence signal is nuclear DAPI counterstain in all images. Scale bars are indicated in lower left corner of each panel.

2.3. Co-Localization Analyses of TNF-α and miR-21

To evaluate co-localization of miR-21 and TNF-α mRNA in budding cancer cells, we performed the combined assay, detecting miR-21, TNF-α mRNA and CK, on four frozen colorectal cancer samples with evident budding characteristics. As a negative control, the scramble LNA probe was combined with dapB RNAscope and cytokeratin (CK) immunofluorescence. We changed the RNAscope detection assay to being HRP-based, using rhodamine substrate, to avoid masking effects of the Fast Red chromogen. In all four cases, TNF-α mRNA was seen in a subset of CK-positive cancer cells in the invasive front, whereas expression in (CK-negative) stromal cells was a rare event (see Figure 4). The TNF-α mRNA ISH signal was often seen in cancer cells with a branching appearance, i.e., cells that show initial outgrowth from the cohesive adenocarcinoma structure (Figure 4g,k), whereas the front-runner budding cancer cells were often TNF-α mRNA negative (Figure 4g,k). miR-21 was as expected seen in stromal fibroblastic cells, and subsets of cancer cells were also positive, including subsets of budding cancer cells (Figure 4f,j). No ISH signal was observed with the scramble probe and the dapB probe (Figure 4d), and the background fluorescence signal was limited to typical endogenous autofluorescence (Figure 4d). The combined stained slides were scanned using a confocal slide scanner with a 20× objective. Examples from the other three cases are shown in the Supplementary Materials. At 20× magnification, the TNF-α mRNA signal appeared as a rather diffuse signal. Hence, some areas of interest were scanned using a 40× water objective. At this magnification, the individual spots from the RNAscope assay were clearly visible (Figure 5).

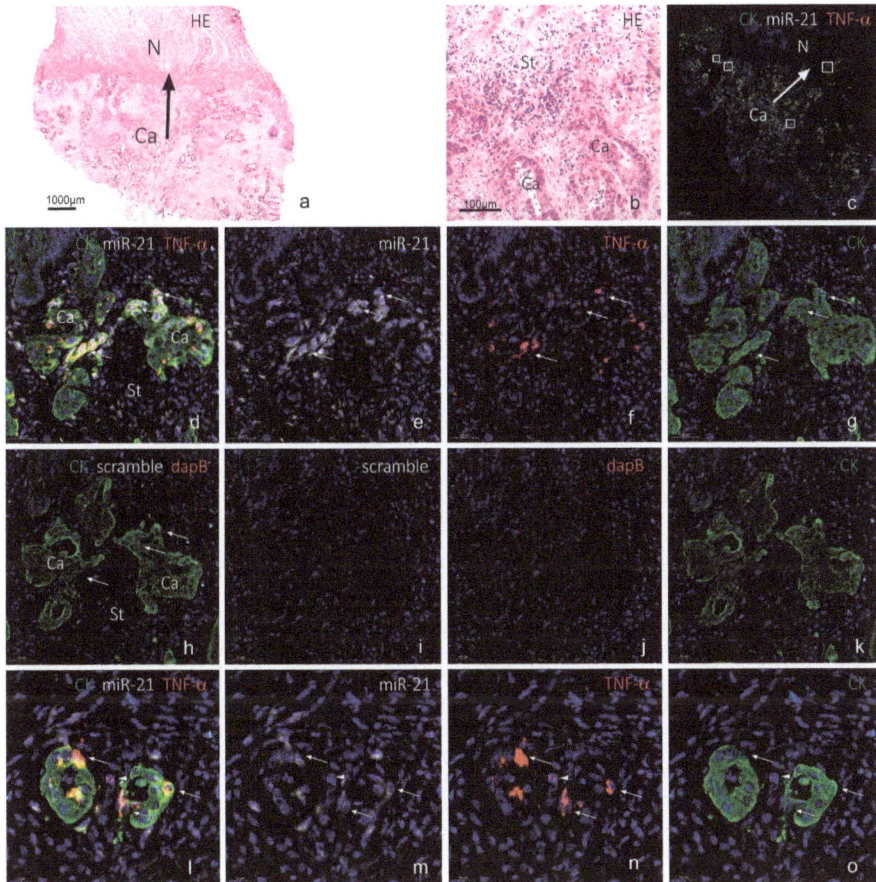

Figure 4. TNF-α mRNA and miR-21 in budding cancer cells in colorectal cancer (HRP procedure). The colon cancer tissue sample (cryo-embedded) is shown after H&E staining at low (**a**) and high magnification (**b**), and after combined fluorescence staining and subsequent confocal slide scanning microscopy at low magnification (**c**). The tissue section shows miR-21 (white in (**c,d,e,l,m**)) or scramble (white in (**h,i**)), TNF-α mRNA (red in (**c,d,f,l,n**)) or dapB mRNA (red in (**h,j**)) and CK (green in (**c,d,g,h,k,l,o**)). Probes to miR-21 and scramble probe were incubated on serial sections (**d–g**, and **h–k**) respectively. Arrows in (**d–h**) indicate the same cancer cell population in the two sections that are positive with the miR-21 and TNF-α probes and negative with the scramble and dapB probes. Arrows in (**l–o**) indicate miR-21 and TNF-α positive cancer cells and the arrowhead indicate a TNF-α positive stromal cell. Sections were scanned using a Pannoramic confocal slide scanner at 20× to give the overview image (**c**) and the framed areas are shown in (**d–g,l–o**) at high magnification (digital zoom), and two frames are shown Figure 5. Arrows in (**a,c**) indicate direction of invasion. Ca, cancer area; N, normal area; St, stroma area. Scale bars are indicated in lower left corner of each panel.

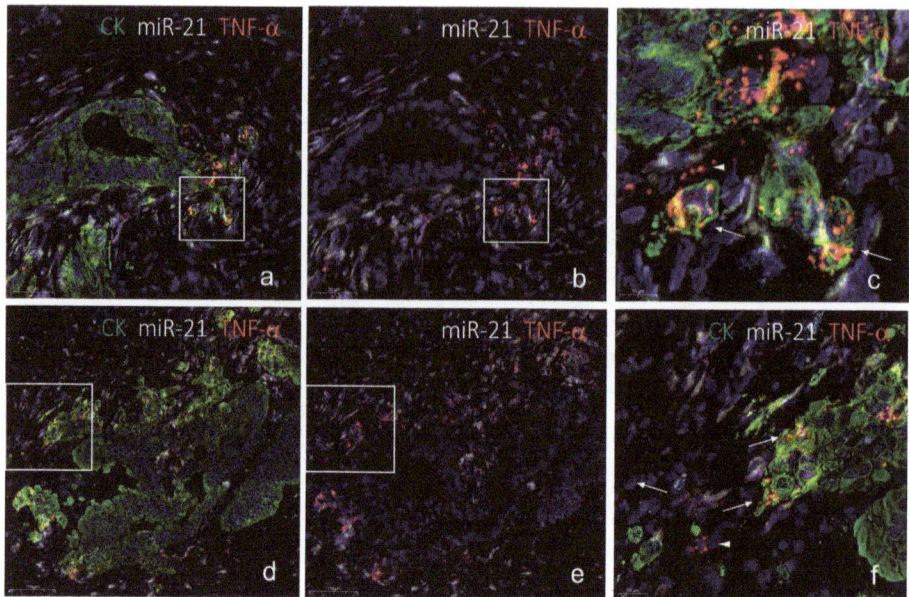

Figure 5. Confocal slide scanning at 40×. The same section as in Figure 4 here scanned using a 40× water objective. The two areas (**a–f**) show typical TNF-α mRNA expression at cancer cell branching with miR-21 expression in TNF-α-positive cells and neighboring TNF-α-positive signal in the stroma (arrowheads in (**c,f**)). Note that miR-21 positive stromal cells are TNF-α-negative. The RNAscope ISH signal is here evident as single dots representing individual mRNA molecules (arrows in (**c,f**)). Scale bars are indicated in lower left corner of each panel.

Since expression of TNF-α mRNA was very restricted compared to miR-21, we evaluated the TNF-α positive cancer cells in the four cases for miR-21 positivity within the same cell. In two cases, virtually all TNF-α positive cells were also miR-21-positive, whereas in the other two cases only 20% or less were miR-21 positive. TNF-α positive cells located in the stroma were generally miR-21 negative. In one of the cases, we noted that a TNF-α mRNA positive cancer cell was often associated with miR-21 in neighboring cancer cells and stromal cells. However, stromal fibroblastic cells located in the vicinity of TNF-α positive cancer cells did generally not show differences in the miR-21 expression patterns compared to areas without TNF-α mRNA signal. These observations suggest that miR-21 is not co-regulated with the TNF-α mRNA, neither within the same cell nor in neighboring cells.

3. Discussion

This study was undertaken to address if in situ localization analyses can help to clarify interplay between miR-21 and the pro-inflammatory cytokine TNF-α during colon cancer progression. We developed a fully automated combined in situ hybridization and immunohistochemistry assay that allowed use of LNA probe technology and RNAscope technology together with immunofluorescence on the same tissue section. Using CK immunofluorescence, all cancer cells at the invasive front of the colorectal cancers were visualized. Co-localization of the TNF-α mRNA in the CK positive cells indicated that this potent pro-inflammatory cytokine is expressed in a small subset of cancer cells located at the invasive front. We found that miR-21 also co-localizes in some, but not all, of these cancer cells, suggesting that miR-21 is not co-regulated with TNF-α mRNA, and that the presence of TNF-α mRNA does not lead to suppression of miR-21 expression. In normal mucosa, TNF-α mRNA was noted as a few single dots representing very low expression and miR-21 was only occasionally seen in normal mucosa [7], indicating that the cancer or associated inflammation may be required to induce expression of the two transcripts. We speculate that miR-21 in TNF-α mRNA positive cancer cells may suppress an autocrine effect of TNF-α to mediate cell death, e.g., via PDCD4 [40].

We previously developed a microRNA ISH method on a Tecan Genepaint instrument [7], which provides highly reproducible staining results; however, we found that RNAscope was not feasible for this instrument due to reagent consumption and consequently high cost. On the other hand, fully automated RNAscope assay procedures, e.g., on a Ventana instrument [30], have been developed by the vendors (ACDbio and Roche) and implemented in our Lab. To combine microRNA ISH with RNAscope, we therefore first established a microRNA ISH procedure on the Ventana instrument. Using the AP-based staining method, and similar reagents developed previously on the Tecan instrument, we obtained identical results with a slightly improved sensitivity and signal-to-noise ratio. Because pre-treatment procedures, required for microRNA ISH and RNAscope assays, were not compatible for FFPE samples, we developed the combined assay on frozen tissue, which also has the advantage of having better RNA integrity if processed for cryo-protection immediately after resection and appropriate storage. The combination of the three techniques into a triple fluorescence assay can potentially have many applications, including co-localization studies of a microRNA and its target mRNA and protein.

We investigated four cases of colorectal cancer that all included the invasive tumor front and prevalent cancer cell budding. In each of the four cases, multiple events of TNF-α mRNA expression and miR-21 expression in cancer cells could be evaluated. Because the same localization pattern was noted for TNF-α mRNA in the four cases, and also that the four cases showed the similar miR-21 localization pattern, we assume that our findings are representative for such adenocarcinomas despite the limited number of cases. The invasive growth pattern, however, varied with respect to the cancer cell structures, the presence of inflammation and fibrosis. In all four cases, the localization of TNF-α mRNA was similar, thus TNF-α positive cancer cells were often located at branching points of the cohesive adenocarcinoma structures located toward the invasive front or in detached de-differentiated budding cancer cells or in larger islands of cancer cells. TNF-α mRNA positive cancer cells were only occasionally seen in the central areas of the tumors. The characteristic TNF-α mRNA expression pattern was seen in both paraffin and cryo-embedded samples, and the same pattern was obtained with both the AP- and HRP-based RNAscope kits. Differences in staining intensity and prevalence of positive cells may be related to biological differences or differences in mRNA integrity. For specificity evaluation of the RNAscope assay, we included reference probes, of which, most importantly, the negative control probe resulted in the lack of ISH signal except from background. RNAscope probes to mRNAs for MMP9, PPIB and IL-1β all showed the expected expression patterns. Taken together, the observations suggest that the ISH signal obtained with the TNF-α probe is representing the genuine TNF-α mRNA. To our knowledge, this is the first study to show expression of TNF-α mRNA in budding and branching colon cancer cells in situ. The particular localization of the TNF-α mRNA expressing cells suggests a role in aggressive cancer cell invasion. In support of this observation, Li et al. [41] measured increased TNF-α mRNA levels in total RNA isolated from micro-dissected colorectal cancer budding cells and the surrounding microenvironment. In addition, TNF-α has been found to be involved in branching morphogenesis in in vivo models, e.g., during rat mammary gland development facilitated by MMP9 [42,43], suggesting that the cancer cells re-establish cellular mechanism used during early organ development. It is also of particular interest that TNF-α was found to induce an invasive phenotype in ovarian epithelial cystic structures [44]. In the latter study, TNF-α was added to the growth medium with the cystic structures, suggesting a paracrine mechanism inducing invasion, which would be expected if the source of TNF-α was inflammatory cells, including TNF-α-loaded neutrophilic granulocytes that may not carry the TNF-α mRNA. The observations in our study suggest that TNF-α also mediate an autocrine mechanism that contributes to cancer invasion. A larger sample set would be required to address if the expression of TNF-α mRNA by ISH or the co-expression with miR-21 is of prognostic relevance.

Digital slides obtained by confocal slide scanning microscopy allowed systematic evaluation of individual cells. We evaluated TNF-α mRNA and CK positive cells in each of the four cases for the presence of miR-21. The miR-21 positive fraction varied from 20% to 100%, suggesting no general correlation. In one of the cases, we noted that a TNF-α mRNA positive cancer cell was often associated with miR-21 in neighboring cancer cells and stromal cells. This observation could suggest that TNF-α stimulates miR-21 in various cells in the local neighborhood independent of the cell type. TNF-α has been reported to upregulate miR-21 expression in a variety of cell types, including human renal epithelial cells [45] and in Caco-2 cells used in an intestinal barrier model [46]. In a study by Xu et al. [47], it was observed that the expression of miR-21 increased gradually with low concentrations of TNF-α, while being suppressed at high concentrations. Thus, TNF-α is likely to have an effect on miR-21 expression, but in a complex tissue with many different cells and different signaling pathways interacting, distinct correlation patterns can be difficult to decipher. This may explain why no obvious correlation between TNF-α mRNA in cancer cells and stromal miR-21 was observed. Such patterns, may be better characterized in early stage cancers or in appropriate in vitro cell models. Conversely, miR-21 may control TNF-α expression. Ando et al. [48] found that a miR-21 mimic upregulated TNF-α in T-cells and Zhang et al. [49] reported that miR-21 directly targets TNF-α in bronchial epithelial cells. Future studies may disclose whether miR-21 and TNF-α are cooperating, competing or completely independent during cancer progression.

In conclusion, we report that miR-21 and TNF-α mRNA both are expressed at the invasive front of colon cancers and are co-localized in a subset of budding cancer cells and cells located at branching points and in clusters of cancer cells. Our attempts to clarify potential interactions between miR-21 and TNF-α did not result in consistent co-expression or converse expression patterns. If miR-21 expression is suppressed by TNF-α, therapeutic use of Infliximab would increase miR-21, which would promote tumor progression. Therefore, better understanding of the TNF-α/miR-21 interplay is highly warranted.

4. Material and Methods

4.1. Tissue Material

Three FFPE and four frozen colon adenocarcinoma tissue samples (Table 1) were obtained from Asterand (BioIVT, W Sussex, UK). The tissue samples were selected according to resection year (2015–2017), a well-defined invasive area and, for the frozen samples, also evident tumor cell budding and high RNA integrity number (RIN) values (>7). Evaluation of tissue sections stained with hematoxylin and eosin showed that all cases were moderately differentiated adenocarcinomas. The Asterand/BioIVT tissue samples were obtained according to the ethical principles defined in the Nuremburg Code, on the recommendations to consider when sourcing human biospecimens (https://info.bioivt.com/biospecimen-sourcing-white-paper).

Table 1. Tissue samples.

Case	Procurement	Type	Differentiation	Budding	ISH Method
1	FFPE-A	Ad	Mod	1 (BD1)	Chromogen
2	FFPE-B	Ad	Mod	2 (BD1)	Chromogen
3	FFPE-C	Ad	Mod	10 (BD3)	Chromogen
4	Cryo-A	Ad	Mod	12 (BD3)	Fluorescence
5	Cryo-B	Ad	Mod	6 (BD2)	Fluorescence
6	Cryo-C	Ad	Mod	6 (BD2)	Fluorescence
7	Cryo-D	Ad	Mod	5 (BD2)	Fluorescence

Ad, adenocarcinoma; BD, budding density (according to guidelines issued by International Tumor Budding Consensus Conference, April 2016); FFPE, formaldehyde fixed, paraffin embedded; Mod, moderately differentiated. Budding scores: BD1: 0–4, BD2: 5–9, BD3: >10 buds per 0.785 mm^2 field of view.

4.2. Probes for in situ Hybridization

Double-labeled locked nucleic acid (LNA™) probes were obtained from Qiagen (Exiqon-Qiagen, Hilden, Germany) as double digoxigenin-labeled LNA:DNA chimeric oligos. The microRNA antisense probes had approximately 30% of DNA replaced by LNA to increase the binding affinity (RNA T_m): miR-21-5p (TCAACATCAGTCTGATAAGCTA, RNA T_m = 83 °C), miR-17-5p (TACCTGCACTGTAAGCACTTT, RNA T_m = 89 °C), miR-126-3p (CATTATTACTCACGGTACGA, RNA T_m = 84 °C), and scramble (ATGTAACACGTCTATACGCCCA, RNA T_m = 86 °C). The microRNA probes recognize both the mature forms as well as precursor forms. RNAscope probes, or branched DNA (bDNA) probes, are designed to have high specificity and high sensitivity (www.acdbio.com). The high specificity is obtained by design of two antisense DNA oligonucleotides, also called double-Z probes, to bind adjacent sequences as pairs on the target sequence. The high sensitivity is partly based on design of typically up to 20 pairs for individual mRNA targets. The following RNAscope probes were obtained from ACD, Biotechne (Newark, CA): TNF-α (Tumor necrosis factor, target region: 67–1079, 13 zz pairs), IL-1β (Interleukin-1, target region: 2–1319, 20 zz pairs), MMP9 (matrix metalloproteinase-9, target region: 93–1422, 30 zz pairs), dapB (a *Bacillus subtilis* gene, 414–862, 10 zz pairs), and PPIB (Cyclophilin B, 139–989, 16 zz pairs).

4.3. Automated Chromogenic LNA ISH

MicroRNA ISH was performed using LNA probes on 5-μm-thick FFPE sections essentially as described previously [7], in which the specificity of the ISH staining was also analyzed. In brief, sections were deparaffinized and proteinase-K treated. Probes were added to the sections and incubated at concentrations and hybridization temperatures described earlier [7,39]. The probes were detected with alkaline phosphatase (AP) conjugated anti-DIG antibodies and stained with NBT-BCIP substrate for 30–60 min. In the Ventana instrument (Roche, Basel, Switzerland), protease-1 was used for tissue pre-treatment and probes were incubated at 2nM (miR-21) or 10nM (miR-17 and miR-126).

4.4. Automated Chromogenic RNAscope

RNAscope ISH was performed on 5-μm-thick FFPE sections essentially as described by Anderson et al. [30] using the AP (red) kit at recommended experimental conditions for the Ventana Discovery Ultra instrument. In this setup, we used Amp5 for 32 min (MMP9, IL-1β) or 60 min (TNF-α, dapB, PPIB).

4.5. Automated Combined ISH and IHC

Ten-micrometer-thick sections were obtained and fixed overnight in 4% paraformaldehyde. Staining of microRNA in frozen tissue sections has previously been described [35], using fluorescence detection with horse-radish-peroxidase (HRP)-conjugated anti-DIG antibodies and TSA-Cy5 substrate [34] (Roche). RNAscope ISH was performed as recommended by the manufacturer using the AP kit (Fast red substrate, Figure 3) or the HRP kit (TSA-rhodamine substrate, Figure 4, Figure 5 and Figure S1) in a Ventana Discovery Ultra instrument (Roche) [30]. Blocking of HRP in between the LNA assay and the RNAscope assay was done using the Discovery inhibitor for 10 min (Ventana, Roche Diagnostics). For cytokeratin immunofluorescence, the AE1/3 mouse monoclonal antibody (Dako-Agilent, Glostrup, Denmark) was used at 1:500 and detected with Alexa-488 conjugated anti-mouse Ig (Jackson Immunoresearch, West Grove, PA, USA). The negative control slides used as reference included replacement of the miR-21 probe with the scramble probe and replacement of the TNF-α probe with the dapB probe, whereas the anti-cytokeratin was retained, and was performed on cryo-samples A, B and C. Here, the miR-21 probe was incubated at 10 and 20 nM and the scramble probe at 20 nM. Despite a difference in staining intensity, there was no difference in the cellular staining pattern of the miR-21 signal when comparing the miR-21 probe concentration of 10 and 20 nM. In Figure 4, Figure 5 and Figure S1, the images were acquired from slides incubated with 20 nM probe concentration equal to the scramble probe. Sections were mounted with a DAPI-containing anti-fade solution, ProLong Gold (Thermo Fisher Scientific, Waltham, MA, USA). The detailed procedure is proprietary of Bioneer A/S (Bioneer, Hørsholm, Denmark).

4.6. Slide Scanning and IMAGE Acquisition

All chromogen stained slides were scanned on a Zeiss Axioscan equipped with a 20× objective. All images in Figures 1 and 2 were obtained from such digital slides. Confocal slide scanning was performed using a Pannoramic confocal scanner (3DHISTECH Ltd., Budapest, Hungary), which provides confocal images made by LED light and structured illumination technology. The use of this slide scanner has previously been described in detail for use on sections stained for miR-21 and two subsequent immunoperoxidase stainings [9]. The scanner was equipped with a 20× objective (NA = 0.8, Zeiss, Oberkochen, Germany) and a 40× water immersion objective (NA = 1.2, C-Apochromat (W), Zeiss, Oberkochen, Germany). For the current study, the following LED light sources were applied in the excitations: DAPI 390/22 nms, 520 mW, Cy3 555/28 nms, 370 mW, Cy5 635/22 nms, 510 mW, FITC 475/28, 530 mW. We used one dual pass filter (for FITC and Cy5, custom designed by 3D HisTech) and two single pass filters (TRITC/rhodamine and DAPI, both Semrock, New York, NY, USA). Digital slides were obtained with manually adjusted settings with regards to exposure time, digital gain and excitation intensity. The image acquisition settings for all 4 fluorophores were set at full excitation intensity if not otherwise stated. Image acquisition parameters were: miR-21 (Cy5, exposure time varying from 20–40 ms with digital gain 1), TNF-α mRNA (rhodamine, 20 ms, with 15–30% of full excitation intensity), CK immunofluorescence (FITC, 74–174 ms, digital gain 2), and nuclear counterstain (DAPI, 176–325 ms, digital gain 2). The settings applied to the individual cases were also used on the respective negative control sections incubated with scramble LNA probe and dapB mRNA RNAscope probe (cryo samples A, B, and C). Seven confocal layers of 1 μm were obtained and assembled into one extended focus layer, from which we obtained all fluorescence images in this paper.

5. Conclusions

In this study we developed a combined microRNA and mRNA ISH assay that allows the use of LNA probes and RNAscope probes on the same section as well as enables immunohistochemical detection of cellular markers. We explored expression of miR-21 and TNF-α mRNA at the invasive front of four colon cancers in order to elucidate expression interplay between miR-21 and TNF-α, and found that they are both expressed at the invasive front of colon cancers and are often co-detected in budding cancer cells and cancer cells considered to be part of branching events, however, the co-localization analysis of the four cases did not show consistent co-expression or converse expression patterns.

Supplementary Materials: Supplementary materials can be found at http://www.mdpi.com/1422-0067/20/8/1907/s1. Figure S1. Confocal slide scanning of sections submitted to automated combined staining of miR-21, TNF-α mRNA and CK. The three panels in this supplementary figure show miR-21 and TNF-α mRNA expression in 3 different colorectal cancer cases with varying invasion patterns (cases cryo-B, cryo-C and cryo-D, Table 1). Examples are shown at low (a–d) and high (e–h) magnification, and in a, the framed area is depicted in e–h. In example cryo-B, TNF-α mRNA is seen in foci of cancer cells also positive for miR-21. In example cryo-C, TNF-α mRNA is prevalent and seen in multiple cancer cells that are generally weakly stained for miR-21 or miR-21 negative, whereas miR-21 is prevalent in the stromal cells. In example cryo-D, miR-21 is seen in a few TNF-α mRNA expressing branching cancer cells. For Cryo-B and Cryo-C, in a serial tissue section, the miR-21 probe was replaced with an LNA scramble probe, and the TNF-α mRNA probe was replaced with the dapB mRNA RNAscope probe, that both show virtually no staining (magnification in e–h is identical to i–l).

Author Contributions: Data curation, J.P.J., F.B.S., J.L. and B.S.N.; Formal analysis, F.B.S. and J.L.; Funding acquisition, K.H.; Investigation, T.M. and B.S.N.; Methodology, T.M. and B.S.N.; Project administration, B.S.N.; Resources, T.M.; Validation, B.S.N.; Visualization, J.P.J. and B.S.N.; Writing—original draft, B.S.N.; and Writing—review and editing, J.P.J., K.H., F.B.S. and J.L.

Funding: This research received no external funding.

Acknowledgments: This study was supported by The Danish Agency for Science and Higher Education.

Conflicts of Interest: The authors declare no conflict of interest.

References

1. Ambros, V. microRNAs: Tiny regulators with great potential. *Cell* **2001**, *107*, 823–826. [PubMed]
2. Winter, J.; Jung, S.; Keller, S.; Gregory, R.I.; Diederichs, S. Many roads to maturity: microRNA biogenesis pathways and their regulation. *Nat. Cell Biol.* **2009**, *11*, 228–234. [PubMed]
3. Bartel, D.P. MicroRNAs: Genomics, biogenesis, mechanism, and function. *Cell* **2004**, *116*, 281–297. [PubMed]
4. Krichevsky, A.M.; Gabriely, G. miR-21: A small multi-faceted RNA. *J. Cell Mol. Med.* **2009**, *13*, 39–53.
5. Volinia, S.; Calin, G.A.; Liu, C.G.; Ambs, S.; Cimmino, A.; Petrocca, F.; Visone, R.; Iorio, M.; Roldo, C.; Ferracin, M.; et al. A microRNA expression signature of human solid tumors defines cancer gene targets. *Proc. Natl. Acad. Sci. USA* **2006**, *103*, 2257–2261. [PubMed]
6. Schetter, A.J.; Leung, S.Y.; Sohn, J.J.; Zanetti, K.A.; Bowman, E.D.; Yanaihara, N.; Yuen, S.T.; Chan, T.L.; Kwong, D.L.; Au, G.K.; et al. MicroRNA expression profiles associated with prognosis and therapeutic outcome in colon adenocarcinoma. *JAMA* **2008**, *299*, 425–436. [PubMed]
7. Nielsen, B.S.; Jorgensen, S.; Fog, J.U.; Sokilde, R.; Christensen, I.J.; Hansen, U.; Brunner, N.; Baker, A.; Moller, S.; Nielsen, H.J. High levels of microRNA-21 in the stroma of colorectal cancers predict short disease-free survival in stage II colon cancer patients. *Clin. Exp. Metastasis* **2011**, *28*, 27–38. [PubMed]
8. Nielsen, B.S.; Balslev, E.; Poulsen, T.S.; Nielsen, D.; Moller, T.; Mortensen, C.E.; Holmstrom, K.; Hogdall, E. miR-21 Expression in Cancer Cells may Not Predict Resistance to Adjuvant Trastuzumab in Primary Breast Cancer. *Front. Oncol.* **2014**, *4*, 207. [PubMed]
9. Knudsen, K.N.; Lindebjerg, J.; Kalmar, A.; Molnar, B.; Sorensen, F.B.; Hansen, T.F.; Nielsen, B.S. miR-21 expression analysis in budding colon cancer cells by confocal slide scanning microscopy. *Clin. Exp. Metastasis* **2018**, *35*, 819–830. [PubMed]
10. Zlobec, I.; Lugli, A. Epithelial mesenchymal transition and tumor budding in aggressive colorectal cancer: Tumor budding as oncotarget. *Oncotarget* **2010**, *1*, 651–661.
11. Hase, K.; Shatney, C.; Johnson, D.; Trollope, M.; Vierra, M. Prognostic value of tumor "budding" in patients with colorectal cancer. *Dis. Colon Rectum.* **1993**, *36*, 627–635.
12. Rogers, A.C.; Winter, D.C.; Heeney, A.; Gibbons, D.; Lugli, A.; Puppa, G.; Sheahan, K. Systematic review and meta-analysis of the impact of tumour budding in colorectal cancer. *Br. J. Cancer* **2016**, *115*, 831–840.
13. Patel, V.; Noureddine, L. MicroRNAs and fibrosis. *Curr. Opin. Nephrol. Hypertens.* **2012**, *21*, 410–416. [PubMed]
14. Asangani, I.A.; Rasheed, S.A.; Nikolova, D.A.; Leupold, J.H.; Colburn, N.H.; Post, S.; Allgayer, H. MicroRNA-21 (miR-21) post-transcriptionally downregulates tumor suppressor Pdcd4 and stimulates invasion, intravasation and metastasis in colorectal cancer. *Oncogene* **2008**, *27*, 2128–2136. [PubMed]
15. Yang, Y.; Ma, Y.; Shi, C.; Chen, H.; Zhang, H.; Chen, N.; Zhang, P.; Wang, F.; Yang, J.; Yang, J.; et al. Overexpression of miR-21 in patients with ulcerative colitis impairs intestinal epithelial barrier function through targeting the Rho GTPase RhoB. *Biochem. Biophys. Res. Commun.* **2013**, *434*, 746–752. [PubMed]
16. Thorlacius-Ussing, G.; Schnack Nielsen, B.; Andersen, V.; Holmstrom, K.; Pedersen, A.E. Expression and Localization of miR-21 and miR-126 in Mucosal Tissue from Patients with Inflammatory Bowel Disease. *Inflamm. Bowel Dis.* **2017**, *23*, 739–752. [PubMed]
17. Chapman, C.G.; Pekow, J. The emerging role of miRNAs in inflammatory bowel disease: A review. *Therap. Adv. Gastroenterol.* **2015**, *8*, 4–22.
18. Carswell, E.A.; Old, L.J.; Kassel, R.L.; Green, S.; Fiore, N.; Williamson, B. An endotoxin-induced serum factor that causes necrosis of tumors. *Proc. Natl. Acad. Sci. USA* **1975**, *72*, 3666–3670.
19. Old, L.J. Tumor necrosis factor (TNF). *Science* **1985**, *230*, 630–633.
20. Arnott, C.H.; Scott, K.A.; Moore, R.J.; Hewer, A.; Phillips, D.H.; Parker, P.; Balkwill, F.R.; Owens, D.M. Tumour necrosis factor-alpha mediates tumour promotion via a PKC alpha- and AP-1-dependent pathway. *Oncogene* **2002**, *21*, 4728–4738.
21. Balkwill, F. Tumor necrosis factor or tumor promoting factor? *Cytokine Growth Factor Rev.* **2002**, *13*, 135–141. [PubMed]
22. Guadagni, F.; Ferroni, P.; Palmirotta, R.; Portarena, I.; Formica, V.; Roselli, M. TNF/VEGF cross-talk in chronic inflammation-related cancer initiation and progression: An early target in anticancer therapeutic strategy. *In Vivo* **2007**, *21*, 147–161. [PubMed]

23. Grimm, M.; Lazariotou, M.; Kircher, S.; Hofelmayr, A.; Germer, C.T.; von Rahden, B.H.; Waaga-Gasser, A.M.; Gasser, M. Tumor necrosis factor-alpha is associated with positive lymph node status in patients with recurrence of colorectal cancer-indications for anti-TNF-alpha agents in cancer treatment. *Cell. Oncol.* **2011**, *34*, 315–326.

24. Ham, B.; Fernandez, M.C.; D'Costa, Z.; Brodt, P. The diverse roles of the TNF axis in cancer progression and metastasis. *Trends Cancer Res.* **2016**, *11*, 1–27. [PubMed]

25. Qu, Y.; Zhao, G.; Li, H. Forward and Reverse Signaling Mediated by Transmembrane Tumor Necrosis Factor-Alpha and TNF Receptor 2: Potential Roles in an Immunosuppressive Tumor Microenvironment. *Front. Immunol.* **2017**, *8*, 1675. [PubMed]

26. Cottonham, C.L.; Kaneko, S.; Xu, L. miR-21 and miR-31 converge on TIAM1 to regulate migration and invasion of colon carcinoma cells. *J. Biol. Chem.* **2010**, *285*, 35293–35302.

27. Qiu, Y.F.; Wang, M.X.; Meng, L.N.; Zhang, R.; Wang, W. MiR-21 regulates proliferation and apoptosis of oral cancer cells through TNF-alpha. *Eur. Rev. Med. Pharmacol. Sci.* **2018**, *22*, 7735–7741.

28. Chen, X.; Zhao, L.; Xing, Y.; Lin, B. Down-regulation of microRNA-21 reduces inflammation and podocyte apoptosis in diabetic nephropathy by relieving the repression of TIMP3 expression. *Biomed. Pharmacother.* **2018**, *108*, 7–14.

29. Player, A.N.; Shen, L.P.; Kenny, D.; Antao, V.P.; Kolberg, J.A. Single-copy gene detection using branched DNA (bDNA) in situ hybridization. *J. Histochem. Cytochem.* **2001**, *49*, 603–612.

30. Anderson, C.M.; Zhang, B.; Miller, M.; Butko, E.; Wu, X.; Laver, T.; Kernag, C.; Kim, J.; Luo, Y.; Lamparski, H.; et al. Fully Automated RNAscope In Situ Hybridization Assays for Formalin-Fixed Paraffin-Embedded Cells and Tissues. *J. Cell. Biochem.* **2016**, *117*, 2201–2208.

31. Wang, H.; Su, N.; Wang, L.C.; Wu, X.; Bui, S.; Nielsen, A.; Vo, H.T.; Luo, Y.; Ma, X.J. Dual-color ultrasensitive bright-field RNA in situ hybridization with RNAscope. *Methods Mol. Biol.* **2014**, *1211*, 139–149.

32. Thomsen, R.; Nielsen, P.S.; Jensen, T.H. Dramatically improved RNA in situ hybridization signals using LNA-modified probes. *RNA* **2005**, *11*, 1745–1748. [PubMed]

33. Jorgensen, S.; Baker, A.; Moller, S.; Nielsen, B.S. Robust one-day in situ hybridization protocol for detection of microRNAs in paraffin samples using LNA probes. *Methods* **2010**, *52*, 375–381.

34. Nielsen, B.S.; Holmstrom, K. Combined microRNA in situ hybridization and immunohistochemical detection of protein markers. In *Target Identification and Validation in Drug Discovery*; Moll, J., Colombo, R., Eds.; Humana Press: New York, NY, USA, 2013; pp. 353–365.

35. Nielsen, B.S.; Moller, T.; Holmstrom, K. Chromogen detection of microRNA in frozen clinical tissue samples using LNA probe technology. *Methods Mol. Biol.* **2014**, *1211*, 77–84.

36. Sempere, L.F.; Preis, M.; Yezefski, T.; Ouyang, H.; Suriawinata, A.A.; Silahtaroglu, A.; Conejo-Garcia, J.R.; Kauppinen, S.; Wells, W.; Korc, M. Fluorescence-based codetection with protein markers reveals distinct cellular compartments for altered MicroRNA expression in solid tumors. *Clin. Cancer Res.* **2010**, *16*, 4246–4255. [PubMed]

37. Leber, T.M.; Balkwill, F.R. Regulation of monocyte MMP-9 production by TNF-alpha and a tumour-derived soluble factor (MMPSF). *Br. J. Cancer* **1998**, *78*, 724–732.

38. Nielsen, B.S.; Timshel, S.; Kjeldsen, L.; Sehested, M.; Pyke, C.; Borregaard, N.; Dano, K. 92 kDa type IV collagenase (MMP-9) is expressed in neutrophils and macrophages but not in malignant epithelial cells in human colon cancer. *Int. J. Cancer* **1996**, *65*, 57–62.

39. Knudsen, K.N.; Nielsen, B.S.; Lindebjerg, J.; Hansen, T.F.; Holst, R.; Sorensen, F.B. microRNA-17 is the most up-regulated member of the miR-17-92 cluster during early colon cancer evolution. *PLoS ONE* **2015**, *10*, e0140503.

40. Fluckiger, A.; Dumont, A.; Derangere, V.; Rebe, C.; de Rosny, C.; Causse, S.; Thomas, C.; Apetoh, L.; Hichami, A.; Ghiringhelli, F.; et al. Inhibition of colon cancer growth by docosahexaenoic acid involves autocrine production of TNFalpha. *Oncogene* **2016**, *35*, 4611–4622. [PubMed]

41. Li, H.; Zhong, A.; Li, S.; Meng, X.; Wang, X.; Xu, F.; Lai, M. The integrated pathway of TGFbeta/Snail with TNFalpha/NFkappaB may facilitate the tumor-stroma interaction in the EMT process and colorectal cancer prognosis. *Sci. Rep.* **2017**, *7*, 4915. [PubMed]

42. Varela, L.M.; Ip, M.M. Tumor necrosis factor-alpha: A multifunctional regulator of mammary gland development. *Endocrinology* **1996**, *137*, 4915–4924. [PubMed]

43. Lee, P.P.; Hwang, J.J.; Murphy, G.; Ip, M.M. Functional significance of MMP-9 in tumor necrosis factor-induced proliferation and branching morphogenesis of mammary epithelial cells. *Endocrinology* **2000**, *141*, 3764–3773. [PubMed]

44. Kwong, J.; Chan, F.L.; Wong, K.K.; Birrer, M.J.; Archibald, K.M.; Balkwill, F.R.; Berkowitz, R.S.; Mok, S.C. Inflammatory cytokine tumor necrosis factor alpha confers precancerous phenotype in an organoid model of normal human ovarian surface epithelial cells. *Neoplasia* **2009**, *11*, 529–541. [PubMed]

45. Zarjou, A.; Yang, S.; Abraham, E.; Agarwal, A.; Liu, G. Identification of a microRNA signature in renal fibrosis: Role of miR-21. *Am. J. Physiol. Renal Physiol.* **2011**, *301*, F793–F801. [PubMed]

46. Zhang, L.; Shen, J.; Cheng, J.; Fan, X. MicroRNA-21 regulates intestinal epithelial tight junction permeability. *Cell Biochem. Funct.* **2015**, *33*, 235–240. [PubMed]

47. Xu, K.; Xiao, J.; Zheng, K.; Feng, X.; Zhang, J.; Song, D.; Wang, C.; Shen, X.; Zhao, X.; Wei, C.; et al. MiR-21/STAT3 Signal Is Involved in Odontoblast Differentiation of Human Dental Pulp Stem Cells Mediated by TNF-alpha. *Cell Reprogram.* **2018**, *20*, 107–116. [PubMed]

48. Ando, Y.; Yang, G.X.; Kenny, T.P.; Kawata, K.; Zhang, W.; Huang, W.; Leung, P.S.; Lian, Z.X.; Okazaki, K.; Ansari, A.A.; et al. Overexpression of microRNA-21 is associated with elevated pro-inflammatory cytokines in dominant-negative TGF-beta receptor type II mouse. *J. Autoimmun.* **2013**, *41*, 111–119.

49. Zhang, X.; Ng, W.L.; Wang, P.; Tian, L.; Werner, E.; Wang, H.; Doetsch, P.; Wang, Y. MicroRNA-21 modulates the levels of reactive oxygen species by targeting SOD3 and TNFalpha. *Cancer Res.* **2012**, *72*, 4707–4713. [PubMed]

MDPI

St. Alban-Anlage 66

4052 Basel

Switzerland

Tel. +41 61 683 77 34

Fax +41 61 302 89 18

www.mdpi.com

International Journal of Molecular Sciences Editorial Office

E-mail: ijms@mdpi.com

www.mdpi.com/journal/ijms

www.ingramcontent.com/pod-product-compliance
Lightning Source LLC
Chambersburg PA
CBHW041217220326
41597CB00033BA/6002